AF457860

Your Creative Brain and AI

Your Creative Brain and AI

How We Learn and Consciously Experience Art, Music, and Meaning

Stephen Grossberg

Great Clarendon Street, Oxford, OX2 6DP,
United Kingdom

Oxford University Press is a department of the University of Oxford.
It furthers the University's objective of excellence in research, scholarship,
and education by publishing worldwide. Oxford is a registered trade mark of
Oxford University Press in the UK and in certain other countries.

Published in the United States of America by Oxford University Press
198 Madison Avenue, New York, NY 10016, United States of America

British Library Cataloguing in Publication Data
Data available

Library of Congress Control Number: 2025940693

ISBN 9780198965374

DOI: 10.1093/9780198965367.001.0001

Printed and bound by
CPI Group (UK) Ltd, Croydon, CR0 4YY

The manufacturer's authorised representative in the EU for product safety is
Oxford University Press España S.A. of Parque Empresarial San Fernando de Henares,
Avenida de Castilla, 2 – 28830 Madrid (www.oup.es/en or product.safety@oup.com).
OUP España S.A. also acts as importer into Spain of products made by the manufacturer.

Dedicated to my family

Gail

Deb

Noah

Zoe

and

Brian

with much love

Contents

Preface

Neural Networks, the Father of AI, and Autonomous Adaptive Intelligence

I have been a passionate student of how our brains make our conscious minds since I took Introductory Psychology as a seventeen-year-old Dartmouth College freshman in 1957. In that course, I discovered, and rapidly began to develop, the paradigm of using neural network models to achieve the seemingly impossible dream of solving the classical mind/body problem.

That early discovery changed the course of my life forever, and the impossible dream has, in the intervening sixty-eight years, become a rigorous scientific reality. My brain exploded with ideas that year, and I have been blessed to make breakthrough discoveries until the present time about essentially all the main brain processes that enable our brains to consciously experience and learn about the world.

Another way to describe this type of research is as a program to explain *autonomous adaptive intelligence*, of which human intelligence is the best example. We are *autonomous* because we can do so many tasks on our own. We are *adaptive* because we can learn on our own about the world. And we are *intelligent* in at least the sense that our autonomous adaptation often enables us to successfully realize our goals in a rapidly changing, and sometimes dangerous, world.

Because I have discovered and led so many scientific breakthroughs for over sixty years, I have been called by many colleagues the **Father of AI**, and the **Newton and Einstein of the Mind**. This book will clarify how these names summarize the significance of my scientific work.

A Lucky Life at Boston University: A Supportive Administration and Gifted Students

I should add another blessing that made my productivity possible: I was lucky to be able to attract and train over one hundred gifted PhD students and postdoctoral fellows with whom I collaborated to make these discoveries. I was no less lucky to get administrative and financial support to create the academic infrastructure that made such training possible. I did all this at Boston University (BU), whose president, John Silber, and provost, Dennis Berkey, supported my efforts to found, first, the Graduate Program in Cognitive and Neural Systems and then, two years later in 1991, the Department of Cognitive and Neural Systems (CNS).

Dennis once told another faculty member, in my presence, why BU was willing to take such big risks with me. After all, letting a faculty member start a new department does not happen every day in academe. He said: When Steve makes a promise, he always keeps it. And I had promised BU that I would bring in the funding to make my dream possible.

Before CNS began, I had been a professor at BU since 1975. In 1989, I was awarded an endowed chair that was created for me, when I became the Wang Professor of Cognitive and Neural Systems. This promotion acknowledged my leadership of the wonderful productivity of our research team. This productivity was measured in at least three ways:

The sheer number of discoveries and archival scientific articles that we published.

The international fame of our group. Many people told me that our team helped to put BU "on the map" in all the areas to which we contributed, which included psychology, neuroscience, mathematics, computer science, and engineering.

Money Talks

Last, but not least, the big research grants that we won from multiple funding agencies in Washington, including the Army Research Office (ARO), Air Force Office of Scientific Research (AFOSR), Defense Advanced Research Projects Agency (DARPA), National Science Foundation (NSF), National Institutes of Health (NIH), and Office of Naval Research (ONR). In all, we won over $85,000,000 in grants over the years, at a time when a million dollars was still "real money."

You might immediately wonder: Why did so many agencies in the Defense Department, of all places, support our research on mind and brain? That was because all of our models were rigorously defined by systems of mathematical equations, in particular nonlinear differential equations, and algorithmic approximations of them. We had shown, using mathematical theorems and systematic computer simulations, that our models actually work. By "work," I mean, using the jargon of engineering and technology, they do all sorts of intelligent things, like image and speech processing, adaptive pattern recognition and prediction, cognitive and emotional information processing, value-based decision-making, navigation, and adaptive sensory-motor control and robotics. These were processes that the Defense Department wanted to use to develop cutting-edge technologies for keeping our Nation safe.

How do our models keep our Nation safe? As one example that this is not just a personal delusion, many years ago, after I finished giving a plenary lecture at a major scientific conference, a man came up to me to say that he was from the National Security Agency (NSA) and that "ART [Adaptive Resonance Theory] is protecting the Nation." He then just as quietly walked away. Because what the NSA does is usually classified, he did not go into further detail. I will explain what ART is, and why the NSA might have found it useful, later in this book.

This money from multiple grants paid the salaries of all our PhD students and postdocs, the summer salaries of our faculty, most if not all of the salaries of our administrative staff, a big chunk of my endowed chair salary, and the costs of all of our computers and laboratory equipment. The grant money that was budgeted for "direct costs" is the money that we used to pay for all the things that I just listed. The direct costs did not include the overhead on these grants, often 60% or so in addition to the direct costs budget. The overhead paid, at least indirectly, the costs to support the building that we used, including its office and classroom space, furniture, repairs, heat, light, and so on. In fact, for quite a few years, I was told that we brought in more grant money than any other unit at BU, including our medical school and law school.

In brief, I kept my promise to Dennis: CNS paid its own bills.

We did not, of course, continue to be the top earner in BU for all the years until I retired in 2019. Through vigorous faculty hiring and other smart administrative choices, notably investing in new buildings, and strong administrative support for stand-alone research centers, BU rapidly became a world-class educational *and* research university, with grants to match across multiple academic units.

Neural Networks and AI Hype, and the Newton and Einstein of the Mind

Neural networks and AI are now very much in the news. They are once again a hot topic. I write "once again" because AI was over-hyped many years ago, by people like Marvin Minsky and Seymour Papert at MIT, until that AI bubble burst due to computational weaknesses of their hyped AI algorithms. I was then a professor at MIT, so I could witness this spectacle at close range.

The period that followed this bubble burst in the 1970s is often called the AI Winter (https://en.wikipedia.org/wiki/AI_winter). Indeed, in 1973, Sir James Lighthill, who then held Isaac Newton's position as Lucasian Professor of Mathematics at Cambridge University, wrote a scathing and highly influential review of AI, called the Lighthill Report (https://en.wikipedia.org/wiki/Lighthill_report). Lighthill's report caused a drastic reduction in AI funding, particularly in England.

Despite these setbacks of the symbolic AI that was then being developed, Minsky and Papert argued against funding of neural network research, and made sure to include me in their criticism. This was thus a period when AI was emphatically not the same as neural networks. The opportunistic appropriation by AI of neural networks happened only later on.

I have often thought that Minsky's motivation to attack my work was due, in part, to the fact that his own early neural network research had failed, whereas mine had flourished. His ego could not tolerate such an outcome.

My neural networks research at MIT, although copious and published in distinguished journals like the *Proceedings of the National Academy of Sciences* (see sites.bu.edu/steveg), thus came under fire when I came up for tenure. This happened despite the fact that I was promoted from Assistant Professor to Associate Professor in two years, which was quite rare at that time. As part of my tenure review, Sir James Lighthill was asked to write a recommendation letter for me, with the thought that he would trash my work like he did the rest of AI. Instead, Sir James wrote a long letter full of praise for my work, saying that I was doing what AI should have been doing all along. I was lucky to get to know Sir James personally some years later when I was lecturing at Cambridge.

Just as during those days at MIT, the people who are most vocally hyping AI in the news these days are not usually the people who have made the most important discoveries in our field. The people in the news tend to promote one or two models that do not resemble how humans learn and become intelligent. These models also have serious computational problems that I will discuss in this book, including the fact that they are untrustworthy, unreliable, and literally do not know what they are talking about. However, by functioning as a closed clique, if not a cult, and spending a lot of effort marketing the ideas of their fellow members, the people pushing these models have attracted a lot of public attention.

Despite all of this marketing noise, people who know the broader neural network literature often call me the **Father of AI**, because of my 1957 discovery of the neural networks paradigm and of the main equations that are still used to explain how our minds work. Because I have discovered and led research for over sixty years about neural network models that explain the main processes whereby our brains make our conscious and unconscious minds, in both healthy individuals and clinical patients, I have also been called the **Newton and Einstein of the Mind**.

I retired from Boston University in 2019, just before the COVID pandemic rapidly began to spread and kill in 2020. I then became an Emeritus Professor of Mathematics and Statistics, Psychological and Brain Sciences, and Biomedical Engineering. I also retain the title of Wang Professor of Cognitive and Neural Systems, as well as my office in the building that once housed CNS.

I still have a romantic attachment to this building, partly because I was invited to design it by the BU administration, including its entire floor plan of offices, classrooms, auditorium, library, mailroom and lounge, vestibule, hallways, staircases, elevator, windows, lighting... even the bathrooms! I designed the building to facilitate the human endeavors that I hoped it would encourage. Because I was redesigning a building with a fixed footprint, I made the vestibule small and the hallways narrow to make the offices and other spaces where people actually lived and worked as large as possible. I have treasured memories of the many happy scholarly and social activities that these spaces enabled over the years.

Retirement, My Magnum Opus, and This Book

The Law of Unexpected Consequences reared its head big-time after I retired. I have a great respect for this Law, which has had major consequences for the trajectory of my life on several occasions. For the first time in over forty years, I did not have to think of research ideas, write, win, and manage multiple research grants, and financially pay and train the students and postdocs with whom to do the research that the grants supported, as well as the administrative staff to support all these activities.

I had so much free time that I wrote and published my Magnum Opus, *Conscious Mind, Resonant Brain: How Each Brain Makes a Mind*, in 2021. This book was written for a general audience in a self-contained and non-technical way. I worked very hard to make it as easy and as enjoyable to read as possible. I did my best to do this, constrained by the reality that the human brain is perhaps the most complex system in Nature that humans can hope to scientifically understand. My book explains how our models provide unifying and principled explanations of the data reported in hundreds of psychological and neurobiological experiments, while also reviewing and unifying the work of many other scientists. I also mentioned some of my predictions that were confirmed by subsequent data.

It was heartwarming to me when my book got wonderful reviews from many distinguished psychologists, neuroscientists, and technologists. You can read both the pre-publication and post-publication reviews at https://www.amazon.com/Conscious-Mind-Resonant-Brain-Makes/dp/0190070552. It was also a delightful surprise to learn that the book won the 2022 PROSE book award in Neuroscience of the Association of American Publishers.

As soon as I submitted my Magnum Opus for publication, I continued to do research on topics that interested me. I no longer had collaborators, but I had over fifty years of archival research as a foundation upon which I could confidently build. Without having to worry about grants, or what students and postdocs wanted to study, I was able to think and write about some of the topics that have always interested me, notably visual art, music, how language acquires meaning, and the differences between AI that focuses on Deep Learning and ChatGPT versus AI that develops biological neural networks and applications thereof. This book is the result of that research.

I have written this book, as I did my Magnum Opus before it, to be as self-contained and non-technical as possible. I also wrote it so that the chapters can be read independently of each other. Because of that, I sometimes repeat facts in more than one chapter, if they are needed to understand that chapter independently of all the others.

Read This Book Like a Story

I hope that a first reading of the book will be approached like a story for which you would like to get the gist. Because our brains are perhaps the most complicated systems in Nature, even a "self-contained and non-technical" exposition will necessarily include some challenging facts and ideas. I have tried to present them in an accessible way, partly by stating some of the most basic ones early in each chapter, and partly by including a lot of pictures to illustrate key ideas and results. I have also included enough details to satisfy readers who seek a deeper understanding. Archival articles in which hundreds of supportive psychological and neurobiological experiments using these processes are cited, explained, and quantitatively simulated in my Magnum Opus. I hope that I have found a good balance for the maximum number of readers.

The Blessing of a Happy and Loving Family

Chapter 2 of my Magnum Opus provides an extensive summary of all of the people and institutions who supported me and my work over the years. I am still grateful to all of them. Here I will just mention the people who are dearest to me:

Gail Carpenter, my brilliant, wise, generous, adventurous, and loving best friend, wife, and closest collaborator for the past fifty years.

Deborah Grossberg Katz, our wonderfully creative daughter, who has become a distinguished architect who co-owns, with her husband Brian Phillips, the architecture firm Interface Studio Architects (ISA) LLC, whose many prize-winning designs and built projects I like very much: http://www.is-architects.com/, http://www.is-architects.com/press.

Last but not least, our beautiful, kind, thoughtful, and gifted grandchildren, Noah (15) and Zoe (12), whose many evolving interests and talents are always a delight to witness and hear about.

In all, I feel very blessed.

1

Overview

My personal odyssey and discoveries that led to this book

1 From a Shy Kid in Jackson Heights to the Newton and Einstein of the Mind

How did a shy kid like me get interested in understanding how our brains make our minds? And how did that interest eventually lead me to scientifically explain how our brains carry out, and consciously experience, art, music, and language meanings?

One reason is that I grew up in New York City, where little of Nature was on display except for tall buildings, pets, and people. I was always interested mostly in the people, including why so many well-meaning people from my own childhood did not seem to be living happy lives. Let me illustrate what I mean with some examples.

From the time that I was five years old, in 1945, until I left for college at age seventeen, I grew up in Jackson Heights, Queens, New York, as part of a lower-middle-class Jewish family. I lived with my family in a small apartment in which I shared one bedroom with my two brothers, dormitory-style, while my grandma, Gisella, or Gussy for short, slept on a fold-out couch in the small foyer that everyone walked through to get to the kitchen or living room.

Those of you who have seen the Woody Allen film *Annie Hall* will not be surprised that my family members, all of Hungarian Jewish descent, did not hesitate to yell and scream at any possible provocation. A lot of this yelling happened in the kitchen preparing a meal, or even while eating the meal, which made everything about food an emotionally charged subject.

The yelling when my grandma's three sisters got together with her to cook a Seder dinner in our aunt's small kitchen was a high point in the annual screaming sweepstakes. It was all absolutely worth it because they were master cooks and always made a memorable feast for our assembled extended family, ranging from my great grandparents at one end of a long Seder table that spanned three rooms, to me and the other little kids at the other.

I was incredibly lucky to have a brilliant mother, who was born Elsie Leitner. Elsie was the first person in our family to attend college, against the advice of these same busy-body aunts, who wanted her to immediately go to work, marry, and have babies. Instead, before

Your Creative Brain and AI. Stephen Grossberg, Oxford University Press. © Oxford University Press (2026).
DOI: 10.1093/9780198965367.003.0001

she married for the second time when I was five years old and we moved to Jackson Heights, my mom commuted by train from Woodside, in Queens, New York, where she lived with my grandparents Gussy and Eugene—who everyone called Mishka—to Hunter College in Manhattan, from which she graduated with a bachelor's degree in history when she was nineteen years old.

During this time, my mom also fell in love with the neighborhood ladies' man, Irwin Gottesman, whom she married when she graduated from Hunter. Irwin is my genetic father and that of my brother, Mitchell, or Mitch for short, who was born two years before me. Irwin never went to college. Instead he owned a neighborhood grocery store before he died from Hodgkin's lymphoma when I was one year old and he was twenty-eight. My mom thus became a widow with two small children at age twenty-four.

While my mom spent every day for a year caring for my dying father, his brothers, who also owned a nearby grocery store, agreed to manage his store for them. After my father died, and my mom returned to manage the store, she discovered that his brothers had stolen all the stock from the store for themselves, leaving her with nothing with which to support two small children. When I saw the 1964 film Zorba the Greek, I was reminded of this family disgrace by the scene in which the villagers enter Madame Hortense's bedroom right after she dies to steal all her belongings.

One might have thought that the need to support her two small children might have concerned my father's brothers. But this was actually par for the course for my paternal relatives. For example, when my mom was pregnant with Mitch, Irwin's mom wished that Elsie's blood would turn to oil before she gave birth. When she was pregnant with me, she wished that my mom would be hit by a car before she gave birth. The reason for this shocking, self-defeating, cruelty seems to have been that my paternal grandmother loved my handsome father far too much, in the manner of an incestuous Jocasta complex, and could not bear the thought that my mom would take him away from her, even though they lived just a few blocks from us.

Although my mom tried to keep Irwin unaware of these threats, when he found out about them, he totally broke off contact with his parents. It was shortly after this that my father was diagnosed with Hodgkin's lymphoma, which, although usually treatable today, was then a death sentence.

The loss of our father greatly traumatized my older brother, Mitch, who was three when he died. Since I was only one, I have no memory at all of my father except an indirect one, a vivid snapshot memory of the black bag of the doctor, at the level of my face, as he visited Irwin at home. But Mitch, being three, did remember our father and intensely felt his loss. It was perhaps this trauma, and the attendant collapse of a sense of security that it entailed, that motivated Mitch to beat me up almost daily, typically without any provocation from me except that I made him feel insecure. These beatings may have contributed to me becoming a shy and introspective person for quite a few years. If I had to put a positive spin on this abuse, it would be that I quickly learned the power of language to express and to defend myself by hitting Mitch back where it hurt, if not with fists, then with the greater cutting force of words.

As a young widow with no means of support near home, Elsie went to work in Manhattan as a bookkeeper while Gussy babysat, cooked her wonderful Hungarian food for us, and took me to lots of matinees while Mitch was in school. Some of these movies were highly inappropriate for an impressionable little boy to see, including Frankenstein and Dracula. For years, I slept with my sheet and blanket carefully wrapped around my neck so that Dracula could not feast on me during the night.

My grandma also took me to at least one gangster movie in which a gangster kills a rival by hiding a gun in an accordion camera. When the picture is taken, the camera's flash is the flash of the gun shooting and killing his rival. This traumatic event also had consequences at home. At a big family party, one of my uncles had us all line up in multiple rows in front of the dining room table to take our picture with ... you guessed it ... an accordion camera. I had never seen one before outside of that movie. I became terrified and scrambled to hide under the table, screaming: "I don't want to die!," over and over. My mom was none too pleased and ended my tearful disruption, not by asking what frightened me, but by dragging me into the bedroom and shutting me in there all alone in pitch darkness. It seems hard to believe that such a devoted and intelligent mother could be so thoughtless, but I still have a framed copy of the picture that my uncle took that night, and I am not in it.

While working in Manhattan and bringing up two boys with the help of my grandma, my mom studied hard to pass the then-rigorous, standardized tests to become a public school teacher in New York City. She passed the tests with flying colors and was for many years a much loved fourth grade teacher in public schools that were attended by needy kids from diverse backgrounds. While teaching full time and taking care of her sons, my mom also managed to continue her education, eventually earning her PhD equivalency in education, primarily to earn more money to help pay the family's bills.

It was from my mom that I acquired my life-long love of learning. At first, learning was the only way that I could find to cope with my rather dreary day-to-day life, and to escape into worlds that I could then only imagine by reading about them in books.

As the son of a father who died young, I was all too aware, ever since I was a young boy, of how short life can be. So learning became my lifeline to trying to get a glimpse of the awesome beauty and harmony of the world before vanishing from it as suddenly as I joined it.

I was also aware from an early age of the fact that our minds actively construct our experiences of the world. The earliest awareness of this that I can remember happened when I was relaxing on my side in my crib in our house in Woodside, so I must have been less than five years old. As I lay there, I stared through the crib's vertical bars at the wall across the bedroom. As I gazed at the wall, I must have relaxed the accommodation of my eyes, because the distance of the wall from me seemed to change, back and forth. This experience startled me into vaguely realizing that we construct our perceptual experiences. They are not rigidly fixed independently of our minds. Rather, they are perceptual illusions that are actively constructed in our minds, and that happen to work very well in allowing us to get on with our lives. Due to experiences like these, I have been fascinated by visual illusions all my life, and have worked hard to scientifically explain why we have them, and how we consciously perceive the most important ones.

I cannot rationally explain why I have always felt the need to discover a unifying harmony behind even mundane experiences or, more generally, to discover the underlying principles and meaning of all my experiences in life. The alternative, that life is meaningless and accidental, was unbearable to me. Trying to discover this harmony has always made learning feel like a spiritual quest for me. I have always thought of it as a way to glimpse, albeit briefly, the handiwork of God.

I am still working as hard as I can to better glimpse that harmony through my research in science, which I have pursued with passion during the last sixty-eight years, ever since I made my first major scientific discoveries in 1957 as a freshman at Dartmouth College.

As I have just noted, perhaps because I grew up in New York City, where the most salient objects in the world were buildings, pets, and people, I was always fascinated to learn more about people. I first wanted to try to understand why so many nice and well-meaning people in our family seemed to be living unhappy lives. Over the years, I gradually became passionately committed to understanding all aspects of how our brains make our conscious and unconscious minds, in both healthy individuals and clinical patients.

This passionate quest and its expression in multiple scientific discoveries led the president of Boston University, John Silber, to introduce me as the "Newton and Einstein of the Mind" before I gave the annual 1989 University Lecture to a Boston-area audience. I had been a full professor at BU since 1975 when I came there from MIT, where I was an assistant and associate professor since 1967. Silber was a famously tough-minded leader. He was nonetheless willing to introduce me in such a shockingly positive manner after checking me out with various leading scholars.

Since 1989, I have continued to make a rapid series of breakthrough discoveries about how our brains make our minds, often with the collaboration of over one hundred gifted PhD students, postdoctoral fellows, and faculty collaborators. Most of these discoveries took the form of principled and unifying explanations of data from hundreds of psychological and neurobiological experiments, using the neural network models that I discovered and developed with the help of my colleagues. Others were mathematical discoveries about the design principles, mechanisms, and architectures that made these neural network models work so well.

These discoveries include a proposed scientific solution of the classical Mind-Body Problem. That is the problem of *how*, *where* in our brains, and *why* from a deep computational perspective humans can *consciously see, hear, feel, and know* about objects and events in a changing world that is filled with unexpected events. This revolutionary discovery also showed that we are not conscious just to platonically contemplate the world. Rather, we use our conscious representations to *plan and act to realize valued goals.* In brief, our consciousness is necessary for our survival in a changing, and often dangerous, world. It is the link between neural models of our minds, and how they support conscious and unconscious actions in the world, that constitutes my proposed solution of the Mind-Body Problem.

One might immediately ask: How does one discover such models? I always began by studying data from scores, or hundreds, of psychological experiments that fascinated me. To know what combinations of experiments all probe similar brain mechanisms is, in itself, a skill, especially since you need to group the data before you understand the brain mechanisms that they reflect. This skill depends upon studying lots of data and thinking about how it could have arisen, moment-by-moment, by a performing individual through time.

This is perhaps the hardest thing to teach students: How to understand how psychological data arise from *individuals behaving and adapting in real time.* This is hard to understand because psychological data was, at least then, usually described in terms of graphs of how one variable varies as a function of another variable. For example, how performance accuracy on a given task increases as a function of the number of learning trials. This cumulative summary of the effects of multiple learning trials was pretty far removed from how individuals learned the task in real time.

By studying lots of data for a long time, I gradually began to be able to think fluently in real time, and to develop an intuition about the underlying design principles that the data reflect. I then tried to write down the simplest mathematical model that could embody these principles. I was led in this way, as a freshman at Dartmouth College taking introductory psychology, to discover neural networks whose dynamics could explain how individuals learn in real time.

At first, because I knew nothing about neuroscience, I did not even realize that my networks were *neural* networks, as opposed to merely formal ones. However, by talking to friends of mine who were pre-medical students, I soon realized that the variables in my models corresponded to well-known processes in our brains. Other properties of my model were not well-known, and corresponded to predictions that have been supported by subsequent experiments.

I hereby had an experience that I can best describe as being ecstatic. It is hard to capture the thrill of having used simple psychological concepts to derive brain mechanisms and dynamics of cell activation, modeled by short-term memory (STM) traces; activity-dependent habituation, modeled by medium-term memory (MTM) traces; and learning and memory, modeled by long-term memory (LTM) traces, and then to use these laws to explain the human learning data that I was reading about in class. That thrill imprinted in me the passionate desire to learn as much as I could during my life about how our brains make our minds.

This conceptual and mechanistic foundation has been supported by lots of psychological and neurobiological data, and has enabled many of my subsequent discoveries. As one of many examples, later modeling results explained how specific breakdowns in the normal functioning of my learning models mimic behavioral symptoms of mental disorders that afflict millions of people, including Alzheimer's disease, autism, amnesia, schizophrenia, post-traumatic stress disorder, attention deficit hyperactivity disorder, auditory and visual agnosia and neglect, and disorders of slow-wave sleep. I never dreamed that I would have any insights to offer about this topic.

I published my My Magnum Opus, called *Conscious Mind, Resonant Brain: How Each Brain Makes a Mind* (https://www.amazon.com/Conscious-Mind-Resonant-Brain-Makes/dp/0190070552), in 2021 to provide a self-contained and non-technical summary and synthesis of many of my discoveries over the years about how our brains make our conscious and unconscious minds, in both healthy individuals and clinical patients. One of the reviewers of my book, Daniel Franklin, explained in his review why various scientists have called me the Newton and Einstein of the Mind over the years:

> "In 1989, Stephen Grossberg gave the annual University Lecture at Boston University to a Boston-area audience. The University President, John Silber, introduced Grossberg as the Newton and the Einstein of the Mind. Silber had done his homework before his startlingly laudatory introduction. Grossberg had already introduced the field of biological neural networks as a college Freshman in 1957 as well as the then-revolutionary paradigm of using nonlinear systems of differential equations to model how brains make minds. These equations included the laws for neuronal activation, or short-term memory (STM); activity-dependent habituation, or medium-term memory (MTM); and adaptive weights for learning and memory, or long-term memory (LTM) that still form the foundation of all biological models of how brains make minds. Grossberg then rapidly began introducing and developing an astonishing range and depth of neural models whose topics today, 64 years later, include essentially all the most important processes of conscious and unconscious human intelligence.
>
> "By discovering the foundational laws of how brains make minds, and using them to build unifying theories that explain so much important data, Grossberg has played a role similar to that played by Isaac Newton in his great work on celestial mechanics. This may be why John Silber mentioned Newton in his introduction.
>
> "But why Einstein? Albert Einstein was famous for using thought experiments to derive fundamental laws of physics, including both Special Relativity Theory and General Relativity Theory. Thought experiments use familiar facts from daily life to derive profound scientific conclusions. They are a way to make scientific ideas accessible to many people.
>
> "Grossberg is a master of using thought experiments to derive profound conclusions about learning, cognition, and emotion. His thought experiments use facts that are familiar to everyone because they represent ubiquitous environmental pressures on the evolution of our brains. In particular, Grossberg published a famous thought experiment in 1980 to derive Adaptive Resonance Theory, or ART, which is currently the most advanced cognitive and neural theory of how our brains learn to attend, recognize, and predict objects and events in a changing world. All the foundational hypotheses of ART have been supported by subsequent psychological and neurobiological experiments, and it has also provided principled and unifying explanations of hundreds of additional experiments.
>
> "Nowhere during the thought experiment are the words mind or brain mentioned. ART design principles and mechanisms are thus a universal solution to a learning problem that Grossberg called the stability-plasticity dilemma because it asks how any system can learn quickly without experiencing catastrophic forgetting. Remarkably, ART also explains how humans consciously see, hear, feel, and know things about the world, and use these conscious states to effectively plan and act to realize valued goals.
>
> Grossberg's proposed solution of the mind-body problem thus arises from a computational analysis of how humans and other animals autonomously learn in a changing world. Because the words mind and brain are never mentioned in his derivations, they can be used as a blueprint for designing autonomous adaptive intelligent algorithms and mobile robots for technology and AI.

Returning to when I was a young boy: I developed a terrifying sense of impermanence, probably because my father died when I was just one year old. My mom remarried my second father, Solomon Grossberg, when I was five years old. It was then that our family moved to our crowded apartment in Jackson Heights and added another brother, Kenneth, to cram with Mitch and me into a single bedroom.

Sol had a hard early life. He was orphaned as a boy growing up in a slum in Brooklyn, New York, and, after recovering from infantile paralysis, limped to a nearby street corner every day to sell newspapers to try to keep himself and his younger siblings out of the orphanage. He eventually went to night school to become an accountant, and thereby become a better provider for his younger siblings, but he was so emotionally damaged by his childhood experiences that he let his bosses run all over him at work for his entire life.

Because Sol had no childhood to speak of, he had no idea how to be a father. He was a naturally kind man, but had no concept of the kinds of passions and needs for transcendence that were boiling up in me from an early age. We never managed to have a conversation during all the years that I knew him before he died at age ninety-one.

Without a father who could help me to figure out how to become a man, I began to carefully observe all the men in my life, as well as the women, to figure out what kind of a person I wanted to be. This, too, contributed to me gradually finding myself fascinated by how our minds work.

Elsie and Sol devoted most of their energies to helping me, Mitch, and Ken have the best life that they could afford. As just one example, after Elsie's teaching year ended, both she and Sol would work through the summers to be able to afford to send the three of us to summer camp, out of the heat of a New York summer, at a time when no one had air conditioning. We typically went to camps in the Pocono mountains of Pennsylvania. As soon as I was old enough, I worked at camp to pay my own way, first as a waiter and then a counselor, among other duties. Figure 1.1 shows a picture of me at age sixteen in camp during a family visit. Elsie and Gussy stood beside me as Sol took the picture.

FIGURE 1.1 Me at age sixteen with my mom, Elsie, and my grandma, Gussy, who were visiting me at summer camp in Pennsylvania.

Camp was a wonderful experience in many ways. I learned how to swim and play sports like basketball and baseball. My best strokes were breast stroke and back stroke. I even won races! Even better, I got to do a lot of art, including designing and painting sets for campus plays with a beautiful girl who was studying art at the High School of Music and Art in New York. I also made big strides in overcoming my shyness and in even becoming a leader among my camp friends. I am not the only shy kid who found out that he or she could become quite the ham when performing a definite role in a play or musical on stage. It's hard for me to imagine today that I once even acted and sang the lead in a camp musical comedy!

2 The Jackson Heights Public Library and Language

I was lucky to live in Jackson Heights, partly because Public School 69 (PS 69) was just a few blocks away from our apartment and had excellent teachers and lots of bright and highly motivated students.

I worked incredibly hard in PS 69 and was always at the top of my class. I loved to learn, but was just as often motivated by a fear of failure. Even as a boy, I could see the barriers that society constructs at multiple stages of one's life, barriers that could easily prevent me from finding a meaningful life.

My friends and I were regularly reminded of these barriers at school. For example, in the eighth grade at PS 69, we all took a city-wide test whose highest score was 12.0+, meaning the equivalent of at least a high school graduate finishing grade 12. Our eighth grade teacher, Miss Farrell, who was a caricature of a sour old spinster, with even a pencil stuck in the tight bun of her hair, rather sadistically read out loud the scores of each student to the entire class. Because ours was an unusually gifted class, student names followed by 12.0+, 12.0+, 12.0+, ... were read out monotonously until, to everyone's shock, someone got an 11.9.

That poor student, a friend of mine named Harvey, was publicly shamed and felt like a failure for a long time after that, despite the fact that, in any other school, 11.9 would be viewed as a great success. Harvey's sad example has always reminded me to emotionally support and guide my own students and postdocs with lots of advice and information that I never received about what to expect at the next stages of their lives, including advice about how to prepare, practice, and give lectures and posters at scientific conferences, how to seek and apply to schools and jobs where they could fulfill their talents and, no less important, how to try to find environments where their work would be appreciated and their lives could be fulfilled.

Just a few blocks beyond PS 69 on 37th Avenue was a public library. Since I was already accustomed to walking to school in that direction, it seemed natural to walk a few more blocks to the library.

I was lucky to live in a time when public libraries were in abundance due to the generosity of Andrew Carnegie. Carnegie was an immigrant from Scotland with humble beginnings before earning his place as one of the richest tycoons in history. His own early education was enabled by the generosity of a wealthy man who allowed him to use his private library, since there were no public libraries widely available at that time.

Carnegie paid this debt forward by financing many free public libraries across the United States, as well as in Scotland, Ireland, Great Britain, and Canada. I am one of the millions of beneficiaries of Carnegie's generosity.

While at the Jackson Heights public library, I would wander among the book shelves to discover exciting new worlds of the imagination, described vividly in famous novels by authors like Charles Dickens, Robert Lewis Stevenson, and Mark Twain.

Reading these books triggered in me, and so many other readers, intense emotions and visual imagery of imagined worlds and adventures in them. This was my first sustained exposure to *how humans can learn and consciously understand language meanings*, by linking perceptual and affective experiences to language utterances. My recent work on neural models of how children learn and consciously experience language meanings from their parents and other teachers is one topic of this book. See Chapter 4.

3 The Jackson Heights Public Library and Music

At the Jackson Heights public library, I could also borrow records for free! I felt such a thrill bringing home music by composers whom I had never heard of before, including Bach, Scarlatti, and Vivaldi; Beethoven, Brahms, and Mozart; Chopin and Liszt; Debussy, Faure, and Ravel; Rachmaninoff, Schubert, and Schumann; Franck, Grieg, and Satie; and on and on ... such inexhaustible riches! My parents bought a simple phonograph so I could listen to all this incredible music.

Partly because I spent so much time listening to music, my parents splurged when I was in the eighth grade and bought an upright piano for me to play. I loved to lose myself in the music, and was playing works by Bach, Beethoven, Chopin, Debussy, and Gershwin, among other composers, by the end of the year. After that initial year of playing the piano a lot, I could only play it intermittently once I went to Stuyvesant High School and then on to Dartmouth College and Stanford University.

4 The Fastest Typist in New York! And Thinking about What Kind of Future I Wanted

Playing the piano helped me to acquire another skill that has been useful, indeed critically important, to me in my later life. We had typing classes in school, and my flexible, arpeggio-playing fingers learned to type one hundred words a minute. This skill comes in handy for me even now, because I can type as fast as I can think, so typing my manuscripts and books is not a challenge for me.

At that time, this skill also helped me to gain insight about what I wanted to do with the rest of my life. One summer, in order to help bring in more money to our household, I went to a typing pool in Manhattan to try to get work as a temporary typist. At that time, temporary typists would be farmed out to businesses around Manhattan that needed them. At first I was told that they already had plenty of typists, but I pressed the issue and informed them that I typed really fast, so they let me take their test.

The whirring sound of me typing from their test text was heard in the next room by the boss of the establishment, who came in to see who was making this unexpected sound. As he watched me clicking away, he said that he thought that *he* was the fastest typist in New York. Suffice it to say, I was working the next day.

One establishment that I worked for was a big export–import company. In addition to typing everything in sight, my youthful energy drove me to help bring the entire office into order. The owner of the company took notice. He did not have a son, so he offered to train me to take over his company when he retired.

That offer forced me to think about the fact that, perhaps twenty or so years after I took over, I would be rich enough to retire. But what would I then do for the rest of my life?

This thought experiment forced me to realize that I wanted to find a passion in life that would not diminish as I grew old. Luckily, I found it in science, and am still clicking away at my computer keyboard to write down my scientific discoveries at age eighty-five.

5 Jackson Heights to Manhattan: Stuyvesant High School

The rite of passage from public school to high school required that my classmates and I take another competitive city-wide exam. We needed to score a high enough grade on this exam if we wanted to attend one of the special high schools that New York City proudly offered its brightest students.

I managed in this way to attend Stuyvesant High School, which, then just as now, is one of the top academic high schools in the country. Stuyvesant was, and remains, a powerful argument for investing in public education that any meritorious student can attend.

Famous graduates of Stuyvesant include James Cagney, Thelonious Monk, Joshua Lederberg, Richard Axel, Eric Lander, Robert Fogel, Roald Hoffmann, Paul Reiser, Tim Robbins, Paul Cohen, Bertram Kostant, Lewis Mumford, David Axelrod, Eric Holder, Joseph Mankiewicz, Jerrold Nadler, and Lucy Liu. A list of famous Stuyvesant alumni can be found at https://en.wikipedia.org/wiki/List_of_Stuyvesant_High_School_people. The astonishing range of accomplishments of these graduates just goes to show that hardly anyone has a clear idea of what their life's work will be while they are going to high school.

Attending Stuyvesant required more than a good enough score on an entrance exam. One also had to be able to get there. This was not a foregone conclusion if only because I lived in the borough of Queens, and Stuyvesant is in the borough of Manhattan, which is separated from Queens by the East River. Fortunately, for both my mom's education and my own, the New York subway system, which played a big role in the growth of New York, could take me in less than one hour from a station near our apartment on 75th Street in Jackson Heights, Queens, to within a block of the old Stuyvesant building on East 15th Street off First Avenue in Manhattan.

Mentioning the East River reminds me of an only-in-New-York story about my delightful great aunt, Frieda Yankus, who lived half a block from us in Jackson Heights.

Frieda commuted by subway five days a week, for decades, to and from Manhattan for her job, where she worked as a brilliant bookkeeper who basically ran the small company that employed her. The subway is not, however, the only way to get between boroughs in New York. For example, the 59th Street Bridge also connects those boroughs over the East River.

One day, my father Sol was planning to drive our entire immediate family to Manhattan for a meal at a nice restaurant, Frieda included. My father casually said that we would take the 59th Street Bridge over the East River. Frieda was dumbfounded and asked: "What river?" Having taken only the subway between Queens and Manhattan for her entire life, she had no idea that her thousands of subway trips went through tunnels under the East River!

Frieda's astonishment should give everyone pause who studies how humans form cognitive maps of their world. Even a whip-smart, life-long New Yorker like Frieda had no idea that New York boroughs were separated by water.

6 Taking the Subway to New York's Great Art Museums

Once I got in the habit of taking the subway to go to Manhattan for school, I could easily travel to all of Manhattan's wonderful art museums and theaters, notably the Museum of Modern Art, the Metropolitan Museum of Art, and the theaters where world-class ballet and opera were being performed.

Ballet and opera performances by the New York City Ballet and the New York City Opera were first offered in the New York City Center on West 55th Street, before both moved to the newly built and visually stunning Lincoln Center for the Performing Arts a little further uptown on West 65th Street. The inaugural concert of the New York Philharmonic was performed in Lincoln Center in 1962, conducted by Leonard Bernstein. I could easily travel to both venues for ballet and opera performances by taking the subway from Jackson Heights.

Just as Andrew Carnegie funded free libraries all over America, another wealthy patron, Abbey Aldrich Rockefeller, was a prime mover in getting the Museum of Modern Art (MOMA) built and funded. The Metropolitan Museum of Art was initiated by another wealthy person, John Jay, a lawyer, diplomat, and grandson of a Chief Justice of the United States Supreme Court.

The old MOMA building, on West 53rd Street, right off Fifth Avenue, was also easy to reach by subway from Jackson Heights. I can still remember the thrill I felt when I first walked into the large room filled with masterpieces of Impressionist and post-Impressionist painting, including some of the most famous and beautiful paintings by Cézanne, Degas, Gauguin, Monet, Renoir, Seurat, and van Gogh.

Since I was a little boy, I always loved to draw and paint. I remember my grandfather bending down to look over my shoulder when I was less than five years old to see the picture that I was drawing as I lay on my belly on the floor. I know that I was less than five, because that floor was in Woodside, Queens, where we lived before we moved to Jackson Heights when I was five.

My love for drawing and painting has continued throughout my life. In fact, when I was in high school, I won a city-wide competition that let me take a class held at MOMA with art students from all over New York. These students really knew what they were doing. I did not.

I learned about this opportunity just the day before students were supposed to present their portfolios to the people judging who could take the class. I worked feverishly late into the night and did as much drawing and painting that I could in such a short time. When I showed the results to the people judging student work, I apologized and explained that I had to create the entire portfolio in less than a day. My work was clearly not at the level of some of my classmates, but my passionate motivation and latent talent won me a spot in the class, where I learned a lot and met some other students whose love of art resonated with my own.

These early beginnings looking at and creating art are one foundation for my wanting to write this book.

7 The Subway: Gateway to Ballet, Opera, and Musicals

My early experiences listening to and playing music, and viewing and creating visual art, led me, while I was still in high school, to take the subway to watch ballet performances at the New York City Center. I always felt that ballet combines delights for both the eyes and the ears as few other art forms do. To make things even more exciting, George Balanchine was then in his prime as the artistic director and choreographer of the New York City Ballet.

I did not then realize how extraordinary, and historic, it was to have the privilege of watching the premiers of often more than one brilliant new ballet by Balanchine in a single evening, danced by some of the greatest ballet dancers ever to live, who Balanchine personally trained. Some of the wonderful dancers whom I remember from those early years include Maria Tallchief, Gelsey Kirkland, Patricia McBride, Melissa Hayden, Violette Verdy, Edward Villela, Arthur Mitchell, Jacques d'Amboise, and Andre Eglevsky. Even from the cheap seats at the back of the auditorium, these performances were thrilling to experience.

Other musical experiences, such as performances by the New York City Opera at the New York City Center, were also affordable for me, years before I started to attend performances at the pricier Metropolitan Opera and Broadway musicals. All of these experiences have combined to make me want to understand how humans can learn, consciously perform, and have deep feelings about musical lyrics and melodies with variable rhythms and beats, another topic of this book. See Chapter 3.

8 Art Collecting: "Don't buy it if you don't need it"

Although I did not have time to create my own art once I left for college, I very much wanted, and needed, to live with art. I therefore began collecting art when I had enough money to do so, starting small as a graduate student at the Rockefeller Institute for Medical Research in Manhattan between 1964 to 1967, when I bought my first mask, which was made in Haiti. One of my passions later in life was to enjoy and, when I could afford, to buy plein air paintings by artists who work mostly in Provincetown, Massachusetts.

Provincetown has an honored place in the history of American plein air painting, which began under the leadership of Charles Hawthorne in 1899. Hawthorne, who founded the Cape Cod School of Art, held painting classes on the beach in Provincetown (Figure 1.2), Massachusetts, to capture the extraordinary light there, which is due to the fact that Provincetown is surrounded by water on three sides.

Our family managed to buy a beach house in Truro, Massachusetts. Truro is the town just south of Provincetown, around a twenty-minute car ride away. Our house contains some beautiful plein air paintings by Cape Cod artists that capture the wonderful light here at different times of the day. The painting by John Dowd called *West End Twilight* is illustrative (Figure 1.3). John's work is often compared with that of the great American painter Edward Hopper,

FIGURE 1.2 Charles Hawthorne teaching painting in plein air in 1899 on the beach in Provincetown, Massachusetts. Courtesy of the Provincetown Art Association and Museum.

FIGURE 1.3 A painting by John Dowd, called *West End Twilight*, from my personal collection. John is a highly respected painter in Provincetown whose work has been compared with that of Edward Hopper.

who lived between 1882 and 1967. Although Hopper mostly lived in New York City, he and his wife Jo spent forty summers in a house just a short walk down the beach from our house in Truro. Hopper created many beautiful paintings there of scenes around Truro.

I also love the many statues, masks, and shields of African, Asian, and New Guinea art with which my family and I have happily lived in our primary home in Newton Highlands, Massachusetts, with a few pieces sprinkled around our house in Truro as well.

I unexpectedly started to collect African art many years ago when I walked past a little sliver of a store front between 68th and 69th Street on Eighth Avenue in Manhattan. The store front was packed with African masks and statues to which I had never given much of a thought before then. These pieces called out to me with an intensity that drove me to enter the store.

I was later told that one goal of African artists is to "release the spirit" from the wood that they are carving. On that particular day more than fifty years ago, and every day since, the spirits have been calling out to me, loud and clear.

The store was owned by Abe and Anna King, who were already in their eighties when I met them. Although Abe was diminutive and stooped over in his straight-back chair when we met, he had run a marathon from the Northern tip of Manhattan to its Southern tip when he was a young man.

How Abe began buying and selling African art is a story in itself. He earlier made a living by selling automobile tires for the fancy cars owned by famous actors and actresses working in the legitimate theaters that were then on 42nd Street. These performers often lived in apartment hotels on the upper West Side during long runs of their plays. For reasons that he never explained to me, Abe got advice from his theatrical clients to sell African art after the legitimate theaters on 42nd Street slid into sleazier days.

I thought that the art in Abe's window was wonderful, and probably too expensive for me to afford. As it turned out, the art was affordable, even for me, because, although beautiful, it did not have the provenance that would send the prices through the roof.

I have always found it odd that two essentially identical pieces of African art, one that was used in a village ceremony on a certified date, and the other not so used, can differ in price by several orders of magnitude.

This price difference is, at least to my mind, quite different than the price difference between a painting by, say, Rembrandt, and a copy of that painting. The unused piece of African art is not a copy. It is an original work of art because many African carvers will carve multiple examples of a traditional type of mask or statue throughout the year. Only a few of these originals can be used in ceremonies.

Abe taught me something that I still think of whenever I consider buying a work of art. While I was deciding whether or not to buy a statue or a mask, he liked to say: "Mr. Grossberg, don't buy it if you don't need it." I suspect that all passionate collectors of art know what Abe meant by "needing" to live with a work of art!

Through Abe's shop, I began to meet African art traders, who used to bring hundreds of pieces of African art out of their villages to the now dilapidated apartment hotels that Abe's theatrical customers stayed in years before when he sold them tires. When the legitimate theaters left 42nd Street, the performers who lived on the upper West Side did too, so the hotels where they stayed lost their high-paying customers. An African art trader could then rent a lovely apartment, with full living room, dining room, two bedrooms, and one or two bathrooms for around $50 a week.

I learned to make a deal by buying art from these African traders. That knowledge has come in handy multiple times during my leadership experiences as a department Chairman, research center Director, journal Editor-in-Chief, research society President, conference organizer and chairman, and other academic leadership roles over the years.

Each African trader might have a few great pieces within the large pile of pieces on the living room floor in their rented apartment. You could only buy a great piece if you also bought quite a few lesser pieces, so that they could sell all the pieces that they had before returning home. Choosing an acceptable group of pieces, and arriving at a mutually satisfactory price for them, required a lot of decision-making and bargaining. It also forced me to ask questions about art that are just as important to ask about romantic relationships:

"Is this true love or just an infatuation?"
"Will I still want you when I wake up with you tomorrow morning?"

Buying art that one loves really helps to get in touch with one's feelings, or at least it did so for me.

The first thing that I learned about how to make a deal was from a serious African art collector named Bruce who I met at Abe's store. Bruce introduced me to some of the African traders who he already knew. Before we met the first trader, Bruce told me to be ready to walk out if a trader was not acting in good faith, even if I was privately dying to buy one or more of his pieces. I actually got up my nerve to do that once, but was called back before I was out of the door.

Suffice it to say that, after getting some experience making deals with several traders, when I knocked on a hotel room door where a trader whom I knew was staying, I was greeted with a big smile and the salutation: "Hello, Brother Steve." Both my wife, Gail Carpenter, and I were invited to visit these traders at their homes in Africa whenever we wanted. You can imagine from this anecdote that I bought a lot of pieces from them that I "needed"!

I still find the variety of images and techniques used to create African art breathtaking. In Figure 1.4, the left panel shows a wooden mask that depicts the serene face of a Baule queen, while the right panel shows a wooden Ngbaka mask with a wonderful yin-yang broken symmetry

FIGURE 1.4 Two African masks, one of a Baule queen (left) and another from the Ngbaka tribe (right) from my personal collection. The carvers of even great African masks are usually unknown.

FIGURE 1.5 A beaded royal throne made by the Bamileke tribe from my personal collection. As with the African masks, the creator of this throne is unknown.

FIGURE 1.6 A bronze head of a Benin queen, or Oni (left), and a bronze statue of a seated Baule king with attendants, both from my personal collection.

in its black-and-white surfaces. The Baule tribe descends from the Akan peoples who lived in Ghana and the Ivory Coast. The Baule tribe now lives in the center of the Ivory Coast. The Ngbaka tribe lives in Northern Democratic Republic of the Congo.

Figure 1.5 shows a royal Bamileke throne. The cowry shells and beads that cover the throne were used to make royal art, because the king or tribal chief often hoarded all cowry shells and beads, and was one of the only people in the tribe wealthy enough to commission major art pieces. The shells and beads were laboriously attached to burlap, which was first wrapped around the wooden sculpture underneath it.

Other African art pieces were rendered in Bronze, such as the Benin head of an Oni, or Congolese queen, as shown in Figure 1.6, left panel, and in Figure 1.6, right panel, a seated Baule king with attendants. This particular head of an Oni dates from the nineteenth century. The most famous Benin bronzes were made in the thirteenth century and thereafter. They were stolen by force from the palace of the Oba, or king, of Benin in a raid by the British in 1897. There is now an active movement to have all the Benin bronzes returned to Africa. The seated Baule king and attendants date from the eighteenth century. As noted above, the Baule tribe lives in the Ivory Coast, whereas the Benin tribe lives East of that in what is now part of Nigeria.

Having seen examples of African art rendered in wood, beads, cowry shells, and bronze, let me end with the example of the ancient terracotta Nok figure in Figure 1.7

FIGURE 1.7 A terracotta figure from the Nok tribe in Africa from my personal collection.

that is estimated to have been made between 500 BC and 200 AD. Historians and archaeologists refer to this piece as an example of Nok art because these artifacts were first discovered near the modern Nigerian town of Nok.

How I began to collect New Guinea art was no less accidental. It began when my wife Gail, our young daughter Deb, and I vacationed on the Great Barrier Reef after Gail and I both gave invited lectures at the University of Adelaide in Adelaide, Australia. After completing our lectures, we first visited Kangaroo Island, which is just South of Adelaide, where we saw an amazing diversity of life, ranging from kangaroos, camels, and koalas to penguins, seals, and emus. I still vividly remember what felt like a personal encounter with an emu on Kangaroo Island, where we stood staring at each other at close quarters for quite a long time.

Gail, Deb, and I then visited Green Island in the famous Great Barrier Reef for a final holiday before returning home. As things turned out, there was nothing much that we could do at the resort on Green Island where we stayed, since none of us was an experienced snorkeler, so when we heard that there was a crocodile farm nearby, we figured that Deb at least might enjoy seeing that.

Little did we suspect that the crocodile farm was owned by George Craig, who went up river in New Guinea after World War II to hunt crocodiles, but also found lots of magnificent art in the villages there. George managed to collect perhaps the world's greatest collection of New Guinea art during these trips, and sprinkled it around his crocodile farm.

When I saw this amazing art, I did what I always do: I asked if it was for sale. Soon George came out to meet the unexpected visitor who was interested more in his art than his crocs. We all got on very well, and George allowed me to buy some wonderful pieces for very little money, shortly before he became famous as the world's greatest collector of New Guinea art, and prices for his collection started to increase accordingly. Here is the URL of a story about George from the web page of Michael Hamson, a well-known contemporary dealer in New Guinea art: https://oceanicart.com/PROVENANCE/George-Craig/1.

One of the most important kinds of New Guinea art is war shields that were used quite often between neighboring tribes who regularly went to war with each other. The risks and costs of these wars can be imagined from the fact that, not so long ago, members of some of these tribes were cannibals and headhunters. New Guinea shields embodied the kind of creative fervor in these tribes that went into the masks and statues that were created in Africa, and included symbolism that was intended to both frighten enemies and to symbolize ancestor spirits who were thought to be protective during battle.

Figure 1.8 shows three of the beautiful shields that I got from another important collector of New Guinea art, Todd Barlin, who currently lives in Melbourne, Australia, where he is the Director of Oceanic Arts Australia https://www.oceanicartsaustralia.com/about/. Suffice it to say, the shields that I was lucky enough to buy from Todd were no longer inexpensive! One reason for this, apart from their beauty and rarity, is that these shields have unimpeachable provenance.

FIGURE 1.8 Three New Guinea shields from my personal collection. See the text for details.

For example, the shield on the left of Figure 1.8 comes from the Wogamush River in the April River Area of the Upper Sepik River in East Sepik Province, Papua, New Guinea. It was field collected by Dr. Fred Gerrits in the 1960s before it became the second shield to be purchased by Todd Barlin. Fred Gerrits was born in 1933 in Bandung, Indonesia. After graduating with a degree in medicine in Holland, he settled in Papua in the 1950s, where he met his future wife Nel. As of 1964, the Gerrits worked in various hospitals, until Fred was appointed as the Tuberculosis and Leprosy Control Officer for the provinces of West and East Sepik, with a base in Maprik at the foot of the Prince Alexander Mountains.

The shield in the middle of Figure 1.8 was collected on the Frieda River by the Australian crocodile hunter, John Pasquarelli, in 1965. Pasquarelli was collecting artifacts for The Museum der Kulturen in Basel, Switzerland. He kept this shield for himself because he knew it was an important piece, more rare and important than the one he sold to Sotheby's, New York, in 2004, which sold for a lot of money at auction.

The shield on the right in Figure 1.8 comes from the April River region in East Sepik Province of Papua. It was made in the early twentieth century and collected by the Australian artist Wallace Thornton when he was working in a studio at Elizabeth Bay House in the 1930s.

Experiences with art and music like the above made me want to scientifically understand the creative processes that enable human brains to create and appreciate art and music. Although I began modeling how brains make minds in 1957, it has taken me several decades to accumulate enough insights to offer a scientific analysis of the processes that are needed to consciously perceive, learn, recognize, have feelings about, and create aspects of art and music.

9 From Stuyvesant to Dartmouth College: Al Hastorf and John Kemeny

The above experiences occurred when I was in college, graduate school, or during the years after that when I was a professor at MIT or BU. Stuyvesant High School gave me a wonderful preparation for what was to come.

Stuyvesant was even more competitive than PS 69. One reason was that lots of the students were Jewish kids who needed scholarships to go to the Ivy League schools that we were all longing to attend, such as Dartmouth, Harvard, Princeton, and Yale. Even before we could hope to get a scholarship, we had to score high enough in our school and city-wide exams not to be struck down by the "Jewish quota" that was quietly observed by all of these colleges and universities. As I recall, there was a quietly shared understanding among these prestigious schools that not more than 2% of their entering classes were allowed to be Jewish students, no matter how smart and accomplished we were.

Again I worked like a dog in Stuyvesant and graduated in 1957, ranked first in my class and the class valedictorian. This was my ticket to the Ivy League!

I should not overlook the fact that City College of New York famously let anyone attend for free and provided a world-class education. You were free to attend, and just as free to flunk out. Some of the most famous Americans were launched by a superb education at CCNY, including quite a few Nobel Laureates. Its reputation was such that it was sometimes called the "Harvard of the proletariat."

When I was in Stuyvesant, I had never heard of CCNY. And although I love New York, and have regularly returned to it to "recharge my batteries" over the years, I then needed to get away from home to begin to learn how to live independently. I chose Dartmouth partly because Dartmouth offered me the largest scholarship, among all the Ivy League schools to which I applied, and used it to supplement the National Science Foundation scholarship that I also won.

A generation of lucky students like me could try to win these national scholarships because Sputnik 1 was launched by the Soviet Union in October 1957, just as I was graduating from high school: https://en.wikipedia.org/wiki/Sputnik_1. As a result, the US government started to pour money into scholarships and other programs aimed at rapidly training a new generation of American scientists and engineers who could out-compete the Russians.

I also wanted to go to Dartmouth because I noticed that it had a Senior Fellowship Program. If a student was lucky enough to become a Senior Fellow, then he could spend his senior year doing research, without the need to take regular courses.

I knew by the time I left Stuyvesant that I could not endure much longer the pressure of working relentlessly in my courses to earn grades high enough to stay first in my class. This pressure did not let up in Dartmouth, because I needed to do well to keep the scholarships that I had already won.

I very much wanted a year off this "rat race" before I left college, so that I would have a chance to figure out what I wanted to do with the rest of my life. I had no idea what "research" entailed but, as it turned out, the kind of mind-brain research that I unexpectedly started to do as a seventeen-year-old Dartmouth freshman was also what I worked on as a Senior Fellow, and still pursue with passion and love even today.

There was also a Dark Side to attending Dartmouth, one which I luckily was able to avoid. At that time, fraternities were the center of the College's social life. The boys at these fraternities often drank a lot, drove recklessly at night, while drunk, over winding country roads to visit nearby girls' schools, and sometimes behaved pretty outrageously toward the girls whom they invited to their fraternity parties. I heard about one such party where the boys all simultaneously "dropped trou" to embarrass the girls whom they invited. These misogynistic parties, notably at the Alpha Delta Phi fraternity, were one inspiration for National Lampoon's 1978 movie "Animal House."

I luckily found a few fellow students in my class at Dartmouth with whom I shared a lot in common. They are my dear friends to this day. I also had a steady girlfriend for several years at Dartmouth, who also lived in Queens, New York.

Dartmouth had two distinguished professors who changed the course of my life. My work may never have gotten off the ground as a Dartmouth undergraduate were it not for Albert Hastorf, who was then chairman of the psychology department (Figure 1.9, left panel), and John Kemeny, who was then chairman of the mathematics department (Figure 1.9, right panel). Al and John created the academic infrastructure whereby I was able to become the first joint major in mathematics and psychology at Dartmouth.

John was a remarkable man in many ways. As a Princeton graduate student in mathematics, he was Albert Einstein's assistant. In this capacity, he helped Einstein with the mathematics that was needed to develop Einstein's Unified Field Theory. Among his many other accomplishments, John co-invented the BASIC programming language, as well as one of the world's first computer time-sharing systems. He did both of these projects with Thomas Kurtz, another mathematics professor at

FIGURE 1.9 Two of my mentors at Dartmouth College, Albert Hastorf (left) and John Kemeny (right). Both pictures were taken from the web: https://www.aip.org/library/ex-libris-universum/the-emilio-segre-collection, AIP Emilio Segrè Visual Archives, Physics Today Collection.

Dartmouth. John strongly believed that students who were gifted in mathematics should go into the social sciences to develop mathematical models for them.

In one of the courses that I took from John, we read his just-published book, an *Introduction to Finite Mathematics*, which includes applications to the social sciences. Although the book focused on discrete mathematical models in the social sciences, it inspired me to continue doing work on the continuous neural network models that I had just discovered.

I took another course from John as a sophomore that read through one of his other just-published books, *A Philosopher Looks at Science*. This book had a profound effect on me. It is still in print on Amazon with a five-star rating: https://www.amazon.com/Philosopher-Looks-Science-John-Kemeny/dp/0442043244.

In this course, I learned from John that science is a highly personal, creative endeavor, always evolving and never finished, and not just something to learn in a cookbook fashion as revealed truth. Part of my experience in the course was to write essays about topics related to John's book. After the class was over, John, like all other Dartmouth faculty, wrote "commendations" for those students who they thought had done exceptional work. John wrote that I had written the best essays that he had ever read. Although I was told that I was awarded more commendations in my courses than any other Dartmouth student ever had, John's high opinion in his commendation helped me to muster the courage to continue my work.

After taking John's course, I read as many books as I could get my hands on about the history of physics so that I could better appreciate that "settled" science is also an evolving, and highly personal, human endeavor. Knowing some of the ups and downs of famous scientists of the past, some of whom even felt driven to commit suicide, helped to prepare me for the many challenges to my own work that I intermittently experienced for the next twenty years.

I was lucky to be studying with John Kemeny then, just as I was lucky to be able to see new George Balanchine ballets at New York City Center at around the same time. John Kemeny went on to become President of Dartmouth, among his many other accomplishments: https://en.wikipedia.org/wiki/John_G._Kemeny. It was in 1972, when John was President, that he announced Dartmouth would admit women.

Al Hastorf was not an intellectual powerhouse like John Kemeny, although he made valuable contributions to social psychology, notably social influences on perception. Al had a different kind of genius, especially as a scientific advisor and leader. In 1961, the year that I graduated from Dartmouth, Al moved to Stanford University where he became a much-loved chairman, dean, vice president, and provost.

Becoming a joint major in mathematics and psychology made sense for me because, as I was taking introductory psychology as a Dartmouth freshman, I became fascinated by how humans and animals learn. I was particularly interested in how humans learn sequences of items, whether they are sequences of nonsense syllables, words, objects, navigational turns, or skilled arm movements.

It was the study of how humans learn lists of things that led me in 1957 to discover and to rapidly develop the paradigm of using neural networks that are defined by systems of nonlinear differential equations to model how brains make minds, including the main mathematical laws for neuronal activation, or STM traces; activity-driven habituation, or MTM traces; and adaptive weights for learning and memory, or LTM traces, that are still used today.

Although I began by modeling how our brains learn sequences of items, this early study rapidly blossomed into a wide-ranging series of discoveries about essentially all the foundational brain processes whereby our brains make our conscious and unconscious minds, in both healthy individuals and clinical patients.

As I earlier noted, it is because I discovered the neural network paradigm and its foundational equations in 1957 that I am often called the Father of AI. Also as I noted above, I am called the Einstein of the Mind because, after achieving a deep understanding of the principles and mechanisms whereby I could model psychological and neurobiological data about a given faculty of biological intelligence, I could rederive these models from thought experiments, just as Einstein did for his Special Theory of Relativity and his General Theory of Relativity. Because of the important role of thought experiments in my own work, John Kemeny's personal connection to both Einstein and me has always felt particularly precious.

10 A Transformational Year: Supplanting Fear with Love

Strong support by Hastorf and Kemeny enabled me to become a Senior Fellow in my senior year at Dartmouth and to devote that year to doing research to further develop the neural network models that I discovered in my freshman year, and that formed the conceptual and mathematical foundation for my life's work.

This was, in more than one way, a transformational year for me. When I no longer had to work for a grade, I learned how to work for love, and would from that time on follow my intuitions and passions to determine what I wanted to understand scientifically. As my later scientific work has explained, positive motivations like love are much better at sustaining creative acts than fear. Chapters 13 and 17 of my Magnum Opus explain why (Grossberg, 2021).

11 From Dartmouth to Stanford: Statistical Learning versus Neural Networks

Recommendations from professors like Hastorf and Kemeny, coupled with my high grades and perfect 800s on each of my national College Board exams, won me a National Science Foundation graduate fellowship and acceptances to three graduate departments at Stanford University. I particularly wanted to get my PhD at Stanford because it was one of the two places in the United States that had a strong program in mathematical psychology. The program at Stanford was housed in the famous Institute for Mathematical Studies in the Social Sciences. Several distinguished professors were working on learning theory at the Institute, including Bill Estes, Dick Atkinson, Gordon Bower, and Pat Suppes.

At the time I met Dick Atkinson, or more formally, Richard C. Atkinson, he was a young man—just ten years older than me—and movie-star handsome. Dick was well-known for being a ladies' man as well as a serious modeler. He went on to have a distinguished career in academic administration, including as the Chancellor of the University of California at San Diego, the President of the University of California system, and the Director of the National Science Foundation. But on the day that I walked into Dick's office, before any of those things happened, the first thing he said to me was: "So you're the guy who Kemeny said is going to save mathematical psychology." I will summarize below why Kemeny thought mathematical psychology needed saving.

Bill Estes, or more formally, William K. Estes, was regarded by many as the father of mathematical psychology. Bill began as an excellent psychological experimentalist, having published a seminal article in 1941 with B. F. Skinner on fear conditioning. This work was part of a time-honored tradition of *associative learning* by many great psychologists (Figure 1.10), including Ivan Pavlov (Pavlov, 1927), Donald Hebb (Hebb, 1949), Clark Hull (Hull, 1943), Edward Thorndike (Thorndike, 1927), B. F. Skinner (Skinner, 1938), Edward Tolman (Tolman, 1948), and Bill Estes (Estes, 1950).

All of the above psychologists were American, with the exception of Pavlov, who was Russian. Learning theory had become a primarily American research topic. Many of the greatest pioneering contributors to visual perception were European, including Hermann von Helmholtz and Ernst Mach in Germany, and James Clerk Maxwell in Scotland (Figure 1.11). All of these European scientists were famous physicists as well as students of perception.

My neural models over the years have provided principled and unifying explanations and quantitative computer simulations of the types of psychological data that each of these pioneers discovered. Many of these explanations are summarized in a self-contained and non-technical way in my Magnum Opus.

Bill realized that just adding variations of experiments, all probing the same learning process, to the already huge literature in experimental psychology needed to be supplemented by some kind of unifying theory. Bill was not, however, trained in theory or mathematics. He did the best that he could by using models of the type that John Kemeny and Laurie Snell wrote about in their 1960 classic book *Finite Markov Chains*.

Bill used finite Markov chains to develop what he called *stimulus sampling theory*. In a typical version of his model, there were two urns, each containing abstract elements. The elements in one urn were "learned" and the elements in the other urn were "unlearned." The process of learning in the model was described by the probability that elements in the unlearned urn moved to the learned urn on a given learning trial. The process of forgetting consisted of the probability that elements in the learned urn moved to the unlearned urn. "Stimulus sampling" was the abstract operation that initiated movement of the elements between the urns.

Variations of this simple model, whose states and probabilities were estimated to fit the conditions of different experiments, provided a useful theoretical unification of the data from a lot of statistical learning experiments.

The problems with the model included the facts that: It was just an abstract Markov Chain that learned in discrete time steps. It provided no explanation of how brains work,

FIGURE 1.10 Pictures of seven leading psychologists who advanced our understanding of how humans and animals learn. See the text for details. Pictures taken from the web: Ivan Pavlov: https://www.britannica.com/biography/Ivan-Pavlov; Donald Hebb: https://psicologiaymente.com/biografias/donald-hebb; Clark Hull: https://www.verywellmind.com/clark-hull-biography-1884-1952-2795504; Edward Thorndike: https://www.verywellmind.com/clark-hull-biography-1884-1952-2795504; B. F. Skinner: https://geniuses.club/genius/burrhus-frederic-skinner; Edward Tolman: https://donaldclarkplanb.blogspot.com/2021/07/edward-tolman-1886-1959-latent-learning.html; William Estes: https://webapp1.dlib.indiana.edu/archivesphotos/results/item.do?itemId=P0033403 Image courtesy: Indiana University Archives.

or learn in continuous time, let alone the different kinds of learning that occur in different parts of the brain, such as perceptual learning, recognition learning, and reinforcement learning. Also, being a statistical learning model, it described probabilities of learning by groups of subjects, not how individuals learn.

One link of Bill's model with brain dynamics was the concept of *stimulus sampling*. In my neural network models, as in our brains (Figure 1.12), one cell body population "samples" another when its activity, or cell body potential, gets large enough to send signals down its axons to the synaptic knobs where adaptive weights can "sample," and thereby learn about, the activities of the postsynaptic cells that the knobs abut. This learning process alters the strength with which future signals down the axons will release a chemical transmitter that activates the postsynaptic cells.

There was another link between Bill's model and my own: Although my models deterministically simulated individual cells and cell populations, their dynamics also had a statistical interpretation. That is because I interpreted the activities, or potentials, of my cell populations to be the *statistical means*, or averages, of the activities of many activated cell sites in the population. These excitable sites could be in an individual cell, in multiple cells in a population, or both. Interpreting cell activities as the means of a deterministic model was possible under the assumption that the variances of cell site fluctuations were small compared with their means.

I failed to arouse the interest of any of the senior faculty in mathematical psychology at Stanford in my work. My work was closest to that of Bill because we both used a concept of "stimulus sampling." But Bill had his hands full trying to explain data with his statistical model, and did not have the training to appreciate the learning dynamics of my nonlinear neural networks. A difference in personalities also contributed to my failure: I was very shy and not experienced at promoting my ideas. And Bill was famously

FIGURE 1.11 Pictures of three great scientists who worked at the end of the nineteenth century in both physics and psychology. See the text for details. Pictures taken from the web: Hermann von Helmholtz: https://en.wikipedia.org/wiki/Hermann_von_Helmholtz; James Clerk Maxwell: https://www.britannica.com/biography/James-Clerk-Maxwell; Ernst Mach: https://archiv.radio.cz/en/static/inventors/mach

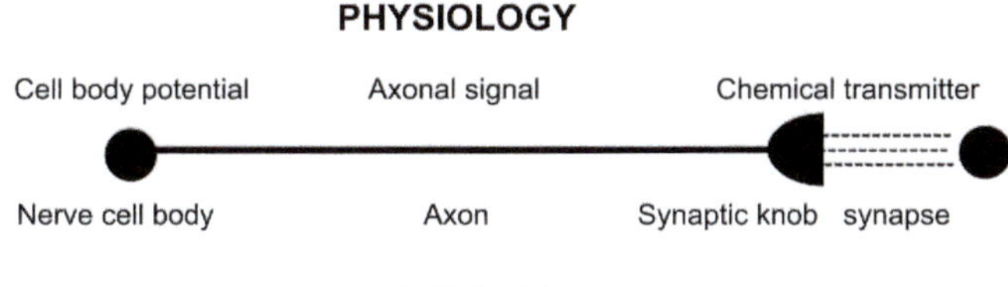

FIGURE 1.12 A schematic of how signals from a neuronal cell body travel along the cell's axon until they reach its synapse, at which time they release a chemical transmitter into the synaptic cleft that alters the activity of a post-synaptic cell. Author created.

reticent. When I went to his office to try to interest him in what I was doing, he would sit, Sphinx-like, not saying a word for what felt like an eternity. I felt like whatever words I could muster were sucked into the Black Hole of his silence.

Bill's reticence in private was hard for me to understand because, when Bill gave a lecture, he was fluent and charming. This side of Bill was most evident to me many years later after Bill left Stanford to become a professor at Harvard, while I was nearby as a professor at Boston University. Bill and his socially delightful and generous wife, Kay, threw wonderful Christmas parties in their home at 95 Irving Street near the Harvard campus in the house that William James had earlier built and lived in for many years (Figure 1.13). William James is sometimes called the Father of American Psychology to acknowledge his foundational contributions to the field, including his seminal book *Principles of Psychology* that was published in 1890. James also let Edward Thorndike (see Figure 1.10), when he was a Harvard graduate student, use the basement of his house to raise the chickens that Thorndike used to study animal intelligence. See https://historycambridge.org/james/James%201.html for more details about this.

Given this history of having the Father of Psychology, and the Father of Mathematical Psychology, both having lived in the same house made me feel that the house was a kind of shrine whenever I went there.

Other works of William James also influenced my own thinking, including the book of his Gifford lectures on *The Varieties of Religious Experience* that was published in 1902. I was thinking of these lectures when I called my major 2017 article about consciousness *Towards Solving the Hard Problem of Consciousness: The Varieties of Brain Resonances and the Conscious Experiences that They Support.*

See https://www.sciencedirect.com/science/article/pii/S0893608016301800.

My concept of stimulus sampling by cells in neural networks was needed to explain that the functional units of learning are *distributed patterns*, or *vectors*, of adaptive weights, or LTM traces, across a *network*, not individual traces between a particular pair of *cells*, as Figure 1.12 might suggest. Indeed, my explanation in 1957 of data about serial verbal learning depended greatly on the dynamics of distributed patterns.

In particular, distributed patterns of cell activation, or STM traces, influenced the learning of distributed patterns of LTM traces across the network. I was driven to consider

FIGURE 1.13 William James and his home in Cambridge, Massachusetts. Pictures taken from the web and assembled by the author: William James home: https://historycambridge.org/james/James%201.html. Picture of William James: https://historycambridge.org/james/James%201.html. Credit: History Cambridge.

pattern learning by properties of classical learning data like the bowed serial position curve (Figure 1.14). This famous curve illustrates that, in many situations where we learn a list of items that occur one after another through time, we can remember the items near the beginning and the end of the list better than items in the middle. Said more colloquially: We often remember how a series of events began and ended, but the middle is a muddle.

Stimulus sampling of distributed STM patterns was needed to explain the bowed serial position curve, because each of the cell populations that was activated by serial presentation of a list (Figure 1.15) could only learn associations with other active cells during the time interval when it was active enough to "sample" the other cells' activities. There are typically more cells that are simultaneously active when the middle of the list is being presented, thereby causing more response interference to learning there. That is why the number of errors is greatest near the middle of the list in Figure 1.14.

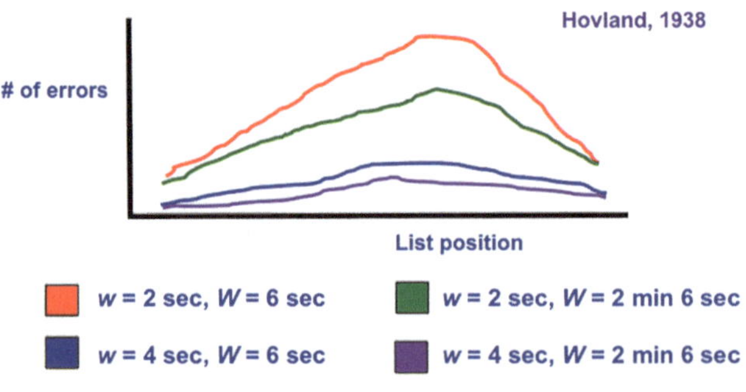

FIGURE 1.14 The bowed serial position curve that describes the errors humans make when learning a list via serial verbal learning. Author created.

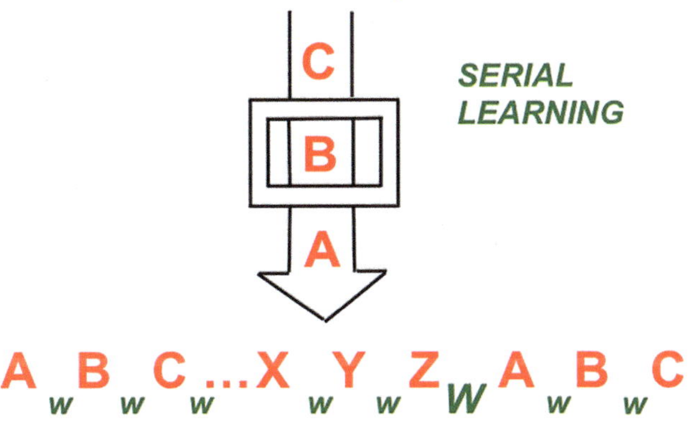

FIGURE 1.15 A schematic of the serial verbal learning paradigm. See the text for details. Author created.

My Magnum Opus (Grossberg, 2021) explains how this happens in greater detail, as well as many other serial learning properties, like backward learning, and the distribution of anticipatory and perseverative errors at different list positions. The combination of all these properties forced me into realizing that a neural *network* that can learn associations between any pair of items in a list, such as from A to B *and* from B to A, was required to explain verbal learning data.

My efforts to explain how patterns of LTM traces learn from patterns of STM traces through time also forced me into a particular associative learning law that I call the *gated steepest descent law*. In particular, in order for a *pattern* of LTM traces to *match*, through learning, a *pattern* of STM traces, some of the LTM traces may need to *increase* to match large STM activities, while other LTM traces need to *decrease* to match small STM activities in a given STM pattern. Thus, my LTM laws combined what some scientists would later call Hebbian (increasing LTM strength) *and* anti-Hebbian (decreasing LTM strength) properties. In summary, gated steepest descent learning enables an LTM trace to increase (Hebbian) as it learns to match a large STM activity, and to decrease (anti-Hebbian) as it learns to match a small STM activity.

My learning law contradicted the famous Hebb Postulate that practice always strengthens associations (Figure 1.16). Donald Hebb (Figure 1.10) introduced this learning law in his famous 1949 book titled *The Organization of Behavior*. Here, Hebb wrote that "when the axon of cell A is near enough to excite a cell B ... A's efficacy, as one of the cells firing B is increased" (Hebb, 1949). The Hebb Postulate is wrong, and it is wrong for a

FIGURE 1.16 The Hebbian learning laws proposed by Donald Hebb. Author created.

basic reason: Hebb chose the wrong computational unit. The unit is not a single pathway, as in Figure 1.16. Instead, the unit of LTM is a *pattern* of LTM traces that is distributed across a network. When one needs to match an LTM pattern to an STM pattern, then both increases *and* decreases of LTM strength are needed.

The Hebb Postulate also fails for a reason that does not depend upon this insight: If an LTM trace could only increase, and if all LTM traces have a finite upper bound, as do all processes in biology, then there could come a time when an LTM trace reached its maximum value. After that time, further learning with that LTM trace would be impossible, even if the world changed. We would then be stuck with learning that ended years ago, and could not adapt to novel future environments. My learning law enables us to learn quickly throughout life without experiencing catastrophic forgetting. It solves what I called the *stability-plasticity dilemma*, as I note below in my discussion of Adaptive Resonance Theory.

12 From Stanford to Rockefeller

I left Stanford with an MA in mathematics after taking 90 credits of graduate mathematics. I did not try to earn my PhD at Stanford because all of the graduate faculty whom I met were developing well-established branches of mathematics, such as functional analysis, harmonic analysis, and probability theory, whereas I wanted to analyze the mathematics of my neural networks, which they knew nothing about. It was one of several times in my life when people who could control my future asked: "What happened to Grossberg? He was so promising."

I did have a wonderful advisor at Stanford, Professor Karel deLeeuw, who could not help me in my quest, but who was a comfort to know. Karel's door was always open to all students who sought him out. Tragically, his door was also open to a troubled graduate student, Theodore Steleski, who murdered him from behind in 1978, when Karel was forty-eight, as he sat at his office desk. This was not the only time that a person who was kind to a mentally ill person entered their twisted fantasies, only to be murdered, or otherwise grievously harmed, by them.

With no future for me there, I left Stanford in 1964 to finish my PhD, which I hoped to do at MIT. I was also attracted to the Rockefeller University, then the Rockefeller Institute for Medical Research, because it had some faculty who might be sympathetic to my academic goals. When I realized that my unhappiness at Stanford was not going to abate, I got up my courage to write a letter to Rockefeller asking about possible opportunities for graduate studies there. To my delight, the powers that be checked me out at Dartmouth and Stanford, and I got a letter inviting me to visit there, all expenses paid, for an interview.

Rockefeller has had a huge impact on medical science. Rockefeller was America's first biomedical institute, and was founded in 1901 shortly after France's Pasteur Institute, which was founded in 1888, and Germany's Robert Koch Institute, which was founded in 1891. Rockefeller was founded as The Rockefeller Institute for Medical Research by John D. Rockefeller, who had also founded the University of Chicago in 1889. Cures for quite a few diseases were discovered by scientists at Rockefeller, many of whom were later honored for their work with Nobel Prizes.

One epoch-making discovery was made in 1944 at Rockefeller by Oswald Avery, Maclyn McCarty, and Colin MacLeod, who showed that DNA is the material in genes that controls inherited traits. A summary of discoveries by Rockefeller scientists over the years can be found at: https://en.wikipedia.org/wiki/Rockefeller_University.

I wanted to transfer to MIT in Cambridge, MA, because I hoped that I could work for Norbert Wiener (Figure 1.17). Wiener was a famous child prodigy who went on to do seminal work in many fields. In addition to his important work in information theory and probability theory, he created the field for which he is perhaps most famous, calling it *cybernetics*. What cybernetics is was explained by title of Wiener's 1948 book on the subject: *Cybernetics: Or Control and Communication in the Animal and the Machine*. As Google notes: "the word cybernetics arises from the Greek word κυβερνήτης (kybernētēs, steersman, governor, pilot, or rudder)." Wiener was one of the first scientists to theorize that intelligent behavior required feedback mechanisms. That was music to my ears because feedback mechanisms were ubiquitous in my neural networks, even in my 1957 model of verbal learning.

Wiener **Norman and Fagi Levinson**

FIGURE 1.17 Photos (from the left) of Norbert Wiener and of Norman and Fagi Levinson. See the text for details about how they influenced my life and career. Pictures taken from the web: Norbert Wiener: https://en.wikipedia.org/wiki/Norbert_Wiener; Norman Levinson: https://www.nasonline.org/wp-content/uploads/2024/06/levinson-norman.pdf; Zipporah ("Fagi") Levinson: Picture taken from the author's personal collection. Courtesy MIT Museum.

My original plan was to stop at Rockefeller in Manhattan, spend some time with my family in Jackson Heights, Queens, and then continue up to MIT in Cambridge, MA.

The Law of Unexpected Consequences struck, however, in a major way as I was traveling across country from Stanford to New York: Norbert Wiener died.

Suddenly, my invitation to visit the Rockefeller Institute felt like a godsend. In addition to its intellectual attractions, Rockefeller at that time was a very elegant place, with a huge endowment from the Rockefeller family. Rockefeller is located East of York Avenue in Manhattan, and has a fancy entrance gate on East 66th Street through which one enters a campus with beautiful gardens and stately buildings. When I first went through those gates, I felt like I was entering an academic heaven, whose effect was amplified by the fact that the name of the guard at the gate house was Angel ... really!

My "interview" at Rockefeller consisted of having lunch with Mark Kac (Figure 1.18, left panel) in an elegant, wood-paneled, lunchroom on the Rockefeller campus. The main advice that I remember Mark giving me during our lunch together was: "Steve, never grow old." Mark himself managed to live to only age seventy.

Mark was a famous mathematician who worked mostly in probability theory, with applications to statistical mechanics and number theory. Mark was one of the scientists with whom I hoped to study at Rockefeller, having very much enjoyed my reading and classes in probability theory at Dartmouth and Stanford. I also realized that my own neural models were a kind of real-time nonstationary statistical mechanics. This connection of my work with a type of physics that I loved to study excited me a lot.

Mark used to talk daily with his colleague and friend, George Uhlenbeck, another famous statistical physicist who had been recently hired by Rockefeller. Uhlenbeck gave meticulously prepared lectures to Rockefeller graduate students. He was particularly famous for his discovery in 1925 with Samuel Goudsmit of electron spin, or angular momentum. I enthusiastically approached Uhlenbeck to tell him about connections that I envisaged between the statistical mechanics of my nonstationary neural networks and statistical mechanics. Uhlenbeck sat quietly until I finished my presentation, after which he passed summary judgment: "Grossberg, physics is closed." Subsequent discoveries like superstring theory and the Higgs boson have shown that George was overly pessimistic.

Mark might never have made it to Rockefeller. After receiving his PhD in Poland, he luckily looked for a position abroad, and in 1938 was granted a scholarship which enabled him to work in the United States. I write "luckily" because his parents and brother, also Jewish, had remained in Poland, and were murdered by the Nazis in mass executions in 1942.

Mark's many scholarly accomplishments are summarized in https://mathshistory.st-andrews.ac.uk/Biographies/Kac/. Perhaps his most famous paper was published in 1966

FIGURE 1.18 Photos (from the left) of Mark Kac, Detlev Bronk, and Gian-Carlo Rota, three faculty at the Rockefeller University who played important roles in my graduate education. Credit: Konrad Jacobs, Erlangen/Wikipedia; Photograph by Donna Coveney/MIT; Mark Kac: https://en.wikipedia.org/wiki/Mark_Kac; Detlev Bronk: https://digitalcommons.rockefeller.edu/faculty-members/14/; Gian-Carlo Rota: https://mathshistory.st-andrews.ac.uk/Biographies/Rota/

while I was a graduate student at Rockefeller. Its title illustrates his ingenuity: *Can You Hear the Shape of a Drum?*

Mark loved to tell jokes and was always checking to be sure that everyone was having a good time. His picture in Figure 1.18 nicely captures the welcoming smile that he offered generously to us all.

After having lunch with Mark, I was ushered into the beautifully furnished office of Detlev Bronk, who was then the President of Rockefeller (Figure 1.18, middle panel). Bronk was part of a distinguished family that could trace its origins to colonial times. He was, for example, a descendant of Pieter Bronck, an early settler in New Netherland. The borough of the Bronx in New York City is named after, Jonas Bronk, a relative of Pieter Bronck.

Bronk made his name as a scientist, educator, and administrator. His scientific work included the seminal discovery with Lord Edgar Adrian of the all-or-nothing nature of the action potential. Together they showed that action potentials, or "spikes," travel in a non-decremental way down axons from their cell bodies (Figure 1.12). Their two landmark papers in 1928 and 1929 showed, in addition, that the changing firing rates of such discrete spike trains in motor nerve axons command target muscles to contract more or less in a continuous analog way. These discoveries helped to establish biophysics as a recognized scientific discipline.

Bronk's educational and administrative roles included being President of Johns Hopkins University from 1949 to 1953, President of The Rockefeller University from 1953 to 1968, and President of the National Academy of Sciences between 1950 and 1962.

The sun was setting and casting long shadows over Bronk's beautiful office when I was ushered into it for an

FIGURE 1.19 (Top row) Pictures of several leading experimental neuroscientists whose data my neural network models explained. (Bottom row) Pictures of me and Christoph von der Malsburg when we first met in the 1970s. The picture of me was taken in 1976, the year that I introduced Adaptive Resonance Theory, and is from my personal collection. The other pictures were taken from the web: H. Keffer Hartline: https://en.wikipedia.org/wiki/Haldan_Keffer_Hartline; Floyd Ratliff: https://digitalcommons.rockefeller.edu/faculty-members/68/ Photo by Ingbert Gruttner; Alan Hodgkin: https://en.wikipedia.org/wiki/Alan_Hodgkin; Andrew Huxley: https://en.wikipedia.org/wiki/Andrew_Huxley; Christoph von der Malsburg; https://en.wikipedia.org/wiki/Christoph_von_der_Malsburg

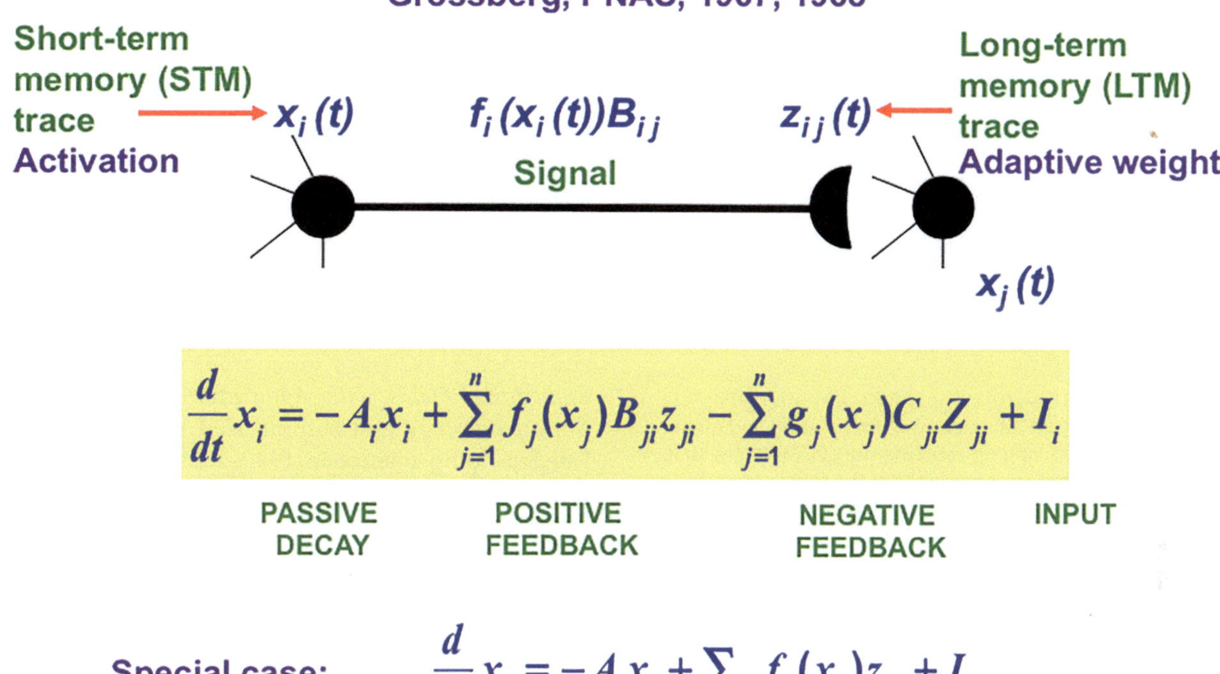

FIGURE 1.20 The neuronal components and mathematical equation that define the neuronal activities, or short-term memory (STM) traces, of the Additive Model. See the text for details. Author created.

interview. My interview with him was as brief as the one with Mark Kac. After settling together into a comfortable sofa, he simply looked kindly at me and asked: "Grossberg, would you like to study here with a full fellowship?" I was exploding with happiness inside when I told him, as calmly as I could, that I certainly would.

There were other scientists besides Mark Kac at Rockefeller with whom I hoped to work. They included some famous neuroscientists, notably H. Keffer Hartline, who worked with Floyd Ratliff (Figure 1.19) to develop an equation to describe the lateral inhibitory interactions within the compound eye of the *Limulus polyphemus*, or horseshoe crab, retina. Hartline won the Nobel Prize in 1967 for his experiments on this process, which he began in the 1930s. It is interesting to read how Hartline adopted whatever technologies were being developed at the time to carry out his experiments over the years.

I was attracted to the work of Hartline and Ratliff in part because my own Additive and Shunting models of on-center off-surround networks also included lateral inhibition. Figures 1.20–1.22 summarize the STM,

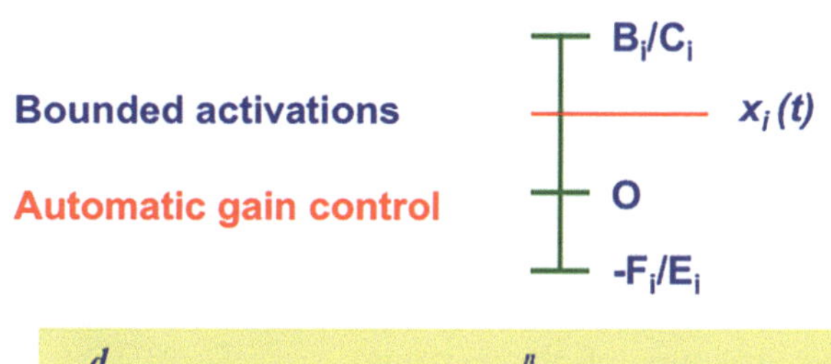

$$\frac{d}{dt}x_i = -A_i x_i + (B_i - C_i x_i)[\sum_{j=1}^{n} f_j(x_j)D_{ji}y_{ji}z_{ji} + I_i] - (E_i X_i + F_i)[\sum_{j=1}^{n} g_j(x_j)G_{ji}Y_{ji}Z_{ji} + J_i]$$

INCLUDES THE ADDITIVE MODEL

FIGURE 1.21 The neuronal components and mathematical equation that define the neuronal activities, or short-term memory (STM) traces, of the Shunting Model. See the text for details. Author created.

MEDIUM AND LONG–TERM MEMORY

MTM: Habituative Transmitter Gate

$$\frac{dy_{ki}}{dt} = H(K - y_{ki}) - Lf_k(x_k)y_{ki}$$

LTM: Gated Steepest Descent Learning

$$\frac{dz_{ki}}{dt} = M_k f_k(x_k)\big(h_i(x_i) - z_{ki}\big)$$

FIGURE 1.22 Typical laws for habituative transmitter gates, or medium-term memory (MTM) traces, and for learning and memory by adaptive weights, or long-term memory (LTM) traces, that are used in my neural network models. See the text for details. Author created.

MTM, and LTM equations that I discovered when I was at Dartmouth. The STM Additive and Shunting models in Figure 1.20 and Figure 1.21 both include lateral inhibitory terms. Figure 1.22 shows the gated steepest descent equation that I used to overcome problems of the Hebb Postulate (Figure 1.18).

Rafael Lorente de Nó was another famous neuroscientist whom I wanted to meet, both because of his own seminal work, and also because he was the most famous student of Santiago Ramón y Cajal of Spain, one of the greatest neuroanatomists to ever live. It was Cajal (Figure 1.23, left panel) who discovered that discrete neuronal cells, or neurons, interact in our brains, rather than the continuous electrical network, or reticulum, that was promoted by Camillo Golgi of Italy (Figure 1.23, right panel).

Cajal and Golgi shared the Nobel Prize in 1906 for their work on the structure of the nervous system. Cajal won for his pioneering work on discovering the Neuron Doctrine, or the fact that discrete neurons exist. Golgi won for his discovery of a revolutionary staining method whereby brain tissue could be visualized. Cajal modified the Golgi method to make his own epochal discovery. Remarkably, Golgi used his Nobel lecture to reaffirm his Reticular Hypothesis and to deny that individual neurons exist, even as Cajal asserted that they do (e.g., Warmflash, 2016)!

13 My First "publication" in 1964: A Monograph on Embedding Fields

Every entering graduate student at Rockefeller was required to write an essay about his or her scientific interests. In response to this requirement, I wrote and typed a 443-page monograph titled *The Theory of Embedding Fields with Applications to Psychology and Neurophysiology* that summarized my discoveries since I began doing science in 1957. Keep in mind that 1964 was before most of the technologies existed that we take for granted these days to prepare our written work, including word processing. Even the Xerox machine had just recently been commercially released. I therefore typed the monograph myself and drew its hundreds of illustrations and mathematical symbols using a pen with black ink.

My Rockefeller professors were surprised to get a monograph instead of an essay. They could not evaluate my work, so they arranged for 125 copies to be made. Some of the copies were given to faculty at Rockefeller. Most of them were mailed to the leading psychology and neuroscience labs in the United States to see if anyone could figure out what I was doing.

FIGURE 1.23 Pictures of Ramon y Cajal (left) and Camillo Golgi (right), who both won the Nobel Prize in 1906 for their work on the structure of the nervous system. Pictures taken from the web: Ramon y Cajal: https://en.wikipedia.org/wiki/Santiago_Ram%C3%B3n_y_Cajal Cajal.PNG/Wikipedia; Camillo Golgi: https://public.websites.umich.edu/~gusb/golgilife.html

I wanted to speak with Lorente de Nó about overlaps between his important experimental discoveries and my neural model explanations of his and related neurophysiological data. When I came into his office, he immediately complained that my monograph did not include anything about the ephaptic interactions whereby action currents within an active nerve fiber can influence neighboring nerve fibers, a topic that he had written about with Amedeo Marrazzi in 1944. He was very surprised when I then showed him where I had discussed such interactions in my monograph. After that, he invited me to join his lab. Although I felt honored by his generous invitation, I declined it as graciously as I could by explaining that trying to discover and develop my neural network models was a full-time job.

14 Gian-Carlo Rota and Theorems about Pattern Learning by Neural Networks

Luckily for me, Gian-Carlo Rota came to Rockefeller as a professor just when I most needed his support (Figure 1.18, right panel). Although Mark Kac might have been a suitable PhD thesis advisor for me, Mark was either working on his own problems in probability theory and statistical physics or traveling to one or another conference or scientific leadership meeting in Washington, D.C., He was therefore rarely around. Gian-Carlo was willing to serve as my PhD advisor, which saved me from being thrown out of school.

That could happen at Rockefeller to even the most gifted students, because Rockefeller was organized as a series of laboratories. The laboratory chiefs behaved like feudal lords with absolute power over their labs. If a Rockefeller PhD student could not find a lab chief to protect him in his first year there, then he could be summarily thrown out of school. Students had to be protected by a lab chief even before they knew what scientific topics interested them. Once in a lab, their research projects, and thus the trajectory of many years of their lives as scientists, were determined for them by their lab chief.

Gian-Carlo protected me from this fate. He was a brilliant, generous, and generally wonderful man who, like an angel from heaven, came from the MIT mathematics department to Rockefeller in 1965, shortly after I arrived, and left in 1967, just after I finished my PhD, to return to his professorship at MIT in the same mathematics department where I was also hired. Gian-Carlo was more than a scientific mentor to his students. He did his best to create a welcoming social environment for us. I still remember fondly when Gian-Carlo treated us to dinner at elegant restaurants in Manhattan, including The Brasserie, where he introduced us to their famous black forest cake, among other delights.

Neither Mark nor Gian-Carlo could evaluate my scientific work. It was too different from the fields of their own expertise. Gian-Carlo worked in a wide range of mathematical fields, including functional analysis and probability theory. He is best known for his contributions to transforming the field of combinatorial analysis, or combinatorics, from a grab bag of computational tricks for counting and other operations on finite structures into one of the most important branches of modern mathematics, with profound connections to other areas of mathematics, physics, biology, and computer science.

In addition to this already remarkable scholarly breadth, Gian-Garlo was also a philosopher who sensed the philosophical importance of what I was doing. He kindly gave me the political protection and academic freedom to do the mathematical research that enabled me to complete my PhD thesis.

Gian-Carlo was the second mathematics professor in my life who was also a serious philosopher. John Kemeny was the first (Figure 1.9, right panel). Both of them provided crucial political support for my research when I really needed it. Whereas John was a philosopher of science, Gian-Carlo was a philosopher of phenomenology. Phenomenology seeks to explain the nature of things through the way people experience them. His background in phenomenology primed Gian-Carlo to appreciate the work of someone like me who was trying to develop realistic brain models of how people experience the world.

15 A Halcyon Time at Rockefeller: Theorems, Pumping Iron, Music, and Piano

I was therefore able to experience a period of time at Rockefeller when I could work in a supportive environment. With my intuitions about how humans and animals learn to guide me, I was able to develop novel mathematical methods to analyze the dynamics of the neural networks that I discovered at Dartmouth. These neural networks were a new kind of nonlinear dynamical system for which I could prove global limit and oscillation theorems about how they learn spatial patterns of information and store these learned patterns in memory. The oscillations of my neural networks described how learning evolved, while the existence of limits as learning settled down described how learned memories were stored through time. *Global* theorems are theorems that are proved for any initial starting point in the system. They are quite rare in nonlinear dynamical systems, where one often uses tricks to linearize the systems and then approximate their properties using various perturbation methods.

During this halcyon period, I benefited in several ways from the largess at Rockefeller. First, I could walk from Rockefeller's Pearly Gates to Bloomingdale's department store at 59th Street and Third Avenue to wander around their huge collection of records, which I could afford to buy, a few at a time, using my Rockefeller fellowship. That was a period when classical music performed by great musicians was recorded by labels like Angel records, whose record jackets were works of art. I still have all of these records in the family room of my home in Newton Highlands, Massachusetts.

From my time at Rockefeller to the present, I have been lucky to feast on performances of great classical pianists like Vladimir Horowitz, Artur Rubinstein, Glenn Gould, Martha Argerich, Sviatoslav Richter, Claudio Arrau, Rudolf Serkin, and Emil Gilels. Some of my favorite great classical violinists are Jascha Heifetz, Isaac Stern, Yehudi Menuhin, David Oistrakh, Fritz Kreisler, and Nathan Millstein.

While I lived in New York and Boston, I have also been lucky to enjoy a lot of world class opera, including favorite singers like Luciano Pavarotti, Plácido Domingo, Piotr Beczala, Beverly Sills, Shirley Verrett, Joyce DiDonato, Renée Fleming, and Anna Netrebko.

Gail and I were regular attendees of the many wonderful concerts of the Boston Celebrity Series. We managed to get season tickets, year after year, in the front row center of the auditorium in Symphony Hall. Such seats were not particularly sought after, because lots of audience members prefer sitting further back in the auditorium where the acoustics might be better. I personally loved to see the fingerings and facial expressions of the musicians up close, sweat and all.

One evening, a young tenor, who was totally unknown to us, walked out on stage for his solo recital with a white handkerchief in his hand. We were thrilled, and a bit overwhelmed, by the beauty, range, and sheer power of his singing. The tenor was Luciano Pavarati in his prime.

I also bought and listened to a lot of jazz, rhythm and blues, and soul music, including music of both men and women. The men included John Coltrane, Louis Armstrong, Charlie Parker, Charles Mingus, Dave Brubeck, John Coltrane, Bill Evans, Ray Charles, Marvin Gaye, James Brown, Otis Redding, Al Green, Stevie Wonder, Michael Jackson, and Thelonious Monk.

My favorite women performers include Ella Fitzgerald, Carmen McRae, Billy Holiday, Nina Simone, Sarah Vaughan, Diana Ross, Whitney Houston, Aretha Franklin, Tina Turner, Dionne Warwick, Roberta Flack, Cher, Joni Mitchell, Lauro Nyro, Julie London, Peggy Lee, and Dinah Washington. My days at Rockefeller were thus full of music as I did my studies, as they have been all my life.

I still remember a magical evening when, as a Dartmouth student home for a holiday, I wandered around Greenwich Village in Manhattan and was drawn to the Village Vanguard, where John Coltrane was playing the saxophone as only he could. During those simpler days, I was able to stand right next to the platform on which Coltrane was playing and get totally enveloped by the sound. That experience was one of the reasons that I started collecting records of jazz performers.

Because the Rockefeller Institute was so rich at that time, there was a room with an open counter where investigators and students could order just about anything. Some of us called it the Gift Shop. One of my fellow students ordered one hundred rabbits there for his experiments. My order advanced science less directly than that:

I ordered dumb bells and a bench press set to be able to work out with weights, another habit that I pursued throughout my life. Thus was established Steve's Gym in the basement right under the elegant office of Detlev Bronk. My little gym was set up outside a practice room that contained an upright piano. One other person whom I knew practiced piano there. Otherwise, I was frequently alone in the basement where I used to stay until I alternated pumping iron, playing the piano, and reading or thinking about something or other until I experienced quite a wonderful endorphin rush.

16 Rota Is Manic-Depressive and No One Believes My Theorems: Los Alamos

Just when I had proved all the global limit and oscillation theorems for my PhD thesis at Rockefeller, a period of drama and anxiety occurred when Gian-Carlo was committed to Paine-Whitney hospital after experiencing an episode of manic-depressive psychosis, a problem that creative people seem to have more frequently than others. Thankfully, he eventually recovered. But while he was ill, my work was exposed to some challenging internal politics at Rockefeller that I fortunately survived.

One problem was that the professors who filled in for Gian-Carlo did not believe my theorems. They had never seen theorems like them, which makes sense because the dynamical systems whose properties the theorems prove were about how our brains learn and remember, which requires different concepts and methods to explain than the methods used to explain more traditional physical processes. One of the mathematicians who kindly filled in for Rota was Jürgen Moser (Figure 1.24, left panel). Jürgen made great contributions to the mathematics of celestial mechanics, which led to him winning several prestigious prizes for his mathematical work, including being awarded the first George David Birkhoff Prize in 1968, and his being named Director of the Courant Institute of Mathematical Sciences in New York. When Jürgen looked at my networks, he predicted properties of them that I knew to be wrong. He made these mistakes because he was using his intuitions about celestial mechanics to try to understand brain learning.

I also had an encounter with the brilliant American physicist and educator, Richard Feynman (Figure 1.24, middle panel), whose versatility and creative research in many branches of physics, including quantum mechanics and quantum field theory, also did not prepare him to understand how our brains work. I first met Feynman before I was scheduled to give a lecture at the Thinking Machines Corporation in Cambridge, MA, near MIT, that had recently been co-founded by Danny Hillis. As usual, I arrived much too early and was led by a staff member to wait in a lovely open space until my hosts could welcome me.

While I was waiting, I was stunned when Richard Feynman, who was then visiting Thinking Machines, came into the room saying: "I was sent here to entertain you." Suffice it to say, I felt highly entertained, indeed honored, to have a chance to chat with him. During our chat, Feynman was totally unpretentious during our lively conversation about vision. As it turned out, he had earlier gotten interested in how our brains see. Feynman discussed the fact that, as he studied how the photosensitive retinas in our eyes work, he soon realized that his training in physics could not explain what he was learning.

In particular, our retinas have a blind spot where the optic nerve forms to carry signals from retinal photoreceptors to our brains. This blind spot is as large as the fovea, where all detailed vision is experienced. The retina is also covered by nourishing veins that occlude light from reaching photodetectors at multiple positions. Feynman could not figure out why we were not consciously aware of these big holes in the retinal mosaic. How did our conscious percepts of visual scenes get completed? Feynman soon realized that he could not find intuitively plausible principles for how our brains do this, so he got out of trying to do vision. My work has explained how this happens using the kinds of perceptual boundaries and surfaces that I will describe in Chapter 2.

The examples of Moser and Feynman illustrate that my neural networks and their properties exemplified a new scientific paradigm, and were not just a ho-hum variation of known mathematics and physics. I have gradually come to understand that my work is a pioneering contribution to the revolutionary paradigm of *autonomous adaptive intelligence*, whose applications will, I believe, transform our societies during the coming century.

FIGURE 1.24 Pictures of Jürgen Moser, Richard Feynman, and Stanislav Ulam, who I had the pleasure of interacting with when I was a graduate student and young scientist. See the text for details. Pictures taken from the web: Jürgen Moser: https://en.wikipedia.org/wiki/J%C3%BCrgen_Moser Klaus Moser/Wikipedia; Richard Feynman: https://www.themarginalian.org/2013/08/08/richard-feynman-on-the-meaning-of-life/; Stanislav Ulam: https://en.wikipedia.org/wiki/Stanis%C5%82aw_Ulam; AIP Emilio Segrè Visual Archives, Ulam Collection.

Mark Kac was, fortunately for me, a friend of Stanislav Ulam (Figure 1.24, right panel), a Polish-American mathematician and nuclear physicist, who was then at Los Alamos Laboratories in New Mexico running simulations on hydrogen bomb explosions. To run the simulations, Los Alamos had the most powerful computers in the United States. Mark therefore asked Stan if, as a favor, he would run computer simulations of my dynamical systems.

Mark had a close professional as well as personal relationship with Stan, if only because both of them were Jewish mathematicians who left Poland for the United States shortly before Poland was overrun by the Nazis. In fact, today there are Mark Kac Applied Mathematics Fellows at the Los Alamos Center for Nonlinear Studies: https://cnls.lanl.gov/external/Kac-fellows.php.

I felt lucky that *any* computers at that time could simulate my neural network models. In fact, when I was working on my models at Stanford, shortly before I left Stanford for Rockefeller, I tried to get the models simulated on the computers in the Stanford Computer Center. I was told that their computers were not powerful enough to simulate the nonlinear systems of fast-slow (STM-LTM) differential equations that defined my neural networks. That was one reason why I worked so hard to prove mathematical theorems in my PhD thesis about learning in my neural networks. The lack of computer simulations thus turned out to be a blessing in disguise.

With supercomputers in our pockets these days, whether in the form of iPhones or another smart phone, young people may not appreciate how rapidly computational power increased during the past few decades. Many people explain this by citing Moore's Law, or the prediction by Gordon Moore, who co-founded Intel with Robert Noyce in 1968, that the number of transistors in a computer chip will double every two years or so, along with computer processing power: https://en.wikipedia.org/wiki/Moore's_law. My colleagues and I have lived through the effects of Moore's Law by being able to simulate far more complex neural networks on our computers as the years unfolded. This was particularly important when we tried to simulate how vision, audition, speech, language, and cognition work.

Back at Los Alamos, Stan himself did not at first believe my results, but he was thankfully curious enough to simulate my dynamical systems on his computers to see what happened. The results saved my life as a scientist because, lo and behold, the Los Alamos computer simulations mimicked my theorems!

I have always found it droll, if not downright strange, that theorems which are arguably the most powerful theoretical method that humans have to achieve some degree of certainty in science—if you are lucky enough to be able to prove them—took second fiddle to computer simulations, which are often beset with problems of numerical approximation. This is the only time in my life when complete global theorems were doubted enough to require their validation by computer simulations, a decidedly inferior type of evidence.

That is not to say that computer simulations are not essential. Theorems may be invaluable to prove general properties of systems. But simulations are necessary to demonstrate their parametric properties, like how the dynamics of specific brain interactions give rise to interactive, or emergent, properties that mimic the detailed properties of observable behaviors. That is why my colleagues always turned to computer simulations to validate our neural network models of how vision, audition, speech, language, and cognition work.

17 From Rockefeller Student to MIT Professor: A Happy Start

Once my theorems were validated at Los Alamos, I got strong recommendation letters from both Mark and Gian-Carlo. In fact, Mark's letter said that I was the most original thinker whom he had ever met. Gian-Carlo's letter was equally glowing. I was therefore hired by the MIT applied mathematics department in 1967 as Assistant Professor, at the same time that Gian-Carlo returned to the same department at MIT. Based on the rapid stream of articles that I published there, I was quickly promoted to Associate Professor, especially because I was starting to get counter-offers from other universities.

Before reminiscing more about my experiences on the faculty at MIT, I want to mention an anecdote that illustrates the continued importance of music in my life then, and to the present day. When I arrived in Cambridge, Massachusetts, as an MIT assistant professor, I luckily found a lovely apartment near Harvard Square, which was the heart of a vibrant Cambridge social scene. I quickly started to hunt for furniture to fill it. I knew that I did not want a secondhand bed, so I bought a new bed at a discount store on Massachusetts Avenue near Central Square.

I also started looking at ads in the local newspaper for used furniture. This was, of course, decades before one could do a Google search for such an item. The first piece of used "furniture" that I bought was a Steinway grand piano! It was being sold at a secondhand furniture store in Somerville as just another piece of secondhand furniture. Apparently it was among the furnishings from the home of an old lady who had recently died, and was priced just like her other old pieces of furniture. This was yet another lucky experience of The Law of Unexpected Consequences. I felt quite at home with just a bed and a Steinway grand, as I started to find other furniture with which to fill my apartment.

That Steinway is still part of my life and is a "member of the family" living downstairs in the music room of our home.

There were several faculty in the MIT applied mathematics department, in addition to Gian-Carlo, who were very supportive of my work. The most distinguished of these faculty was Norman Levinson (Figure 1.17, center panel). Norman wrote his PhD thesis in mathematics under Norbert Wiener. So, although I never got to meet Norbert, I got to know and love the family of his most important student.

Norman was a brilliant mathematician who did important work in Fourier transforms, complex analysis, nonlinear differential equations, number theory, and signal processing. He became Institute Professor at MIT in 1971, shortly before I came up for tenure there. Here is a good summary of Norman's rich and varied life: https://www.nasonline.org/wp-content/uploads/2024/06/levinson-norman.pdf.

Norman married Zipporah (Fagi) Levinson in 1938 (Figure 1.17, right panel). The story of how they met and married is charming in itself, and also involves one of the many stories of Norbert Wiener's famous idiosyncratic behavior. An earlier story involves Norbert traveling to the home of Norman's parents, who lived in Revere, Massachusetts, and who immigrated to America as poor Russian-Jewish immigrants. Norbert made the trip to convince them that their son would not starve as a college professor.

This was not obvious to everyone at that time. Fagi told me that, when they were first married, she had to cut and sew their clothes, and collect discarded wooden boxes from the streets to convert them into bookcases, among other economies. Fagi was, or at least soon became, an accomplished seamstress. I still have two bathroom robes that she made from scratch for me.

Norman met Fagi through her brother, the topologist, Henry Wallman. On their first date, when Fagi excused herself to use the bathroom, a guy came up to Norman to warn him about getting "fresh" with Fagi, because she had apparently recently thrown a "fresh guy" over her shoulder for all to see. Norman proposed to Fagi just two days after they met, and they married within a week.

Fagi told me several delightful stories about how idiosyncratic Wiener could be. Here's one: In the middle of their wedding night, Norbert banged loudly on the door of the bedroom where Norman and Fagi were staying. He wanted to talk math with Norman, honeymoon be damned!

There are many heartwarming anecdotes about Fagi. She used to throw the most wonderful parties at their home for all the mathematicians and their many other friends, and eventually became known as the Den Mother of the MIT mathematics department. See https://news.mit.edu/2010/obit-levinson.

Norman and Fagi welcomed me into their family. They had two daughters about my age, but no sons. They adopted me as a kind of scientific son. Whenever both Norman and I would attend the same mathematics conference, Norman and Fagi would drive me around in a rented car to see the sights with them. After Norman died tragically at age sixty-three in 1975 from a virulent form of brain cancer, my wife, Gail Carpenter, and I asked Fagi if she would be the god-grandmother of our daughter, Deborah. Fagi did this lovingly and generously throughout her life. We always included her in dinner parties and other celebrations in our home, and visited her regularly in the houses where she lived in the Boston area, in Wellfleet on Cape Cod during the summer months, and thereafter in senior housing when she could no longer live independently.

Visiting Fagi in Wellfleet was one of the ways that Gail and I fell in love with the Cape. We also used to visit Provincetown off-season to stay in a bed-and-breakfast while enjoying walks in town and on nearby Cape beaches, eating in the wonderful seafood restaurants, seeking out available theatrical and musical performances, and ... dare I say it ... disco dancing at various bars in town. Our third reason to visit the Cape was to stay with a close friend of ours in his beach house in Truro. That is how we fell in love with Truro and decided to look for a beach house of our own there.

18 Dumped from MIT Because of OPEC: A Promise that Was Not Kept

At most universities, being promoted to associate professor also came with tenure, which guarantees a permanent position until retirement. Tenure enables a faculty member to take intellectual risks, move into new fields, and tackle difficult problems that can take a long time to solve. MIT and Harvard were two of the universities that still did not offer tenure to associate professors because they thought that everyone in their right mind would be dying to work there, even without tenure. I was, nonetheless, given to believe that tenure would be inevitable when I was promoted to full professor within a few years, as had always occurred in the past when a faculty member in the MIT mathematics department was promoted to associate professor.

Unfortunately, the Law of Unexpected Consequences struck in 1973 when OPEC (the Organization of Petroleum Exporting Countries) dramatically raised oil prices and embargoed oil exports to the United States for its support of Israel during the 1973 Yom Kippur war. The original OPEC countries were Iran, Iraq, Kuwait, Saudi Arabia, and Venezuela. They were joined by, in chronological order,

Qatar, Indonesia, Libya, the United Arab Emirates, Algeria, Nigeria, and Ecuador, all by 1973. The United States was hereby thrown into the most severe economic recession in the postwar era, just as I was scheduled for my promotion to a full professor with tenure.

Schools across the country panicked and started throwing out their untenured faculty to save money. Just before this happened, because no one in our department was an expert in the fields in which I did research, I was asked for names of scientists who could write letters about my work to support my tenure. Because my work is so interdisciplinary, it is relevant to experimental and theoretical work in multiple fields, including psychology, neuroscience, mathematics, engineering, and computer science. As a political novice who got no advice from my senior MIT colleagues about how to prepare my tenure case, I was naive enough to suggest fifty names of every well-known scientist and mathematician I could think of, across all these fields, to write tenure letters for me.

Despite the earlier verbal promise of tenure, I was denied tenure, presumably based upon those letters, because I had excellent reviews of my teaching and departmental service. Several of the tenured faculty in the department were so upset by this that they secretly showed me the letters. The distribution of opinions in the letters was bimodal: One group wrote things like I am a genius who deserves the Nobel Prize. The other group questioned my right to try to model the brain at all.

Because of this divide, the department decided to ask someone who was so distinguished that his opinion would surely carry the day. They therefore asked Sir James Lighthill (Figure 1.25, left panel) to write a letter about my work because Sir James had recently written a scathing report for the British Science Research Council, commonly known as the Lighthill Report (https://en.wikipedia.org/wiki/Lighthill_report), about all the things that were wrong with Artificial Intelligence, or AI, as it was being currently practiced. Sir James was then the Lucasian Professor of Mathematics at Cambridge University in England, the Chair that the incomparable Isaac Newton (Figure 1.25, right panel) earlier held at Cambridge.

Sir James' recommendation letter for me was quite different than what he wrote about AI in the Lighthill Report. He wrote a long and glowing letter for me detailing that my work was exactly what the fields of AI and neural networks should be doing. We became quite friendly in later years, and chatted whenever I was invited to lecture at Cambridge University.

Just as John Kemeny's personal connection to both Einstein and me has always felt precious, so too has Sir James Lighthill's connection to both Newton and me through his Lucasian Chair, especially after I began to be called the Newton and Einstein of the Mind after John Silber's introduction of me with that title before my 1989 University Lecture.

After all this effort to get enough recommendation letters to make an informed tenure decision based upon the quality of my work, I was thrown out of MIT anyway for ostensibly economic reasons. As I understand what happened, after the mathematics department supported my case in response to Lighthill's letter, the case was sent up to the MIT Science Council for final approval, which would have been essentially automatic during better economic times. As luck would have it, another colleague from the math department was also coming up for tenure. His research was more routine than my own, but he was in charge of managing the department's computer network, a thankless but necessary task. The Science Council decided that, for financial reasons, it could only approve one case from our department that year, and it was not mine.

This turned out to be a short-sighted decision, even from just an economic perspective. It was not the job of the MIT Science Council to anticipate that, in my next job at Boston University, I was Principal Investigator or Co-Principal Investigator on around $85,000,000 in grants to fund the discovery and development of the artificial and biological neural network models for which I am internationally known. Almost all of this money was used to pay for the salaries of PhD students, postdoctoral fellows, and faculty summer salaries. The overhead from these grants, in effect, also paid for the academic year salaries

FIGURE 1.25 Pictures of Sir James Lighthill (left) and Isaac Newton (right). Sir James played an important role in my career when I was a young professor at MIT and he held the Lucasian Chair in Cambridge that Newton had earlier held. Pictures taken from the web: Sir James Lighthill: https://en.wikipedia.org/wiki/James_Lighthill; Sir Isaac Newton: https://sciencephotogallery.com/featured/1689-sir-isaac-newton-portrait-young-paul-d-stewart.html PAUL D STEWART/SCIENCE PHOTO LIBRARY.

of all faculty in the Department of Cognitive and Neural Systems (CNS), which I founded at BU to enable this kind of training and research to occur. These grants enabled CNS to exist, and thrive, during the many years when I was its founding Chairman.

I also doubt whether the powers that be at MIT would have let a young professor start a new department. BU around that time was eager to support the goals of its gifted young faculty members, especially if they could pay their own bills with grants, as I was able to do. We sometimes likened the BU of the early Silber years to an academic Wild West where anything was possible if you could deliver on your promises. So the Law of Unexpected Consequences struck again, and made it possible for me to create a unique interdisciplinary intellectual community in BU where our training and research on artificial and biological neural networks could prosper.

I know that, when I was Chairman of CNS, if I saw letters like my own tenure letters, I would have strongly supported that candidate. People who do revolutionary work almost never get unanimous support for their work. It is often more frequently the case that they, at least initially, get loud disapproval, as has occurred to even famous visual artists and composers who break with tradition.

19 Some Famous Composers Who Were Not Always Appreciated

Since this book is about how our brains can consciously understand, appreciate, and create art and music, among other topics, let us keep in mind that the work of some famous artists and composers was not always appreciated. Here are a few examples of famous composers who experienced rejections:

Johann Sebastian Bach, a German composer who lived from 1685 to 1750, is today regarded as one of the greatest composers who ever lived. However, after his death, general interest in his compositions declined because they were considered old-fashioned compared to the emerging galant style: https://en.wikipedia.org/wiki/Galant_music. It was not until almost 80 years later that a Bach revival began when Felix Mendelssohn arranged a performance of Bach's Saint Matthew Passion in 1829. Bach's positive reputation then continued to grow unabated until the present time.

Another famous rejection was experienced by Igor Stravinsky after the premiere in Paris of his great orchestral work, The Rite of Spring, on May 29, 1913, with choreography by Vaslav Nijinsky. Rather than being too old-fashioned, Stravinsky's work was too avant-garde for the tastes of many in the audience, who were scandalized by the work, and emitted hisses, boos, and catcalls. Of course, The Rite of Spring is now recognized to be a revolutionary work of genius that redefined the course of twentieth-century music.

More recent musical legends like the Beatles have also had some less-than-positive experiences. When their 1969 masterpiece, Abbey Road, was released, they were already world famous. The record was nonetheless initially panned by critics who worked for *The New York Times*, the London *Times*, and *Life* magazine. These editorial opinions rapidly switched when the record sales were overwhelmingly successful, eventually reaching almost twenty million copies.

It is worthwhile keeping in mind that new work that is too much of this, or too little of that, can get rejected for opposite reasons, as the compositions of Bach and Stravinsky illustrate. Sometimes the most readily accepted musical or scientific work is just a little ahead of the mainstream understanding, which gives mainstream critics pleasure that they can understand the "new" work and may thus write positive reviews of it. Such cautious work may, however, not be the work that is remembered, or that changes the course of history, whether musical or scientific.

These stories remind us that even the most gifted artists have faced problems when their audiences, for one reason or another, were not ready to value their work. Knowing how they turned out will hopefully encourage young people who experience bumps in the road to their goals to persist in doing everything in their power to reach those goals.

20 "Doing Art" in My Lectures and Books Using Evolving Audio-Visual Tools

Once I left home to go to college at Dartmouth, I did not have much time for drawing or painting in my later life. There was one big exception to this loss, which was that, over the years, I created thousands of illustrations for my lectures and books. My illustrations evolved in lock-step with visual display technology.

My proclivity for writing in-depth studies of scientific topics emerged when I was a high school student at Stuyvesant. For example, we had an assignment to write an essay about a topic of our choice in a biology class. I wrote and typed a book of several hundred pages about how our hearts work, and illustrated it with hand-drawn, 3D-shaded illustrations of the human heart. My book attracted so much interest that the Stuyvesant Principal asked to display it in a glass-covered case in the hallway outside his office. I happily complied, but when I went to his office to pick it up at the end of the term, I was told that it was thrown away when they changed the display. Losing my heart book, though a heavy blow, fortunately did not break my heart. It did, though, teach me an early lesson that some honors come with hidden costs.

My Senior Fellow thesis at Dartmouth College was my next long publication. It integrated my discoveries since my freshman year about how my neural network models explain aspects of how brains make minds.

The Senior Fellow thesis provided the foundation for my first "publication" in 1964, which was supposed to be an essay that all entering graduate students at The Rockefeller Institute for Medical Research were required to write. My "essay" turned into the 443-page monograph that I earlier mentioned, which I called *The Theory of Embedding Fields with Applications to Psychology and Neurophysiology.* My monograph summarized my discoveries since I began doing science in 1957. As usual, I typed the monograph and drew its hundreds of illustrations and mathematical symbols using a ballpoint pen with black ink. Interested readers can download this monograph from my web page. Its URL is: https://sites.bu.edu/steveg/files/2016/06/Gro1964EmbeddingFields.pdf.

The 125 copies of my monograph that were sent to leading psychology and neuroscience labs in the United States were made using mimeographs. For those too young to have used them, it was quite an experience! You typed your text on an original copy of a page behind which was a master page into which your typed letters cut holes. The master page was then used to make multiple copies by forcing ink into its letter-shaped holes. Anyone old enough to have used a mimeograph will remember that, if you typed something incorrectly, you brushed a correction fluid over the incorrect holes, and then retyped the corrected text over the dried fluid, hoping for the best.

There also wasn't much in the way of visual technology for professors to use in 1967 when I first started to teach at MIT. I first drew pictures and wrote down equations and key phrases with white chalk on the blackboard, to complement the verbal explanations during my lectures. The next innovation was the use of colored chalk. Chalk was followed by transparencies that I drew on, first with black pens and then with colored pens. The finished transparencies were projected on screens using overhead projectors. It was a big step forward to type the transparencies in color on the computer using a word processor.

The overhead projectors needed bulbs to project images, and bulbs tended to blow out at inconvenient times. I got to the point that I always kept two replacement bulbs handy, just in case. Once when I was giving an invited lecture at an international conference, all three bulbs blew, in rapid succession … really! Someone ran around the building searching frantically for replacement bulbs while the conference ground to a halt.

There came a time when slides were professionally prepared by university audiovisual services. The early slide projectors would jam frequently enough to make using them frustrating, and the devices used to move the slides forward were also flaky. These problems receded when lectures could be created via powerpoint and projected electronically using one's personal laptop, although now you had to be sure that, when you lectured in different countries, you or your host had appropriate voltage converters and plug adaptors to power your computer and project your powerpoint transparencies.

This varied training made it possible for me to lecture effectively both in class and at conferences all over the world. In 2019, just before the first COVID pandemic struck, I gave a lecture series to students in our Department of Cognitive and Neural Systems (CNS) at Boston University on topics related to my work, for which I created around 800 powerpoint transparencies in color. I liked to prepare the transparencies so that "the talk gave itself," in the sense that the images and phrases on the transparencies were so well organized and clear that I could give the lectures effectively even if I was having a bad day. Students could then also study my lectures with comprehension without me around because I had the lectures put on the web. Some of these lectures can still be found on my personal web page sites.bu.edu/steveg/ under the headers *Tutorial Lecture Series*, and *Lectures*. These transparencies became a primary source of the 600+ color figures that I published in my 2021 Magnum Opus.

21 A Life-Changing Year at MIT before Going to BU: Gail Carpenter

As I was being rejected from MIT, a colleague who I knew at Boston University told me that BU was seeking to hire gifted new faculty. Based on my MIT tenure letters, I was then hired as a full professor of mathematics, without tenure, but was told that my tenure case would be decided by the BU President, John Silber (Figure 1.26, left panel), as soon as he returned from a trip in Europe. Silber was then the recently hired BU President, and was vigorously seeking to hire gifted new faculty to convert BU from a regional school of little distinction to a world-class educational and research powerhouse.

My switch from MIT to BU was delayed for a year because of a snafu in setting up my new professorship. It was during that year that Gail Carpenter (Figure 1.27, left panel) joined the mathematics department at MIT as an instructor, having just finished a brilliant mathematics PhD thesis at the University of Wisconsin. Gail's thesis proved mathematical

FIGURE 1.26 Pictures of John Silber (left) and Dennis Berkey (right), who strongly supported the founding of the Department of Cognitive and Neural Systems at Boston University at a time when they were top administrators there. Pictures taken from the web: John Silber: https://horatioalger.org/members/detail/john-r-silber/; Dennis Berkey: https://blueinstitute.org/dennis-berkey%2C-ph-d

theorems about how nerve impulses propagate down axons in discrete activity pulses, or spikes (Figure 1.14). Data about neuronal spikes that were recorded from the squid giant axon, and an empirical model to fit their data, were developed by Alan Hodgkin and Andrew Huxley (Figure 1.19). Their empirical model is usually called the Hodgkin–Huxley model. For their experimental and theoretical work, Hodgkin and Huxley won the 1963 Nobel Prize in Physiology or Medicine, sharing it with Sir John Eccles.

The original Hodgkin–Huxley equations applied only to some kinds of spiking dynamics generated in the squid giant axon. Gail defined a large class of Generalized Hodgkin–Huxley systems that could accommodate spiking data from multiple organisms and different parts of the brain, and proceeded to use geometrical methods to prove and classify the types of spikes that these different systems could generate, ranging from single spike and periodic spiking solutions to bursting dynamics and even chaos. Some of these oscillatory types are summarized in Figure 1.27 (right panel).

I learned a lot from these theorems because Gail's geometric methods made clear how changes in the phase portraits of the neuronal dynamics, which provided a pictorial way to understand how the spikes were caused, generated different spiking patterns. I also realized that her work on single cells, and mine on neural networks, had a lot of overlapping elements, if only because both types of results were based on the dynamics of membrane, or shunting, equations (Figure 1.21).

I met Gail at a faculty social at the beginning of the term. It was soon clear to me that we would have a deep and enduring relationship, indeed that she was the love of my life. In addition to being able to talk with Gail about anything, I loved the combination of her warm, friendly, adventurous, and generous personality, her lovely face and figure, and her brilliant logical and mathematical mind. After spending more and more time together through the Fall and early Winter, Gail and I became

FIGURE 1.27 Picture of Gail Carpenter (left), the most important person in my life, and of some of the types of nerve impulse dynamics that Gail mathematically explained (right). Gail's picture taken from the web: Left panel: Gail Carpenter: http://techlab.bu.edu/members/gail/; Right panel: Computer simulations taken from Carpenter, G. A. (1979). Bursting phenomena in excitable membranes. *SIAM Journal on Applied Mathematics*, 36, 334–372.

lovers on my thirty-fifth birthday, on New Year's eve, 1974. We have been inseparable ever since.

22 Finally a Happy Home at Boston University, After Some More Drama

Before Silber returned from Europe, I was told by the BU mathematics department chairman, who was handling my tenure case, that he had never seen such a strong tenure dossier, so a positive tenure decision was a foregone conclusion. Hearing this was very important to me because I also had a tenured offer from the University of Rhode Island, which I had to decline before Silber returned.

It was therefore quite a shock to me to have my tenure case rejected by the Dean of our college before Silber had a chance to look at it. The Dean told me that he was throwing me out because the mathematics department was already too large. Actually, there were only about twenty faculty in it at the time, to serve very large student body. Now there are over one hundred.

This was my second tenure rejection in a row! If it could not be overturned, my academic career as a research scientist was finished. I therefore appealed this decision to Silber himself.

While we waited for Silber to return, Gail and I were both quite depressed as we faced an uncertain future with the prospect that her rather small salary as a professor in the mathematics department of Northeastern University might have to support us both. It was hard for us to think about anything else, so we would curl up in bed most days and pull our portable TV to the foot of the bed to watch Francis the Talking Mule movies, which were the only thing that was mindless and funny enough to distract us.

When Silber returned, he summoned me to his elegantly furnished office, which reminded me a lot of Detlev Bronk's office at Rockefeller. John got right down to business. He apologized for the delay in awarding me tenure, approved my tenure and promotion to full professor with tenure, and kindly said that, with me now on board, he no longer had to worry about the future of the mathematics department.

In at least two senses that was true: The research that my colleagues and I did during the subsequent years put BU on the map internationally as the world's leading center of research about neural networks and biologically inspired technologies, and our grants were for quite a while the largest throughout the university.

Suffice it to say that the Dean who denied me tenure was not Dean for long.

With the security of tenure in hand, I proceeded to found several academic institutions at Boston University. First, in 1981, I founded the Center for Adaptive Systems (CAS). The goal of CAS was, and remains to this day, to be an interdisciplinary research and training center for postdoctoral fellows to work with me, and later also Gail, to discover and develop principled theories of brain and behavior, notably concerning how individual humans and animals adapt so well on their own to rapidly changing environments that may include rare, ambiguous, and unexpected events. The Center also develops technological applications that are inspired by its biological models. I like to summarize its research goals with the following two questions:

> How does the brain control behavior?
>
> How can technology enable biological intelligence?

These goals were restated in several different ventures that I led at BU. Here is a quote from one of the National Science Foundation (NSF) proposals that I won to fund a large number of distinguished invited speakers, as well as travel fellowships for graduate students from around the world, to attend an annual international conference that I organized and chaired at BU:

> This grant would enable the annual International Conference on Cognitive and Neural Systems to continue on May 24–27, 2000 and in May, 2001. The conference focuses on the two general themes: How Does the Brain Control Behavior? and How Can Technology Enable Biological Intelligence? The conference would also emphasize Learning and Intelligent Systems. Many conferences are covering cognitive or neurobiological data. A few conferences are covering cognitive or neurological models, but not both. Yet other conferences are covering artificial neural networks and other neuromorphic applications. The Cognitive and Neural Systems conference is unique in covering both cognitive and neural experiments and models, as well as a range of intelligent applications, and doing so within a conference program with a single program track. The meeting was prepared to do this because this is precisely what the CNS graduate program at Boston University was created to do. The conference will include invited tutorials and lectures, and contributed lectures and posters by experts on the biology and technology of how the brain and other intelligent systems adapt to a changing world. The conference is aimed at researchers and students of computational neuroscience, connectionist cognitive science, artificial neural networks, neuromorphic engineering, and artificial intelligence.

The CAS enabled me to hire postdoctoral fellows who had received their PhDs in a range of more traditional departments, and to train them, via participation in collaborative research, to be increasingly independent neural network modelers.

As the above NSF grant proposal abstract indicates, CAS worked out so well that I was invited by the BU administration in 1989 to found the Graduate Program in Cognitive and Neural Systems (CNS).

Gail, who was then a professor at Northeastern University, and I became co-Directors of the new Graduate Program. It was also in 1989 that I was awarded my endowed Wang Chair in Cognitive and Neural Systems. Several of the best postdocs whom I had trained were hired as tenure-track assistant professors when CNS was founded. To get any of them hired or approved for tenure, however, they had to be evaluated and voted upon by a cognate department, such as biology, computer science, mathematics, or psychology. Because CNS was just a graduate program, it could not do that itself.

We were initially worried that no students would want to study at CNS. To our surprise and delight, many gifted students wanted to join us because we were, at the time, a unique program. The program showed such progress in its teaching and research productivity that we were invited in 1991 to convert CNS into the Department of Cognitive and Neural Systems, of which I became the founding chairman, and remained in this position until 2017. The founding of CNS was enormously important to us for several reasons:

> First, as a department in our own right, we could hire and recommend our faculty for tenure.
>
> Second, we developed a unique graduate and advanced undergraduate curriculum of eighteen interdisciplinary courses. Each course integrated the psychological, neurobiological, mathematical, and computational information needed to investigate fundamental issues concerning mind and brain processes and the applications of neural networks and hybrid systems to engineering, technology, and AI. The latter training included methods and applications in image processing, signal processing, adaptive pattern recognition, information fusion, data mining, intelligent control, and robotics. Our PhD students were required to take at least ten CNS courses, but could take more of them if they wished.

Our students were also encouraged to acquire disciplinary expertise by taking courses in departments such as biology, computer science, engineering, mathematics, and psychology. Our goal here was to ensure that our students knew enough about traditional fields to get the jobs that they wanted after they earned their PhDs.

After finishing one academic year of courses, our students began to work individually with one or more advisors to learn how to carry out advanced interdisciplinary research, from problem definition through professional presentation. They began this research in their first summer at CNS and then continued to take a lower course load in their second year as they continued their PhD research. CNS rapidly became the leading department in the world for advanced training and research in developing neural network models of how brains make minds, and the application of these models to solve outstanding problems in engineering, technology, and AI. As a result of the breadth, depth, and rigor of this training, CNS graduates have consistently found excellent jobs in both academic and technological professions. We were often told that many other universities looked to CNS as a model for the kinds of interdisciplinary mind/brain/technology departments that they hoped to found.

Conferences that we organized, like the annual International Conference on Cognitive and Neural Systems (ICCNS), added to the training of our graduate students and postdocs, while also creating opportunities for them to meet the many distinguished scientists whom we invited to give 40–60 minute lectures. We hoped that some of these scientists would also offer them jobs after they left us. More junior scientists were invited to give shorter talks at ICCNS, and both scientists and graduate students gave posters.

Mounting ICCNS took money, and that money was part of the grant proposal budgets that I was lucky enough to win over the years. One does not get a roster of twenty to twenty-five senior scientists to lecture at your conference unless you can pay all their expenses to do so. Likewise, my grants included travel and living stipends for around fifty PhD students from around the world to learn at ICCNS.

I used to hold a reception before each conference began in my rather capacious office, where all the speakers and all the funded students could meet and enjoy each other's company while partaking of a modest buffet. I especially wanted the students to have a chance to meet each other before the intense conference program began. They then had someone whom they recognized to sit next to at the conference sessions, and some of them became friends. One measure of the success and longevity of this arrangement is that some of the PhD students who won travel grants at earlier meetings sent their own best PhD students to learn at ICCNS once they became professors.

It was at such a reception in my office that Henry Markram and Wulfram Gerstner, who I invited to speak at the same ICCNS conference, met each other. This encounter eventually led to founding by Henry of both the Blue Brain Project and the Human Brain Project at École Polytechnique Fédérale de Lausanne (EPFL) in Switzerland, to which Henry moved at Wulfram's invitation. I am very proud of having been able to facilitate such productive relationships.

There were also other social and intellectual benefits that some of our speakers derived from ICCNS.

I remember a heartwarming comment to me from one of the invited ICCNS speakers, Richard F. Thompson, a famous neuroscientist who identified and mapped the neural circuits responsible for classical conditioning in the rabbit. Dick generously told me that he loved to listen to all the lectures because they were a unique opportunity for him to efficiently learn about the latest results in so many fields that he did not usually study, or hear about at the more specialized conferences that he attended.

The format of 40–60 minute invited talks made that possible for scientists and students alike. In addition to lecture length, I chose several conceptually important themes each year and grouped four to five lectures together that contributed to that theme to allow a deeper dive into understanding it. Some scientists who were not speaking at ICCNS came to the conference for the same reason as Dick, year after year.

Finally, having our own department allowed CNS to fully participate in the political and intellectual life of the university on multiple dimensions.

The above comments are just highlights of many years of high productivity and happy collaboration at CNS.

23 Adaptive Resonance Theory: Humans Learn Quickly without Forgetting Quickly

After I got tenure, I exploded with research productivity. One of my most important discoveries was published in 1976, the year after I came to BU. It is called Adaptive Resonance Theory, or ART. In brief, ART is a neural network model and theory that explains how humans and other animals can learn quickly without being forced to forget just as quickly. In other words, we can learn quickly without experiencing catastrophic forgetting, which is still a problem for various other AI learning models, including Deep Learning.

I have called this problem the *stability-plasticity dilemma*. ART solves this problem by achieving both the *plasticity* of fast learning and the *stability* of its learned memories. ART does so by using a combination of bottom-up adaptive filtering and top-down learned expectations that can learn to *pay attention* to those combinations of bottom-up features, called *critical feature patterns*, that are the predictively relevant features, while all other irrelevant, outlier features, are suppressed.

Said in another way, ART uses top-down expectations to *match* bottom-up feature patterns and thereby select, attend, and learn predictively relevant feature patterns, while ignoring irrelevant features. This top-down matching process is what solves the stability-plasticity dilemma. The critical features are also the ones that train the adaptive weights, or LTM traces, in the adaptive filters and expectations, and control successful ART outputs and predictions.

While the bottom-up adaptive filters are activating recognition categories, and the recognition categories are reading out top-down expectations, a positive feedback loop closes between distributed features and categories, and continues to cycle through time. This feedback loop synchronizes, amplifies, and sustains the activity of matched features with their chosen categories. This synchronous, sustained activity is a type of *resonance* (Figure 1.28). Because the sustained activity drives fast learning of critical feature patterns in both bottom-up filters and top-down expectations, I called my theory *Adaptive* Resonance Theory.

ART is currently the most advanced cognitive and neural theory of how our brains learn to attend, recognize, and predict objects and events in a changing world that is filled with unexpected events. All the foundational hypotheses of ART have, by now, been supported by subsequent psychological and neurobiological experiments. ART has also provided principled and unifying explanations of hundreds of additional experiments, and ART has made predictions that have been confirmed by subsequent experiments.

No less important, in an oft-cited article that I published in 1980 in the leading psychology journal, *Psychological Review*, I derived ART from a *thought experiment*. Readers who are familiar with the epochal work

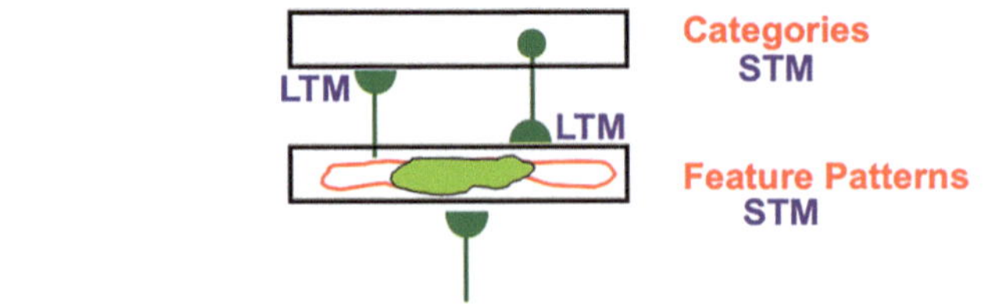

FIGURE 1.28 A two-level neural network as it experiences an adaptive resonance. See the text for details. Author created.

of Albert Einstein may recall that Einstein used thought experiments to derive both his Special Theory of Relativity and his General Theory of Relativity. The power of a good thought experiment is that it is a simple story that develops the logical consequences of a few undeniable hypotheses into a theory that explains profound facts about how Nature works.

My thought experiment demonstrated that ART systems are the *unique* class of solutions of the problem of how *any* system can *autonomously* learn to *correct predictive errors* in a changing world that is filled with unexpected events. The hypotheses from which I derived ART are familiar facts from daily life that we all know. When these facts have acted together as environmental pressures on the evolution of our brains for many thousands of years, ART systems are one result.

The familiar facts from which ART was derived do not mention mind or brain. Thus ART systems are a *universal* class of solutions of this thought experiment. In other words, ART-like systems should be found in any part of Nature that solves the stability-plasticity dilemma. The ability of ART to attend, recognize, and predict objects and events in a changing world must therefore be just one example of a much broader use of ART-like principles in Nature. Indeed, homologs of adaptive resonances seem to occur in multiple organisms and organs, some much more primitive than humans, and not necessarily in brains. I summarize some of these early evolutionary examples, even in single cells, in Chapter 17 of Grossberg (2021).

24 From Solving the Stability–Plasticity Dilemma to Consciousness

The resonances in ART synchronously bind distributed patterns of sensory *features* and the *categories* that learn to selectively fire in response to them (Figure 1.28). I therefore called this kind of resonance a *feature-category resonance.*

After using feature-category resonances to explain detailed properties of objects and events that we recognize in the world, I realized that these resonances subserve our ability to *consciously recognize* these objects and events.

In keeping with the autobiographical flavor of this chapter, Figure 1.29 shows two pictures of me taken in 1976, the year that I began to publish articles about ART. The figure on the right-hand panel was taken by Gail when we were in Paris for a scientific conference.

FIGURE 1.29 Two pictures of me when I was thirty-six years old in 1976, the year that I introduced Adaptive Resonance Theory.

You can see from this picture that I was still pumping iron!

I work out for at least four reasons. One is my bad asthma, which benefits from exercises that encourage regular deep breathing and do not require much endurance. Second, I always felt that working out helped me to think better. I love the rush of endorphins that I experience after exercising. Third, I like to look good for my sweetie at home. And finally, working out is compatible with working on my science. Just as I did at Steve's Gym in Rockefeller University, I alternate each set of pumping iron with a creative activity, whether reading, thinking, writing at my computer, or playing the piano. In fact, right behind me in the home study where I am writing this chapter on my iMac is a Universal Machine home gym. I used to only use it when I could not go to the wonderfully equipped and spacious BU gym. After my retirement and COVID, it is the only gym that I use.

Returning to a discussion of ART, I should mention that, while I was developing ART, I was also developing, starting in 1971, a model about how thoughts and feelings interact to focus motivated attention on causal combinations of valued environmental objects. This research preceded my work on ART, which began in 1975, and provided my first example of an adaptive resonance. I call this model of how cognition and emotion interact the Cognitive-Emotional-Motor model, or CogEM model (Figure 1.30). CogEM learns using an adaptive resonance that I call a *cognitive-emotional resonance.* These resonances support conscious feelings and bind them to the objects and events that cause them. As a result, the critical feature patterns learned by CogEM enable us to plan and act to realize *valued goals.*

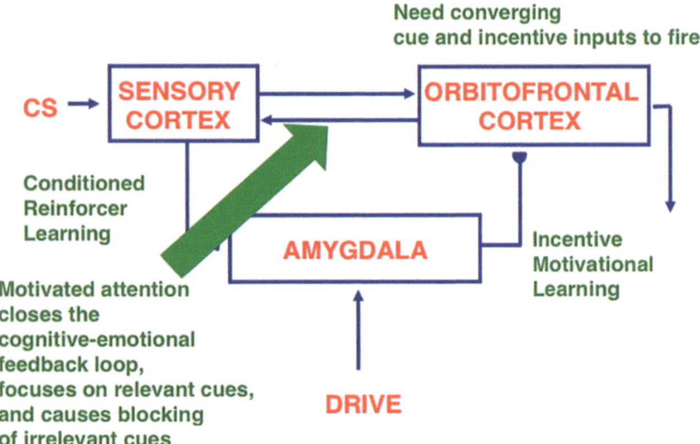

FIGURE 1.30 The Cognitive-Emotional-Motor, or CogEM, neural network. See the text for details. Author's personal property.

25 Six Adaptive Resonances to Consciously Perceive and Recognize the World

As my work continued, I was able to classify six different kinds of conscious awareness, their different functions in behavior, and their different anatomical substrates. In brief, these resonances are:

> *Feature-category resonances* enable us to consciously recognize visual objects and scenes.
>
> *Surface-shroud resonances* enable us to consciously see visual objects and scenes.
>
> *Stream-shroud resonances* enable us to consciously hear auditory objects and streams.
>
> *Spectral-pitch-and-timber resonances* enable us to consciously recognize auditory objects and streams.
>
> *Item-list resonances* enable us to consciously recognize speech and language.
>
> *Cognitive-emotional resonances* enable us to consciously feel emotions and know their sources.

This classification of resonances, and how they work in our brains, proposes a computational solution to the classical mind–body problem. I also explained why not all resonant states are conscious, and why not all brain dynamics are resonant by modeling brain processes that have these properties.

For completeness, let me mention a couple of resonances that do not subserve conscious awareness:

> *Parietal-prefrontal resonances* trigger the selective opening of basal ganglia gates that enable read-out of context-appropriate thoughts and actions. The opening and shutting of basal ganglia gates subserves volitional control. When basal ganglia gates fail to work properly, life-threatening problems like Parkinson's disease may result. Examples of basal ganglia gating are given in Chapters 3 and 4.
>
> *Entorhinal-hippocampal resonances* are predicted to dynamically stabilize the learning of entorhinal grid cells and hippocampal place cells, which play a key role in spatial navigation. The hippocampus plays other important roles as well in learning and memory. Indeed, patients with alcoholic Korsakoff's syndrome can experience a severe form of amnesia that prevents the formation of new memories. These symptoms correlate with hippocampal pathology (Sullivan and Marsh, 2003; Visser et al., 1999). Chapter 3 provides more information about the hippocampus.

Neither of the above resonances is conscious, and thus familiar to us, because they do not include any *qualia*, whether external ones like lights and sounds, or internal ones like hunger or fear. Grossberg (2021) explains how all the above resonances work.

Brain dynamics can also occur that do not include any resonances, notably the ones that control spatial and motor representations. These processes thus do not internally support conscious experiences. These differences can be traced to the computationally complementary organization of our brains.

Figures 1.31 and 1.32 illustrate complementary computing. They show that the perception, object learning,

WHAT	WHERE
Spatially invariant object learning and recognition	Spatially variant reaching and movement
Fast learning without catastrophic forgetting	Continually update sensory-motor maps and gains
IT	PPC

	WHAT	WHERE
MATCHING	EXCITATORY	INHIBITORY
LEARNING	MATCH	MISMATCH

FIGURE 1.31 Properties of complementary computing within the ventral, or What, cortical stream and the dorsal, or Where, cortical stream. Author created.

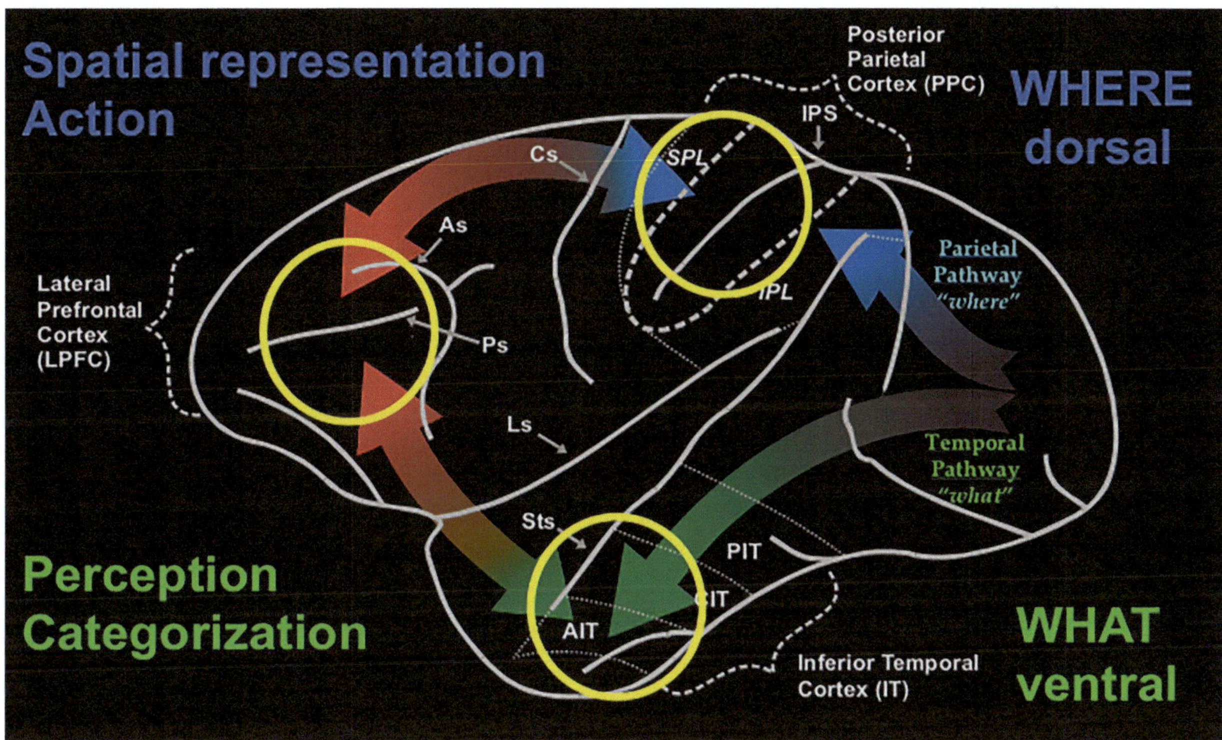

FIGURE 1.32 Anatomical regions where perception and categorization occur in the What stream, and spatial representation and action occur in the Where stream. Adapted from Figure 1 in Two-streams hypothesis, *Wikipedia*, https://en.wikipedia.org/wiki/Two-streams_hypothesis

categorization, and recognition processes that occur within cortical areas like inferotemporal cortex, or IT, in the ventral, or What, cortical processing stream, compute properties that are computationally complementary to the processes that control spatial representation and action within cortical areas like posterior parietal cortex, or PPC, in the dorsal, or Where, cortical processing stream.

Due to complementary computing, IT carries out *match learning* and *excitatory matching*, as in ART, whereas PPC carries out *mismatch learning* and *inhibitory matching*, as in models like the Vector Associative Map, or VAM, model (Gaudiano and Grossberg, 1991, 1992). Due to its inhibitory matching and learning, VAM does not solve the stability-plasticity dilemma. Its learned spatial and motor representations can thereby continually adapt as our bodies change throughout our lives, and thereby effectively control them.

For example, when we consciously see and recognize familiar visual art, surface-shroud resonances and feature-category resonances synchronize in our brains, so we can consciously see and recognize familiar paintings and sculptures.

When we consciously hear and recognize familiar music, spectral-pitch-and-timber resonances, stream-shroud resonances, and feature-category resonances synchronize in our brains, so we can consciously hear and recognize familiar pieces of music.

When we consciously have feelings about the art and music that we are experiencing, the above resonances synchronize with cognitive-emotional resonances.

When we consciously understand language meanings by associating language utterances with the perceptual and affective experiences that they describe, all six of the above resonances synchronize across our brains.

26 The Resonances that Support Art, Music, and Language Meanings

Understanding how several of the above resonances work has made possible my research on the main topics of this book: visual art, music, and language meanings.

27 Laminar Computing: A Canonical Cortical Circuit Supports Intelligence

ART has been incrementally developed over the years with over one hundred collaborators, notably Gail Carpenter, and our many gifted CNS PhD students and postdoctoral fellows. One of the outcomes of this development is a revolutionary computational paradigm that I call Laminar Computing. Laminar Computing was developed in response to the fact that the cerebral cortex, which is the seat of higher intelligence in all modalities, is organized into layered circuits, often with six main layers, that undergo characteristic bottom-up, top-down, and horizontal interactions. Figure 1.33 shows a schematic of this canonical cortical circuit.

Laminar Computing has offered a unified computational explanation of how specializations of this shared canonical laminar circuit design embody different types of biological intelligence. My colleagues and I have modeled how such seemingly different human capabilities as vision and object recognition, speech and language, and cognition all emerge from variations of the same canonical laminar circuit design. Moreover, multiple species use variations of this canonical circuit, including bats (Suga and Ma, 2003; Suga, O'Neill, and Manabe, 1979; Zhang, Suga, and Yan, 1997).

I should emphasize that we developed non-laminar models of each of these human capabilities over a period of years before we gradually began to understand how each of them could be embodied in particular specializations of canonical laminar cortical circuits in an elegant, parsimonious, and predictively more powerful theory.

Remarkably, this canonical laminar design of neocortex realizes the best properties of feedforward and feedback processing, analog and digital processing, and bottom-up data-driven processing and top-down attentive hypothesis-driven processing (Figure 1.33). As part of its top-down attentive processing, this canonical circuit learns to focus attention upon the critical feature patterns that control learning and predictive success, and to dynamically stabilize the memories of previously learned patterns. These important properties depend upon properties of the ART Matching Rule, which is realized by a *top-down, modulatory on-center, off-surround circuit* that is distributed across and within cortical layers by a design that I call *folded feedback* (Figure 1.34). Folded feedback means that top-down signals from a higher cortical region, such as visual cortical area V2, activate layer 6 in a lower visual cortical area, say V1, before layer 6 activates layer 4 in V1.

As Figure 1.34 shows, due to folded feedback, when a top-down ART Matching Rule circuit is combined with a bottom-up data-driven circuit, pre-attentive automatic bottom-up processing and attentive task-selective top-down processing converge in layer 6. This happens in all the brain areas of laminar neocortex (e.g., Figure 1.35).

TOP-DOWN ATTENTION AND FOLDED FEEDBACK

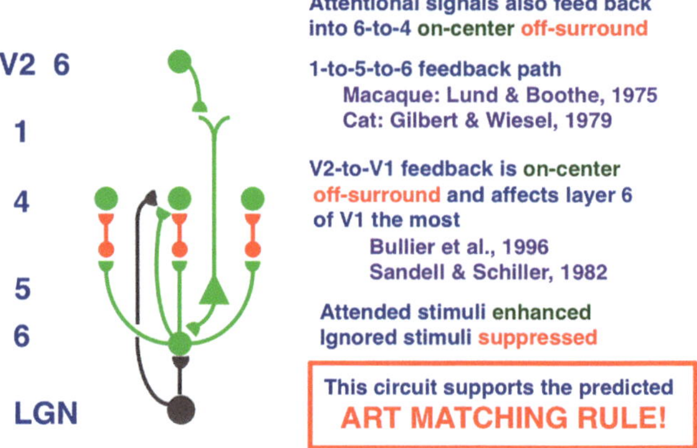

FIGURE 1.34 A schematic of how top-down attention realizes the ART Matching Rule by identified cell types in laminar neocortex using a folded feedback neural network. Author created.

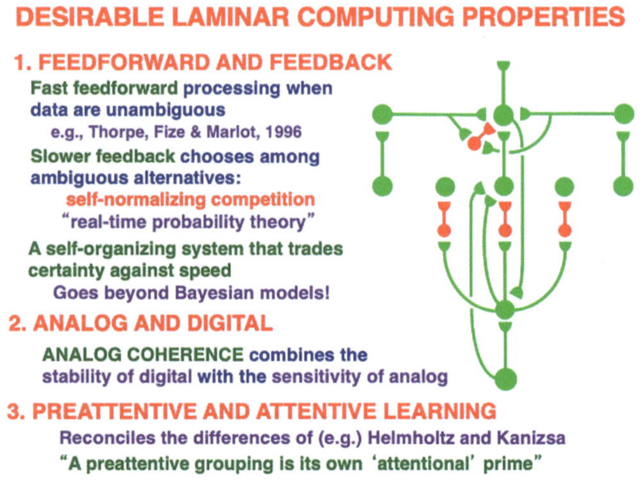

DESIRABLE LAMINAR COMPUTING PROPERTIES

1. FEEDFORWARD AND FEEDBACK
Fast feedforward processing when data are unambiguous
e.g., Thorpe, Fize & Marlot, 1996
Slower feedback chooses among ambiguous alternatives:
self-normalizing competition
"real-time probability theory"
A self-organizing system that trades certainty against speed
Goes beyond Bayesian models!
2. ANALOG AND DIGITAL
ANALOG COHERENCE combines the stability of digital with the sensitivity of analog
3. PREATTENTIVE AND ATTENTIVE LEARNING
Reconciles the differences of (e.g.) Helmholtz and Kanizsa
"A preattentive grouping is its own 'attentional' prime"

FIGURE 1.33 A schematic of the remarkable properties of laminar computing and its realization by laminar neural networks in the neocortex. Author created.

V2 REPEATS V1 CIRCUITRY AT LARGER SPATIAL SCALE

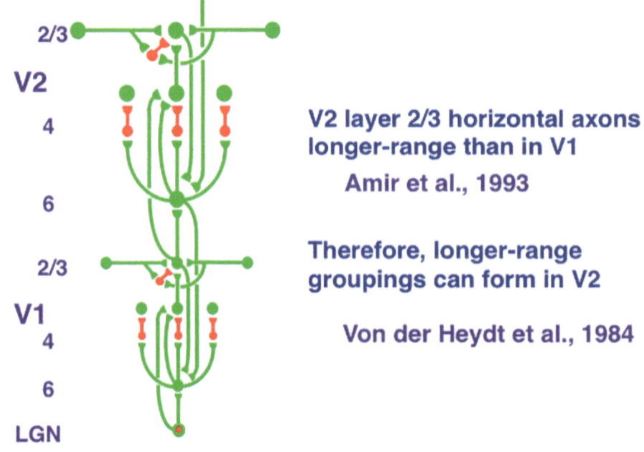

FIGURE 1.35 The canonical laminar neocortical circuit is repeated at multiple cortical levels, where it is specialized to carry out many different processes that realize human intelligence. Author created.

All these brain regions use the competitive interactions within layers 4 and 6 to decide what combinations of bottom-up and top-down signal patterns will survive. Moreover, due to the folded feedback circuit design, a top-down signal from a high level of cortex, say prefrontal cortex, can subliminally *prime* all the layers 4 in lower-level cortices via their modulatory on-centers to prepare for a bottom-up input pattern that may, or may not, match them.

The balance between excitatory processes (in green) and inhibitory processes (in red) during interactions in this circuit also realizes the property of *analog coherence* (Figure 1.36). Analog coherence means that neocortical cells can compute analog, or graded, activities even as they are bound into resonant states by active feedback. In particular, the feedback that supports these resonances does not force cell activities to be either maximal (e.g., 1) or minimal (e.g., 0). Cortical circuits that compute analog coherence can "weigh the evidence" for each input pattern that they process.

More basic research needs to be done to completely understand how laminar cortical designs contribute to all aspects of human intelligence.

Even before that research is done, embodying these designs in variations of a single canonical VLSI computer chip will facilitate the development of increasingly general-purpose, adaptive, autonomous, and intelligent algorithms, machines, and robots for multiple applications. Such a development will be facilitated by the fact that each such chip, by sharing the same canonical design, also shares the same arrangement of input and output pathways that can be used to connect multiple chips into a "silicon brain" capable of AGI, or Artificial General Intelligence.

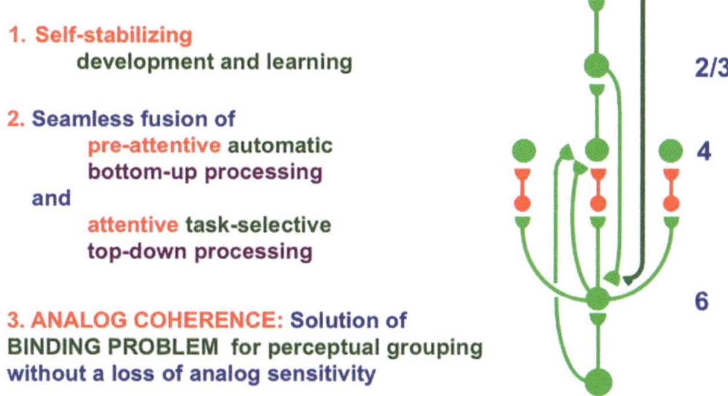

FIGURE 1.36 Some fundamental emergent properties that are carried out by interactions in laminar neocortical circuits, including the crucial property of analog coherence whereby our brains maintain sensitivity to analog neuronal activities while binding them into coherent groupings via feedback interactions. Author created.

28 Complementary Computing: How Brain Specialization Is Organized

As I briefly noted above, another revolutionary paradigm started to emerge from my work over the years and should be embodied in AGI. This paradigm came into view because, as more neural models of various types of brain processes were developed, they collectively began to provide an answer to the basic question: What is the nature of brain specialization? How are the various components of biological intelligence globally organized in our brains?

Many scientists had proposed that our brains possess independent modules, as in a digital computer. The brain's organization into distinct anatomical areas and processing streams shows that brain processing is indeed specialized. However, independent modules should be able to fully compute their particular processes on their own. However, theoretical explanations of much psychological data strongly contradict this possibility.

Anatomical data support these theoretical explanations. For example, Figure 1.37 (lower right image) shows a macrocircuit of the monkey visual system due to Felleman and van Essen (1991), in which multiple parallel processing streams, with multiple, hierarchically organized processing stages, interact via bottom-up, horizontal, and top-down interactions.

Complementary Computing is the name that I gave for my discovery that pairs of parallel cortical processing streams compute *computationally complementary* properties in the brain. Each stream has its own computational strengths and weaknesses, much as in physical principles like the Heisenberg Uncertainty Principle. You can intuitively think of complementary properties by analogies like puzzle pieces fitting together, or a Yin-Yang organization.

Each cortical stream also possesses multiple processing stages. These stages realize a *hierarchical resolution of uncertainty*. "Uncertainty" here means that computing one set of properties at a given stage prevents computation of a complementary set of properties at that stage, and that a hierarchy of interacting processing stages, as illustrated by our model of the neural architecture of the visual system in Figure 1.37, is needed to overcome these uncertainties to compute complete information whereby to interact successfully with changing external environments.

Complementary Computing thus proposes that the computational unit of brain processing that has behavioral

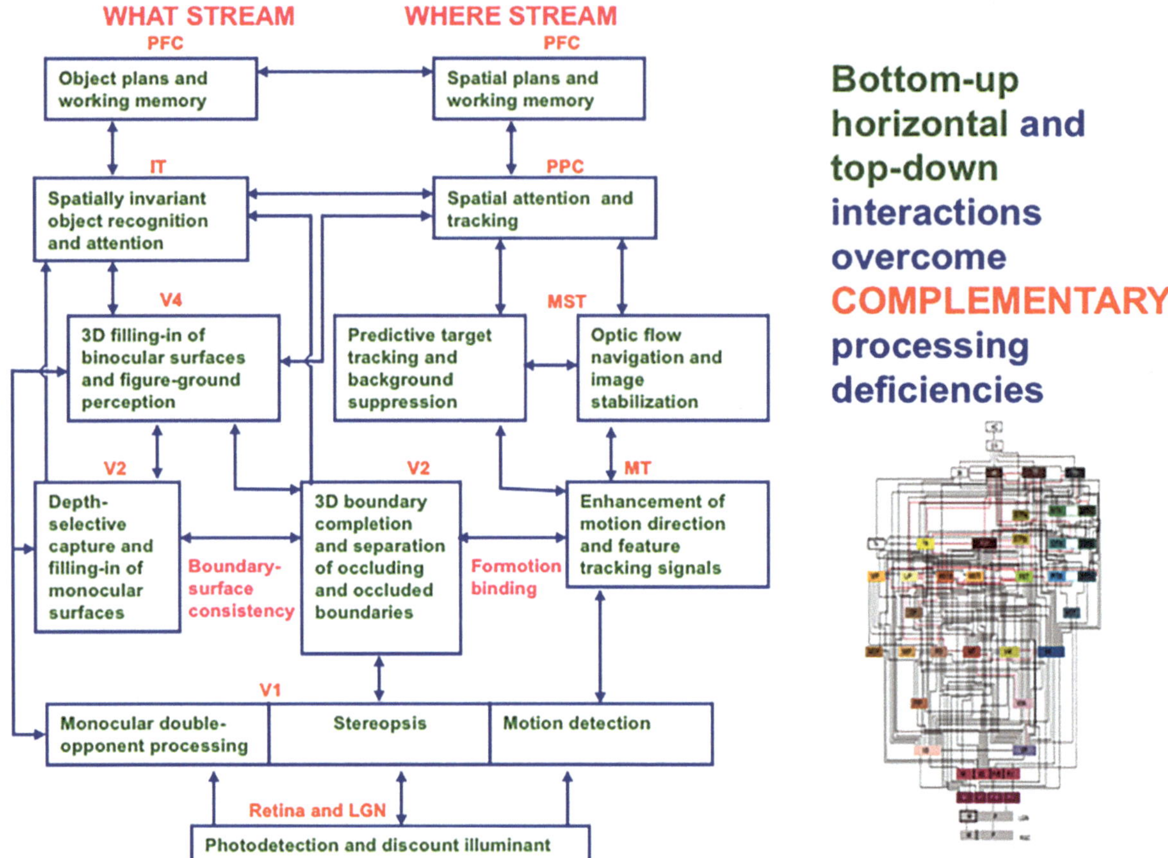

FIGURE 1.37 An emerging unified theory of visual intelligence is carried out by a model macrocircuit of key visual processes (in green) and the cortical areas in which they primarily occur (in red), from the retina to the prefrontal cortex (PFC), including both the What and Where cortical streams. The bottom-up, horizontal, and top-down interactions help each of these processes to overcome computationally complementary processing deficiencies that they would experience without them, and also to read-out top-down expectations that help to stabilize learning while they focus attention on salient objects and positions. Author created.

significance consists of pairs of complementary cortical processing streams with multiple processing stages that interact within and between each other. These interactions include a hierarchical resolution of uncertainty that computes complete information about a particular type of biological intelligence. For example, when we perceive and recognize a great work of art, and act in response to it, our brains use multiple computationally complementary cortical areas (Figure 1.38), including the inferotemporal cortex for perception and recognition, and the parietal cortex for spatial representation and action.

The above discoveries illustrate the fruits of many years of high productivity and happy collaboration at CNS, funded by millions of dollars in research grants that I helped to formulate, write, defend, and lead. My activities at CNS led to many other training and research leadership projects over the years. Two places where highlights of these activities are summarized are my Wikipedia page and an editorial that I wrote when I stepped down as the founder and first editor-in-chief of the journal *Neural Networks*:

https://en.wikipedia.org/wiki/Stephen_Grossberg

https://sites.bu.edu/steveg/files/2016/06/GrossbergNNeditorial2010.pdf

29 Working on Art, Music, and Language Meanings and Zooming the World

During the past few years, I have tremendously enjoyed developing brain models of how humans learn about and consciously experience visual art, music, and language

SOME COMPLEMENTARY PROCESSES	
Visual Boundary	Visual Surface
Interblob Stream V1-V2-V4	Blob Stream V1-V2-V4
Visual Boundary	Visual Motion
Interblob Stream V1-V2-V4	Magno Stream V1-MT-MST
WHAT Steam	WHERE Stream
Perception & Recognition	Space & Action
Inferotemporal and Prefrontal areas	Parietal and Prefrontal areas
Object Tracking	Optic Flow Navigation
MT^- Interbands and MSTv	MT^+ Bands and MSTd
Motor Target Position	Volitional Speed
Motor and Parietal Cortex	Basal Ganglia

FIGURE 1.38 Some computationally complementary pairs of brain processes and their anatomical substrates. Process functions and anatomical substrates are rendered in different colors. Author created.

meanings, the main topics of this book. I have also been invited to contrast our biological neural models with some popular AI models and their serious defects (Chapter 5). These discoveries provide a foundation that I hope may inspire other scientists to learn more about how these foundational processes of the human condition work in our brains.

When COVID struck at the beginning of my studies of art, music, and language meanings, I initially assumed that my life as a conference lecturer had ended. I faced this possibility with gratitude for the many years that I was able to travel to countries around the world to lecture to senior researchers and students alike. I had a really good run, and got to see a lot of the world while doing it.

However, on every long flight to and from a conference where I gave an invited keynote lecture, I typically caught a cold, which triggered an asthma attack. Each asthma attack, as often as not, ended me up in a hospital with pneumonia. Each asthma attack further damaged my lungs. I therefore realized in 2019, just before COVID struck, that I could not fly any more.

It was therefore a pleasant surprise, amid so much tragedy during the terrible COVID pandemic, that many international conferences were held over Zoom. I was honored with invitations to give keynote lectures to conferences organized from places around the world as far-flung as Ankara, Bangkok, Beijing, Boston, Bucharest, Campinas, Edinburgh, Glasgow, Johannesburg, London, Marseille, Melbourne, Mumbai, New York, Newcastle Upon Tyne, Padua, Palo Alto, Porto, San Francisco, Santander, and Tehran.

Just as COVID and Zoom have unexpectedly changed so much about how we run our lives, I have no idea what my own future will bring. I have taken the liberty of writing an Overview of this book that is more personal and autobiographical than I had originally planned, not only to clarify how I came to write a book about art, music, and language meanings but also because I hope that researchers in the future, especially students, will find some of my personal ups and downs instructive in finding their own path to a creative and meaningful life in science.

I deeply resonate with the wonderful song by Stephen Sondheim: I'm Still Here!

https://www.youtube.com/watch?app=desktop&v=3XzlTUgdG6A

Art

How humans consciously see and recognize visual art

"Art is never finished, only abandoned"
Leonardo da Vinci

"The true work of art is but a shadow of the divine perfection"
Michelangelo

"I dream of painting and then I paint my dream"
Vincent Van Gogh

"It took me four years to paint like Raphael, but a lifetime to paint like a child"
Pablo Picasso

"If I could say it in words there would be no reason to paint"
Edward Hopper

Your Creative Brain and AI. Stephen Grossberg, Oxford University Press. © Oxford University Press (2026).
DOI: 10.1093/9780198965367.003.0002

1 Visual Art from the Old Stone Age to the Present

Visual art has been a part of human expression since at least Paleolithic times, when early humans drew primitive pictures on the walls of caves, such as the Cave of Altamira of Northern Spain. The Paleolithic period is also called the Old Stone Age (from Greek: παλαιός *palaios*,[1] "old" and λίθος *lithos*,[2] "stone"), which derives its name from the use of stone tools by the hominins of this period, although they also used wood and bone tools.

Early humans also drew maps as part of wall paintings and rock carvings for thousands of years before they ever drew on papyrus or paper. Maps are useful in every human activity, ranging from depicting the locations of farms, houses, and cities, to navigating to destinations via land, sea, or air. The use of maps and paintings even by early humans illustrates how pragmatic needs of interacting with the world and aesthetic needs of representing it in pictures have always been intertwined.

The longing to pictorially represent the external world has persisted to modern times, as has our need to pictorially represent the inner world of our visual imagination. Our ability to experience the real world through a picture, or a series of pictures, remains one of the foundations of human civilization, whether in the pages of books; the drawings of architectural, medical, and astronomical practitioners; or on our computer, TV, and movie screens.

Every time that we see, or try to create, a picture, our brains are challenged to understand how a few lines or color patches on an approximately flat surface can induce mental representations of objects and backgrounds in a scene that we consciously see in depth. A related challenge is being able to consciously see and recognize objects in pictures that are perceived to be in front of other objects that they partially occlude. No less of a challenge is being able to recognize objects that are partially occluded, even though we cannot see their occluded parts. The paintings of great visual artists expose us to these varied experiences. The sculptures of great artists challenge us in other ways, notably by emphasizing or minimizing features and shapes of naturally occurring, or imagined, objects that we can only see at any given moment from a particular view or perspective.

1 https://en.wiktionary.org/wiki/%CF%80%CE%B1%CE%BB%CE%B1%CE%B9%CF%8C%CF%82

2 https://en.wiktionary.org/wiki/%CE%BB%CE%AF%CE%B8%CE%BF%CF%82

1.1 Paintings, sculptures, and how our brains consciously see them

This chapter first clarifies the brain processes that enable humans to consciously see and recognize two-dimensional (2D) paintings as representations of objects and scenes in a three-dimensional (3D) world. This happens because the same brain processes enable us to consciously see and recognize the 3D objects and scenes in the world that the paintings depict. This unified explanation is a major insight of my neural network models of how humans consciously see. The chapter then shows that these brain processes also enable us to consciously see and recognize 3D sculptures. In this chapter, I mention sculptures of *Michelangelo* di Lodovico Buonarroti Simoni, Gian Lorenzo *Bernini*, and Auguste René *Rodin*.

Whenever an artist manipulates a canvas and experiences conscious percepts of an emerging painting, the artist is performing an experiment that probes different combinations of the brain processes involved in seeing. Artists typically do so without explicitly knowing the brain processes that mediate between their painterly manipulations and their conscious percepts. The particular interests and aesthetic sensibilities of different artists have led them to instinctively emphasize different combinations of brain processes. My models of how these different combinations generate conscious percepts clarify the different styles of individual artists and of entire artistic movements.

This chapter addresses these issues by analyzing paintings by well-known American and European artists. Chapter 1 also mentioned magnificent art that was created in other cultures, notably African, New Guinea, and Asian art.

Illustrative paintings by thirteen artists will be given a unified analysis in the light of neural design principles and mechanisms that have been computationally characterized by neural network models of how advanced brains consciously see. Where appropriate, the structure of the examined paintings will be tied to the artist's intentions or to reviews of the artist's work written by art historians, curators, or critics.

I started developing neural network models of how our brains consciously see the world in the 1980s. As this understanding matured, it became clearer to me how the brain processes whereby we see the world also clarify how we see visual art.

These neural models are used to explain how paintings of Jo Baer, Banksy, Ross Bleckner, Gene Davis, Charles Hawthorne, Henry Hensche, Henri Matisse, Claude Monet, Jules Olitski, Rembrandt van Rijn, Graham Rust, Frank Stella, and Leonardo da Vinci achieved their aesthetic effects. These paintings were chosen to illustrate processes that range from discounting the illuminant and lightness anchoring, to boundary, shading, and texture grouping and classification, through filling-in of surface brightness and color, to spatial attention, conscious seeing, and eye movement control. Variants of this analysis were also provided in Grossberg (2021) and Grossberg and Zajac (2017).

Other valuable books also link the science of vision to the appreciation of art. The books by Margaret Livingstone (Livingstone, 2002) and Semir Zeki (Zeki, 1999) are particularly notable. Both Livingstone and Zeki are leading experimental visual neuroscientists, who described fascinating facts about the brain and visual perception, but did not explain how interacting brain mechanisms give rise to conscious visual percepts. Practitioners of art history and aesthetic theory also do not make the link between brain dynamics and conscious psychological experiences and will thus not be further discussed.

Some writers have questioned whether science can explain how humans appreciate the beauty of artworks, and many others have discounted the possibility that humans can scientifically understand any conscious experience. Conway and Rehding (2013) wrote that "it is an open question whether an analysis of artworks, no matter how celebrated, will yield universal principles of beauty" and that "rational reductionist approaches to the neural basis for beauty...may well distill out the very thing one wants to understand...Its progress in uncovering a beauty instinct, if it exists, may be accelerated if the field were to abandon a pursuit of beauty per se and focus instead on uncovering the relevant mechanisms of decision making and reward and the basis for subjective preferences...This would mark a return to a pursuit of the mechanisms underlying sensory knowledge: the original conception of aesthetics."

This chapter does explain brain "mechanisms underlying sensory knowledge." In addition, Perlovsky (2010) was inspired to write about aesthetic emotions based on some of my neural models. Although the chapter does not "focus on uncovering the relevant mechanisms of decision making and reward and the basis for subjective preferences," the brain processes that control decision-making and reward have been extensively modeled over the years, and used to provide principled and unifying explanations of hundreds of relevant psychological and neurobiological facts; e.g., Brown, Bullock, and Grossberg (1999), Grossberg (1972a, 1972b, 1974, 2018), Grossberg and Gutowski (1987), Grossberg and Levine (1987), Grossberg, Levine, and Schmajuk (1988), Grossberg and Merrill (1992), Grossberg and Schmajuk (1987, 1989).

In Chapter 1, I commented briefly about *surface-shroud resonances* whereby we consciously see objects in the external world, including visual art. I also mentioned the *feature-category resonances* whereby we can consciously recognize familiar art. Feature-category resonances emerge from the brain processes that are modeled by Adaptive Resonance Theory, or ART. I will discuss ART in some detail later in this chapter. In Chapter 1, I also mentioned the CogEM model of how humans learn to experience conscious feelings and link them via *cognitive-emotional resonances* to the objects and events that cause them (Figure 1.33).

The present exposition will summarize brain mechanisms of perception and cognition whereby humans consciously see paintings and whereby painters have achieved their aesthetic goals. Modeling studies that I will summarize clarify how perceptual and cognitive processes interact with emotional processes to create coherent conscious experiences of seeing, knowing, and feeling about familiar objects, including art (e.g., Grossberg, 2013, 2017, 2021). Cognitive-emotional interactions are influenced by reward, so in this sense, the chapter does review "the relevant mechanisms of...reward and the basis for subjective preferences."

An important source of modeling insights clarifies how humans consciously see whatever scenes, familiar or unfamiliar, are before them, including scenes that include paintings. This chapter accordingly analyzes illustrative paintings or painterly theories in light of neural models such as the Form-And-Color-And-DEpth (FACADE) theory of 3D vision and figure-ground perception, including how humans perceive partially occluded objects in a picture (e.g., Grossberg, 1994, 1997; Grossberg and McLoughlin, 1997; Kelly and Grossberg, 2000). The 3D LAMINART model extends this analysis by modeling how identified laminar circuits in visual cortex embody and extend FACADE design principles and mechanisms (e.g., Cao and Grossberg, 2005, 2012; Fang and Grossberg, 2009; Grossberg, 1999; Grossberg, Mingolla, and Ross, 1997; Grossberg and Yazdanbakhsh, 2005). These laminar circuits are variations of the canonical laminar cortical circuit that is used to represent all aspects of conscious perception and recognition (Figures 1.36–1.38).

I am eager to tell you about these models because they have provided unified and principled explanations and predictions of much more psychological and neurobiological data about vision than other available concepts and models. These models are also the only available neural models that propose an explanation of what happens in a viewer's brain when having a conscious visual experience, including when we view a painting. I hereby use our models to link paintings, as visual inputs, to the brain mechanisms that create conscious visual percepts of the paintings. This discussion is organized in a way that may shed new light on artists' aesthetic struggles, and on how humans see the paintings that resulted from these struggles.

I hope that artists who may read about these examples will have a more precise knowledge of how their own painterly manipulations give rise to conscious percepts of their paintings in their own minds. Indeed, an artist who has read some of my earlier discussions on this topic wrote: "As an artist I found Dr. Grossberg's insights into the creative process and vision new and fascinating."

1.2 From gist to scene understanding

When a person first looks at any scene, whether it is a scene in a painting or in the physical 3D world, the information extracted from it most quickly is the scene's *gist* (Friedman, 1979; Intraub, 1999; Oliva, 2005; Potter, 1976; Potter and Levy, 1969). Gist is capable of providing, with a single glance, sufficient information for recognizing what type of scene it is, whether of a city street, forest, mountain, coast, or countryside (Grossberg and Huang, 2009; Oliva and Torralba, 2001). I will explain below how the gist of a scene may be computed from properties of the basic functional units of vision—their *boundaries* and *surfaces*—as can a wide range of other scenic properties, including 3D shape, illusory contours, texture, shading, depth, color, brightness, and object identity. All of these processes typically interact to influence how a 3D scene, or a 2D painting, can generate a context-sensitive 3D representation of the scene in the mind of a viewer.

Gist is computed in our brains from a process that is primarily *bottom-up*; that is, one whereby visual information coming from a scene activates the photosensitive retinas in our eyes, which, in turn, generate signals that propagate ever deeper into our brains. *Top-down* processes also strongly influence the 3D representations that we see. Top-down processes begin at higher brain regions and send signals to brain regions that are ever closer to our retinas. These top-down processes include volitionally controlled *expectations* that focus *attention* upon *critical feature patterns* that embody predictive knowledge of objects and scenes that have been learned from previous experiences. My discussion of ART in Chapter 1 provided an introduction to this vast topic.

Our mechanistic account will need both bottom-up and top-down processes to provide a unified comparative analysis of different paintings. Efforts to understand how top-down and bottom-up processes interact have a long history in visual perception and have triggered an enduring controversy about how we see and recognize the world.

Hermann von Helmholtz, one of the greatest scientists of the nineteenth century, advocated a top-down view when he proposed that we see using *unconscious inferences*, or learned expectations (Figure 2.1, top panel), to see what we expect to see, based on past experiences. I also mentioned Helmholtz as one of three early European vision scientists in Chapter 1 (Figure 1.13).

In contrast, Gaetano Kanizsa (Figure 2.1, bottom panel) provided brilliant counterexamples to Helmholtz's hypothesis using images, such as the one in Figure 2.2 (top right), that we see perfectly well, even though what we see may contradict our expectations (Kanizsa 1955,

FIGURE 2.1 Hermann von Helmholtz (left) and Gaetano Kanizsa (right). Author created. See text for details. The pictures taken from the web Hermann von Helmholtz: https://en.wikipedia.org/wiki/Hermann_von_Helmholtz; Gaetano Kanizsa: https://www.rivistailmulino.it/a/gaetano-kanizsa-br-1913-1993

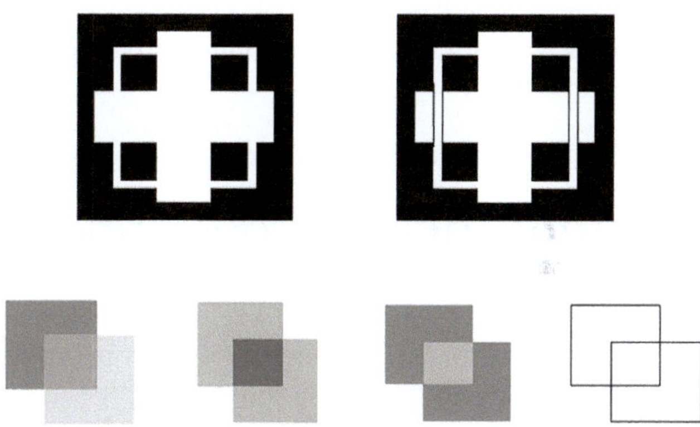

FIGURE 2.2 (top row) Kanizsa stratification. (bottom row) Transparency images. Top row (left column): "Nuove Ricerche Sperimentali Sulla Totalizzazione Percettiva," by G. Petter, 1956, Rivista di Psicologia. 50. Copyright 1956 by Giunti: Gruppo Editoriale. Top row (right column): "Seeing and Thinking," by G. Kanizsa, 1985, Acta Psycologia, 59. Copyright 1985 by Elsevier Science. Bottom row is adapted from Grossberg, S., and Yazdanbaksh, A. (2005). Laminar cortical dynamics of 3D surface perception: Stratification, transparency, and neon color spreading. *Vision Research*, 45, 1725–1743.

1974, 1979). In this figure, we consciously perceive an emergent cross whose horizontal bars lie behind the emergent square, but whose vertical bars are perceived to lie in front of it. This percept should be impossible if our learned expectations about squares usually being flat determined our percepts. Using such cleverly contrived images, Kanizsa emphasized the power of bottom-up visual processes, such as adaptive filtering and perceptual grouping, that act directly on visual scenes and images at early stages of brain processing and can greatly influence our conscious percepts of them.

Both Helmholtz and Kanizsa were partly correct. A more comprehensive understanding has arisen from neural models that explain how bottom-up and top-down processes work together to generate an attentive consensus, or *adaptive resonance*, between what is there in the world and what we expect to see based on our past experiences. Such a consensus typically involves all the processes of bottom-up adaptive filtering, horizontal perceptual grouping, and top-down attentive matching of learned expectations with bottom-up feature patterns (Figure 1.38). ART also explains how we can rapidly learn everything that we know about a changing world throughout life, often without forgetting what we learned many years ago, and how we become conscious of events as we learn about them (Carpenter and Grossberg, 1987, 1991; Grossberg, 1976, 1980, 2013, 2017, 2021).

Helmholtz and Kanizsa could not fully make these connections because they did not have either the intuitive concepts or the mathematical tools needed to express them clearly, which often happens when great pioneers initiate scientific revolutions.

ART concepts will be explained in more detail starting in Section 2.9. Before then, key bottom-up processes first need to be reviewed.

2 Theoretical Introduction

This section reviews some of the basic neural principles and mechanisms that will be used to discuss paintings by several artists. Section 3 will build upon this introduction to provide additional information that will also be needed. Section 4 will apply this information to analyze paintings by Jo Baer, Banksy, Ross Bleckner, Gene Davis, Charles Hawthorne, Henry Hensche, Henri Matisse, Claude Monet, Jules Olitski, Rembrandt van Rijn, Graham Rust, Frank Stella, and Leonardo da Vinci, whose paintings highlight the effects of different combinations of visual properties. Later in the chapter, I will discuss the sculptures of *Michelangelo* di Lodovico Buonarroti Simoni, Gian Lorenzo *Bernini*, and Auguste René *Rodin*.

2.1 Boundary completion and surface filling-in: Visual illusions

The functional units of visual perception are 3D boundaries and surfaces, or more precisely, 3D representations of completed boundary groupings and filled-in surfaces (Grossberg, 1987a, 1987b, 1994). The words "completed" and "filled-in" refer to processes that compensate for the incomplete nature of boundary and surface information that is propagated bottom-up to the brain from each retina. The nature of the incompleteness may be due to the environment, or to the structure of the retina itself.

Figure 2.3 shows a famous image of a Dalmatian in Snow, wherein the incompleteness of boundaries comes from the visual environment. When we first glance at this picture, it may just look like an array of black splotches of different sizes, densities, and orientations. Gradually, however, we can recognize a Dalmatian when boundary groupings form in our brain between the black splotches. These emergent boundaries are *visual illusions* that are not in the image itself. They are created in the visual cortex.

Although they are illusory, these boundary groupings are necessary for us to recognize the dog. This experience of recognition is remarkable because the emergent boundaries are perceptually invisible, or *amodal*: They are not lighter or darker than the white background, nor of a different color, nor are they perceived at a different depth. This percept illustrates that *we can consciously recognize invisible boundaries*. When viewing the Dalmatian in Snow, both invisible and visible boundary groupings cooperate to group image fragments into completed boundaries that we use to recognize the Dalmatian.

The Dalmatian in Snow example illustrates that *some* boundaries are invisible. I predicted many years ago (Grossberg, 1984, 1987a, 1987b, 1994, 1997) that, actually, "*all* boundaries are invisible," at least where they are computed within the boundary formation stream of the visual cortex; see text below. Some other examples of this boundary property are provided in Figure 2.4.

FIGURE 2.3 Dalmation in snow. No discernible source, no credit.

FIGURE 2.4 Four variations of a Kanizsa square. (Left panel, upper row) Kanizsa square, and (Right panel, upper row) reverse-contrast Kanizsa square percepts. The spatial arrangement of pac-men, lines, and relative contrasts determines the perceived brightness of the squares, and even if the illusory square exhibits no brightness difference from its background, as in (Right panel, upper row). These factors also determine whether pac-men will appear to be amodally, or invisibly, completed behind the squares, as in (Left column, bottom row), and how far behind them, as in (Right panel, lower row). Figure in (Right panel, upper row) reprinted from Kanizsa, G. (1976). Subjective contours. *Scientific American*, 234, p. 51. The other figures are variations of that figure drawn by the author.

Figure 2.4 (upper row) includes a Kanizsa square (left panel) and a reverse-contrast Kanizsa square (right panel). The Kanizsa square is a visual illusion that is induced by boundary completion between pairs of collinear pac-man edges. The square is visible because each of the four black pac-men induces brightness contrasts within the interior of the completed square. These contrasts can then spread, or fill-in, surface brightness or color within the illusory square boundary until they hit either the real pac-man boundaries or the illusory square boundaries. This flow of brightness and color behaves like a fluid that diffuses away from a source until it hits a barrier.

Boundaries play the role of a dam that contains the flow of brightness and color and keeps it from flowing outside the contours of the boundary. By acting like barriers, or obstructions, to the flow of brightness and color, boundaries can make themselves visible by causing a different brightness or color to occur on opposite sides of the boundary. Both types of boundaries, both those induced directly by pac-man image contrast and those completed between pairs of collinear pac-men, act as barriers to the spread of brightness and color.

After filling-in is complete, the interior of the Kanizsa square appears brighter than the background around it.

The reverse-contrast Kanizsa square is induced by two white pac-men and two black pac-men on a gray background. It induces a visual illusion that can be *consciously recognized, but not consciously seen.* We consciously *know* that a square boundary is there, but we cannot consciously *see* it. As in the case of the Dalmatian in Snow, when we look at the reverse-contrast Kanizsa square, we consciously recognize a boundary that we cannot see because the boundary is invisible. Why is the illusory boundary that is induced by the reverse-contrast Kanizsa square invisible?

Both the Kanizsa square and the reverse-contrast Kanizsa square are induced by boundary completion between pairs of collinear pac-man edges. In the case of the reverse-contrast Kanizsa square, however, each pac-man in such a pair has opposite contrast (dark/light or light/dark) relative to the gray background. Each black pac-man induces a brightness contrast within the gray square region. Each white pac-man, however, induces a darkness contrast within the gray square region. If the gray square color is chosen appropriately, the filling in of the two brightness contrasts cancels the filling in of the two darkness contrasts inside the square boundary. As a result, the gray color inside the boundary matches that outside the square, so the boundary is invisible.

This explanation works because, despite having opposite contrasts, all four pairs of collinear pac-men can induce an illusory boundary between them. This can happen because our brains add inputs from like-oriented, but oppositely polarized, dark/light and light/dark contrasts at a cell type that can thus respond to opposite polarities. This is accomplished in two processing stages.

The first stage contains cells at each position whose oriented receptive fields are sensitive to just one polarity, either dark/light or light/dark, but not both. These cells are called *simple cells* (Figures 2.5 and 2.6). Next, at each

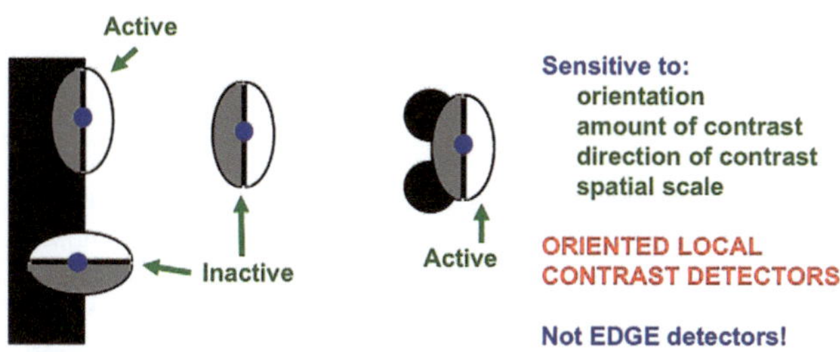

FIGURE 2.5 Properties of a cortical simple cell. As shown in the figure, simple cells that respond to oriented inputs at different positions in a way that is sensitive to the contrast of the image to which they are exposed. Reprinted with author's permission from Grossberg, S., and Zajac, L. (2017). How humans consciously see paintings and paintings illuminate how humans see. *Art & Perception*, 5, 1–95.

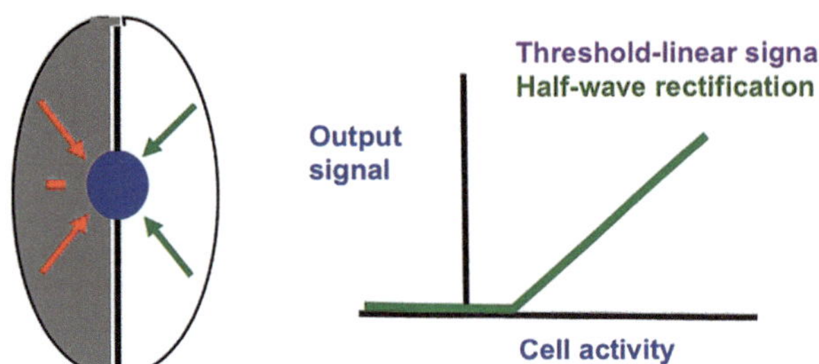

FIGURE 2.6 Receptive field properties of a cortical simple cell. This cell has a vertically oriented receptive field (elliptical region) whose cell body (blue disk) is excited by light that reaches the right half of its receptive field (green arrows) and inhibited by light that reaches the left half of the receptive field (red arrows). When the net cell activity is excited sufficiently, it generates an output signal that grows as it increases above a threshold (see green diagonal line on the right half of the figure). Author created.

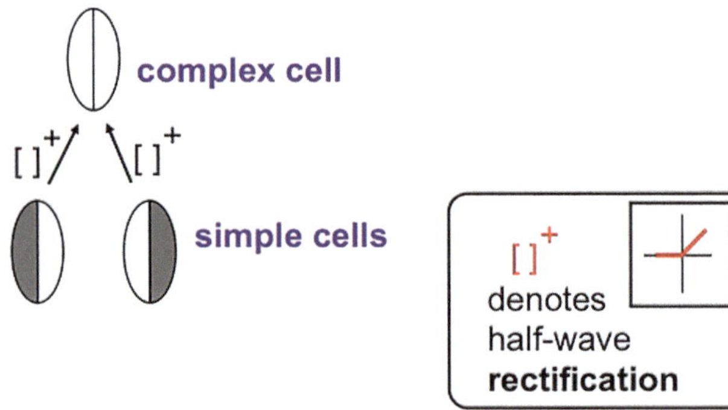

FIGURE 2.7 How output signals from a pair of oppositely polarized but like-oriented, simple cells input to a complex cell. Complex cells pool inputs from simple cells that are sensitive to opposite contrast polarities. Complex cells hereby become contrast invariant, and can respond to contrasts of either polarity. Author created.

position, simple cells with similar orientational tuning, but opposite contrast polarity preference, add their inputs at the next processing stage. The resultant cells are called *complex cells* (Figure 2.7). Both simple cells and complex cells occur in the first region of visual cortical processing, which is called striate visual cortex, or alternatively V1 or area 17.

FIGURE 2.8 David Hubel and Torsten Wiesel at around the age when they won the Nobel Prize for Medicine or Physiology for their experimental work on simple cells and complex cells, among other foundational studies about how the visual cortex works. Image taken from the World Wide Web: https://www.spiegel.de/fotostrecke/hirnforscher-hubel-auf-havard-kongress-geehrt-fotostrecke-114637.html. Associated Press/Alamy Stock Photo.

David Hubel and Torsten Wiesel (Figure 2.8) received the 1981 Nobel Prize for Medicine or Physiology for their discovery of simple and complex cells, as well as for other seminal discoveries about how visual signals are processed by the brains of anesthetized cats and monkeys (Hubel and Wiesel, 1968). Their work triggered research by thousands of neuroscientists who have since used every available technique to study how a wide variety of behavioral stimuli activate processes in multiple parts of our brains, notably in awake-behaving monkeys after those technologies became available. The brains of awake humans have also been studied after non-invasive neuroimaging tools were developed. Invasive recordings have also been done, with patients' permission, with electrodes that are used to help plan brain surgeries with greater accuracy, notably surgeries to treat epilepsy.

2.2 Binocular disparity, laminar computing, and complex cells

My colleagues and I have developed detailed neural network models of the neocortical circuits that create complex cells. These circuits also clarify how complex cells respond selectively to the *depth* of the oriented visual features to which they are selective in external scenes.

Figure 2.9 illustrates how each of our two eyes sees the world from a slightly shifted perspective due to their different positions in our head. *Binocular disparity* is the

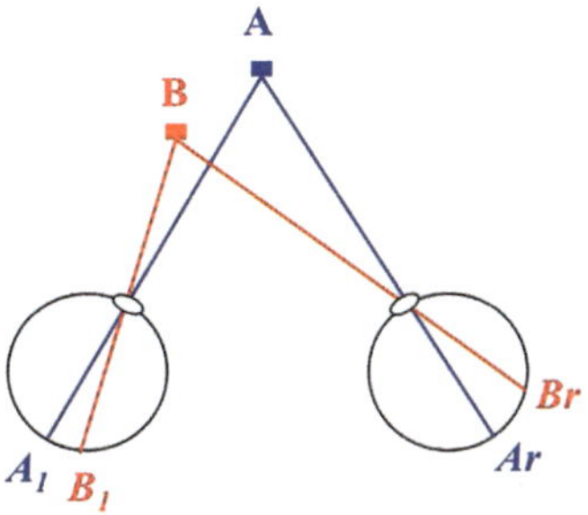

FIGURE 2.9 How binocular disparity occurs on the retinas of our two eyes in response to a visual input, labeled B, in depth while the two eyes fixate, or foveate, another input, labeled A. The images formed on the two retinas in response to an object in the world that is not being foveated are displaced by different amounts with respect to their foveas. This binocular disparity is a powerful cue for determining the perceived depth of the object from an observer. Author created.

term used to describe this difference in image location of an object as it is seen by the left and right eyes. Our brains use binocular disparity as one important cue to extract depth information from the pairs of 2D images that are registered by our left and right retinas. The process that accomplishes this is called *stereopsis.*

Our neural network models explain how simple cells interact to generate complex cell properties. In fact, all neocortical cells are organized into layers, using six main layers and various sublaminae. Remarkably, as I noted in Chapter 1, variations of a *single canonical laminar cortical circuit* carry out multiple psychological functions. Our laminar cortical models have accordingly been able to qualitatively explain, and quantitatively simulate on the computer, challenging psychological and neurobiological data about visual development, perception, recognition, consciousness, cognition, emotion, planning, action, and mental disorders, as well as the analogous processes for audition. As noted in Chapter 1, I have called this computational paradigm *laminar computing* (Figures 1.36–1.38). The following list of references illustrates the range of this modeling work: Berzhanskaya, Grossberg, and Mingolla, 2007; Brown, Bullock, and Grossberg, 2004; Cao and Grossberg (2005, 2012, 2014, 2017a, 2017b, 2018; Chang et al., 2014), Grossberg (1999, 2001, 2003, 2018, 2021a, 2021b), Grossberg and Howe, 2003; Grossberg, Leveille, and Versace, 2011; Grossberg and Kazerounian, 2011; Grossberg and Pearson, 2008; Grossberg and Raizada, 2000; Grossberg and Seitz, 2003; Grossberg and Swaminathan, 2004; Grossberg and Versace, 2008; Grossberg and Williamson, 2001; Grossberg and Yazdanbakhsh, 2005; Leveille, Versace, and Grossberg, 2010; Raizada and Grossberg, 2001, 2003; Yazdanbakhsh and Grossberg, 2004).

Figure 2.10 describes the main cell interactions that transform *monocular simple cells,* that respond to left eye (L) or right eye (R) inputs, but not both, into a depth-selective, like-oriented complex cell. Between these monocular simple cell and complex cell stages is a stage of *binocular simple cells.* A binocular simple cell responds to a left eye (L) monocular simple cell and a right eye (R) monocular simple cell that is sensitive to a shifted position on its retina. The binocular simple cell binocularly fuses such a pair of monocular inputs. It can thereby respond selectively to the binocular disparity of its monocular simple cell inputs. These three cell types occur in identified layers 4, 3B, and 2/3A, respectively, within cortical area V1.

When I first began to model how simple cells interact to generate complex cells in laminar neocortex, I knew that simple cells existed in layer 4 and that complex cells existed in layer 2/3. I also knew that I needed to convert monocular simple cells into binocular simple cells before the binocular simple cells interacted to generate complex cells. What I did not know was where this transformation occurred in the cortical layers. I found that out by searching anatomical data. It was a great relief to me when I realized that a layer 3B exists where this transformation occurs.

Layer 3B is one example of many that I like to call an "acquired taste"; that is, a concept or mechanism that I never dreamed existed until my theoretical method drove me to create it, find its brain correlate, and integrate it into a system-level theory of how humans consciously see.

In order for a binocular simple cell to respond only to approximately equal left-eye and right-eye input signals,

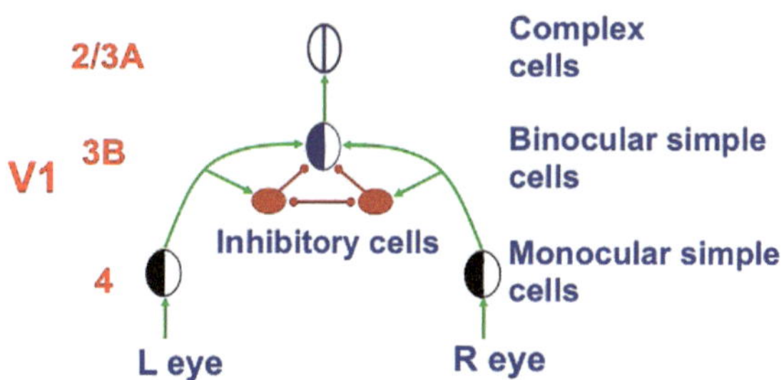

FIGURE 2.10 How interactions among cells in cortical layers 4, 3B, and 2/3A of cortical area V1 give rise to simple cells and complex cells. Author created.

the excitatory inputs that activate it (green arrows in Figure 2.10) are balanced by a network of recurrent inhibitory neurons (red arrows). Cells with this property have been called, by Gian F. Poggio, *obligate* cells (Poggio, 1991).

The obligate property depends on detailed anatomical connections: In addition to inhibiting a target binocular simple cell (red oblique connections), these inhibitory neurons also inhibit each other (red horizontal connections). I have proved mathematically that such a recurrent on-center off-surround network of cells that obey the membrane equations of neurophysiology, also called shunting interactions (Figure 1.21), tends to *normalize its total activity.*

Thus, if only a left-eye monocular simple cell is active, its excitatory input to its target binocular simple cell is balanced by one inhibitory input. Because the excitatory and inhibitory inputs cancel, the binocular simple cell is not activated (one-against-one). The same thing happens if only a right-eye monocular simple cell is active.

However, if both the left-eye and right-eye monocular simple cells are simultaneously activated by approximately equal inputs, then their left- and right-eye excitatory inputs add at their target binocular simple cell. In contrast, their inhibitory neurons, by also inhibiting each other, normalize their *total* activity across both cells. This normalized total activity is roughly the same as it was when a single inhibitory neuron was activated. When both inhibitory neurons are active, they thus send a *normalized total inhibitory signal* to the binocular simple cell (two-against-one). The binocular simple cell is hereby activated and the obligate property is realized.

2.3 Boundary completion and grouping using bipole cells

Boundary grouping and completion occur at a later processing stage than the complex cells. Because boundary completion cells receive their inputs after the stage of complex cells (Figure 2.11), boundary completion can occur between inputs with opposite contrast polarities, as in the reverse-contrast Kanizsa square (Figure 2.4, top right image). Unlike the percept of the Kanizsa square (Figure 2.4, top left image), the interior brightness of the reverse-contrast Kanizsa square percept does not look significantly different from that outside the square. As I noted above, this is because the two white pac-men induce enhanced darkness within the square, whereas the two black pac-men induce enhanced brightness. When these opposing contrasts fill-in within the interior of the square, they tend to average across space and cancel out.

Figure 2.12 shows a computer simulation of the Kanizsa square, and Figure 2.13 shows a simulation of the reverse-contrast Kanizsa square (Gove, Grossberg, and Mingolla, 1995). The input stimuli are in panel A and the simulated percepts are in panel D of each figure. Part B in each figure shows the simulated *feature contours*, whereas part C shows the simulated *boundary contours* in response to each input stimulus.

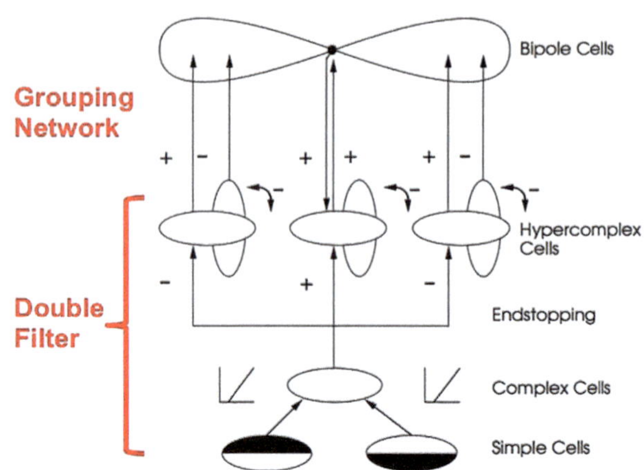

FIGURE 2.11 A double filter and grouping network models how simple, complex, hypercomplex, and bipole grouping cells interact in the visual cortex. Author created.

Boundary contours are the real and illusory boundaries that are formed in response to brightness contrasts of the pac-man inducers with their backgrounds. They are shown in panel C in both Figure 2.12 and Figure 2.13, including the illusory contours that are generated between pairs of pac-men that are like-oriented and positionally aligned.

Feature contours are also induced by brightness contrasts in each stimulus. They result from the process of *discounting the illuminant*, which enables processing of the stimulus contrasts within an image or scene without contamination by illumination gradients. I will explain how and why discounting the illuminant occurs in Section 2.7. Discounting the illuminant helps us to see the true colors in paintings, even when we view them under variable illumination conditions.

The simulated feature contours are shown in panel B of each figure. In Figure 2.12B, all the feature contours are brighter than the background because they are induced by black pac-men on a white background.

In contrast, Figure 2.13B shows that, although the black pac-men induce feature contours that are brighter than the background, the white pac-men induce feature contours that are *darker* than the background.

The final simulated percepts in panel D of each figure are due to filling-in of the feature contours in panel B within the boundary contours in panel C. Note that the filled-in Kanizsa square in Figure 2.12D is both *seen and recognized* because the four bright inducers are averaged

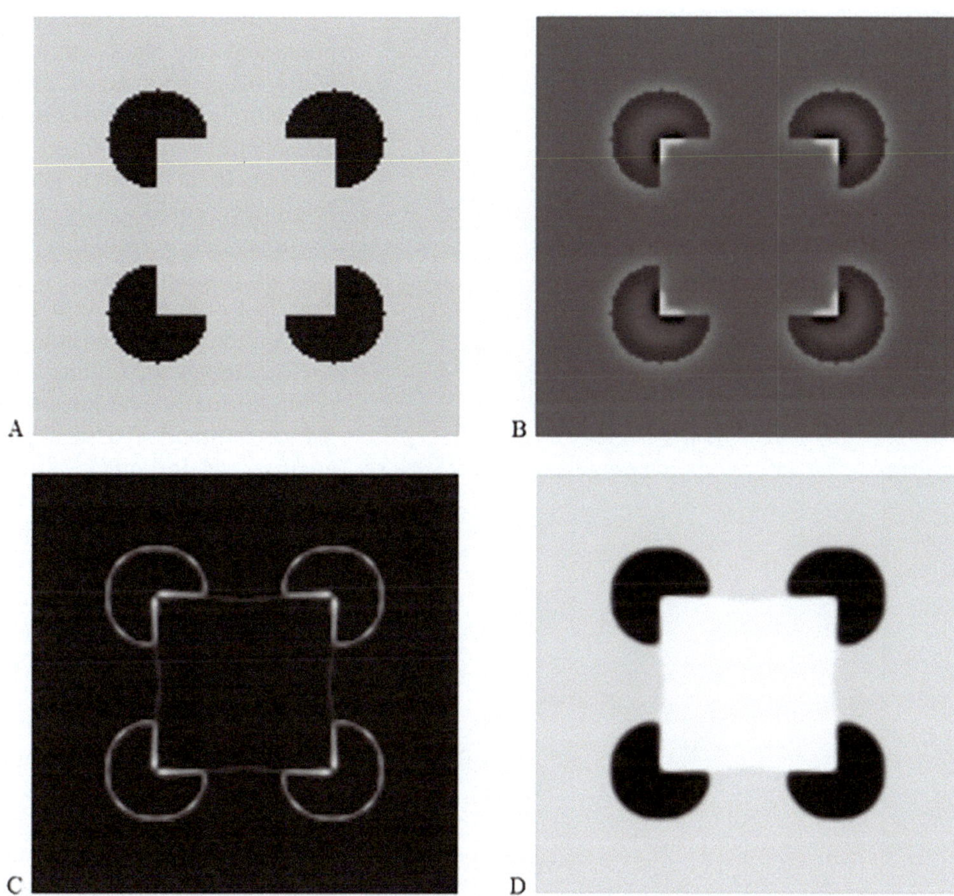

FIGURE 2.12 Computer simulation of how a Kanizsa square image (panel A) gives rise to a conscious percept of a bright Kanizsa square (panel D). See the text for details. Reprinted with author's permission from Gove, A., Grossberg, S., and Mingolla, E. (1995). Brightness perception, illusory contours, and corticogeniculate feedback. *Visual Neuroscience*, 12, 1027–1052.

within the Kanizsa square. In contrast, the filled-in reverse-contrast Kanizsa square in Figure 2.13D is *recognized but not seen* because filling-in of the two brighter and two darker inducers within the square leads to a gray color that is the same both inside and outside the square boundary. This latter simulation shows, just as in the case of the Dalmatian in Snow, that we can *consciously recognize invisible objects*.

The reverse-contrast Kanizsa square percept thus provides another example of invisible boundaries. These boundaries are invisible for two different reasons:

First, because complex cells pool opposite contrast polarities at each position (Figure 2.7), they cannot distinguish between dark/light and light/dark contrasts; hence "all boundaries are invisible" within the boundary formation system. Complex cells, in fact, pool inputs from opposite polarities in response to both achromatic and chromatic inputs (e.g., Thorell et al., 1984).

Although "all boundaries are invisible" within the Boundary Contour system, the positions of boundaries are often indirectly seen if the surfaces that form on opposite sides of the boundary have different brightnesses or colors. In the example of the reverse-contrast Kanizsa square, the effects of the pairs of bright and dark inducers cancel out after surface filling-in occurs, at least in the simulated example, so the filled-in surface brightness inside the square equals that outside the square.

As a result, in response to a properly balanced reverse-contrast Kanizsa square stimulus, surface brightness cannot demarcate the positions of the invisible boundaries using a visible filled-in brightness difference across the boundary. This is the second reason for recognizing, but not seeing, the boundaries of the reverse-contrast Kanizsa square. The *salience*, or strength, of the boundary signals enables us to recognize them, even if the boundaries cannot be consciously seen.

The subtlety of the relationship between seeing and recognizing in these examples illustrates how even excellent experimentalists may interpret their own data incorrectly. Indeed, Thorell et al. (1984) believed that their data implied that complex cells "must surely be considered color cells in the broadest sense" (p. 768). In contrast, in Grossberg (1984), I predicted that "all boundaries are invisible" and that complex cells are *amodal boundary detectors* that pool together signals from multiple simple cell detectors in order to build the best possible boundary signals. This conclusion followed from my theoretical analysis of how, despite the invisibility of *boundaries* within the boundary cortical stream, when boundaries interact with surface filling-in within the surface cortical stream, they can generate consciously visible *surface* percepts.

The image in Figure 2.14 illustrates an important implication of the fact that opposite-polarity boundary signals are added by complex cells at each position. The two gray disks in the image lie in front of a black-and-white checkerboard. As the circumference of each gray disk is traversed, the relative contrasts periodically reverse, from black–white to white–black and back again. Because complex cells pool opposite-polarity signals, a boundary can form around the entire circumference of the gray disk.

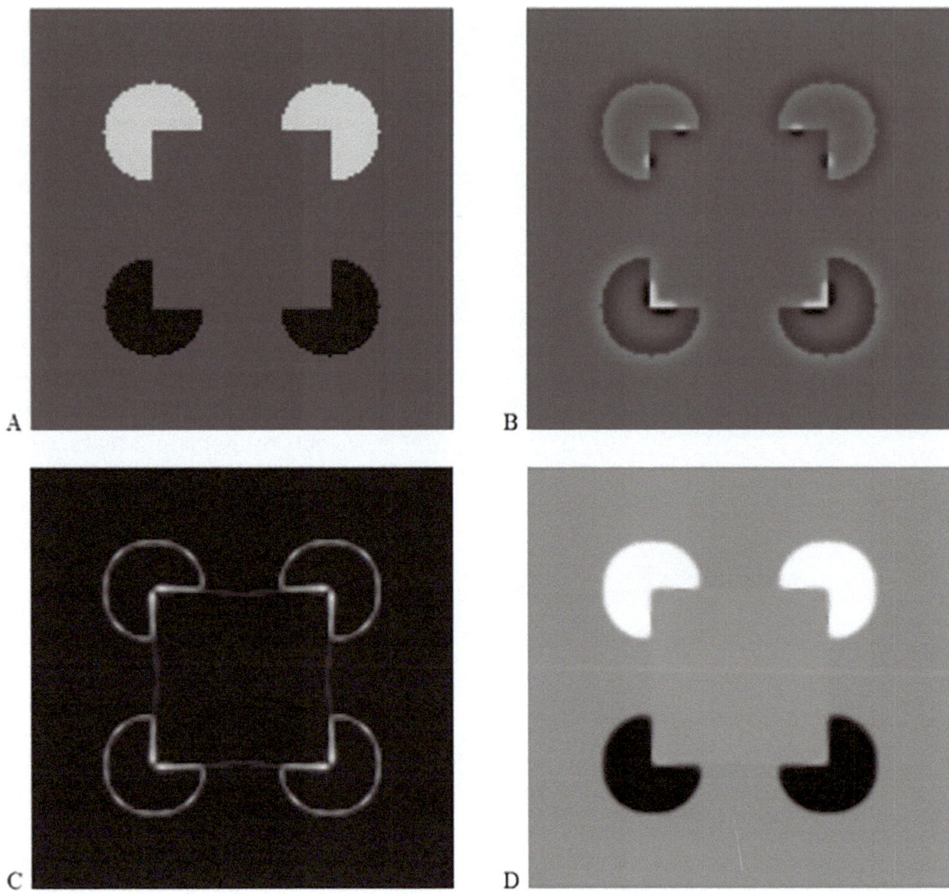

FIGURE 2.13 Computer simulation of how a reverse-contrast Kanizsa square image (panel A) does *not* give rise to a conscious percept of enhanced brightness in the resulting percept (panel D). Reprinted with author's permission from Gove, A., Grossberg, S., and Mingolla, E. (1995). Brightness perception, illusory contours, and corticogeniculate feedback. *Visual Neuroscience,* 12, 1027–1052.

If, instead, the brain computed separate black–white or white–black boundaries using only simple cells, these boundaries would have four big holes in them through which brightness could easily flow. Complex cells prevent this perceptual disaster from occurring, but only at the cost that "all boundaries are invisible."

The image in Figure 2.14 is one of many examples that Kanizsa introduced to argue against Helmholtz's position that all seeing is based on knowledge-based hypothesis testing, or *unconscious inference*, using learned top-down expectations (cf. Figure 2.1). Kanizsa noted that our experiences with regular black-and-white checkerboards should lead us to expect that a white square of the checkerboard is occluded by the gray disk in the lower left of the image, and a black square of the checkerboard is occluded by the gray disk in the upper right of the image. We perceive, instead, an amodally completed black cross behind the gray disk in the lower left, and a white cross behind the gray disk in the upper right. This percept violates our previous experiences with checkerboards, just as our percept of the image in Figure 2.2 (top row, right) violates our expectations of how squares and crosses look.

FIGURE 2.14 A checkerboard that is occluded by two gray disks. The resulting percept contradicts Helmholtz's hypothesis of "unconscious inference." Reprinted with permission from Kanizsa, G. (1979). *Organization of Vision: Essays on Gestalt Perception.* New York: Praeger. Copyright 1979 by Greenwood Publishing Group, Westport, CT.

2.4 Living with your retinal blind spot

The fact that "all boundaries are invisible" has strongly influenced the painterly techniques and theories of many famous artists. Several paintings that illustrate this influence are summarized below. Before describing them, I want to emphasize that the retina itself can create incomplete boundaries out of complete ones that are received from a visual scene. This is because each eye contains a *blind spot* and retinal veins (Figure 2.15).

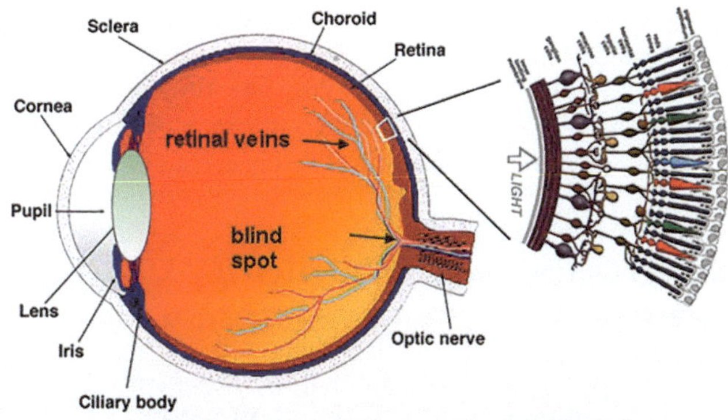

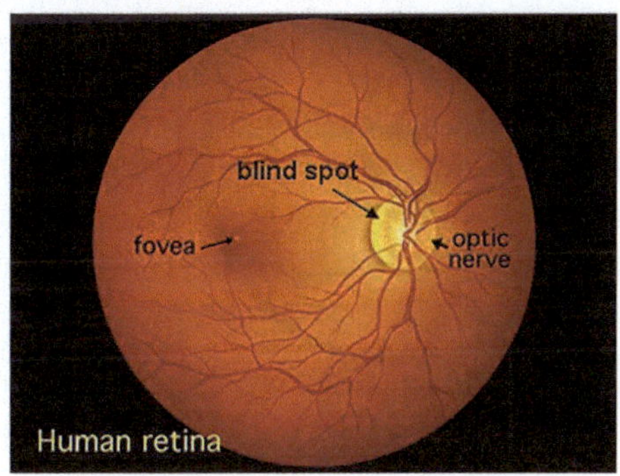

FIGURE 2.15 Cross-section of the eye (top) and top-down view of the retina (bottom), including its blind spot, fovea, and retinal veins. The blind spot and retinal veins can occlude the registration of light signals at their positions on the retina. Reprinted from Part 1, Figure 1, in Kolb, H. Webvision: The Organization of the Retina and Visual System: http:// webvision.med.utah.edu/ book/ part-i-foundations/ simple-anatomy-of-the-retina/.

The blind spot is the place on the retina where multiple pathways, or axons, from photodetectors at other positions on the retina are gathered together to form the optic nerve. The optic nerve sends visual signals from the retina into the brain. Figure 2.15 illustrates that the blind spot is as big as the fovea, where all of our detailed vision occurs. That is why we move our eyes a few times every second using saccadic eye movements to point our foveae at salient objects in a scene.

The retinal veins cover multiple positions on the retina in order to nourish it. The positions that are covered by the blind spot and the retinal veins cannot send visual signals to the brain from objects in the world. Remarkably, we are not aware of these huge holes, or occlusions, in the visual scenes that our retinas receive. That is because our brains complete the boundaries and surfaces at these occluded positions in the same way that they complete boundaries and surfaces in response to Kanizsa squares (Figure 2.12) and reverse-contrast Kanizsa squares (Figure 2.13). I will discuss below in greater detail how our brains do this.

2.5 Boundaries and surfaces are computationally complementary

If "all boundaries are invisible," then how do we see anything, including a beautiful painting? I have predicted that "*all conscious percepts of visual qualia are surface percepts*" (Grossberg, 1984, 1987a, 1987b, 1994, 1997). To accomplish this, every visual scene is processed by two parallel processing streams in the visual cortex. That fact explains why, in the simulations summarized in Figures 2.12 and 2.13, feature contours and boundary contours are computed in parallel before they interact together to determine a final percept. As often occurs in the neocortex, these parallel streams compute computationally *complementary* properties.

Figure 2.16 summarizes what I mean when I claim that boundaries and surfaces exhibit complementary properties: Boundaries are completed *inwardly* between pairs or greater numbers of inducers, in an *oriented* fashion, and are *insensitive* to contrast polarity because they can be generated by inducers with the same contrast polarity or with opposite contrast polarities. On the other hand, surface filling-in proceeds *outwardly* in an *unoriented* fashion and is *sensitive* to contrast polarity because we can see the difference between light and dark. These properties are manifestly complementary: inward vs. outward, oriented vs. unoriented, insensitive vs. sensitive. They fit together like a key and lock, or like yin and yang.

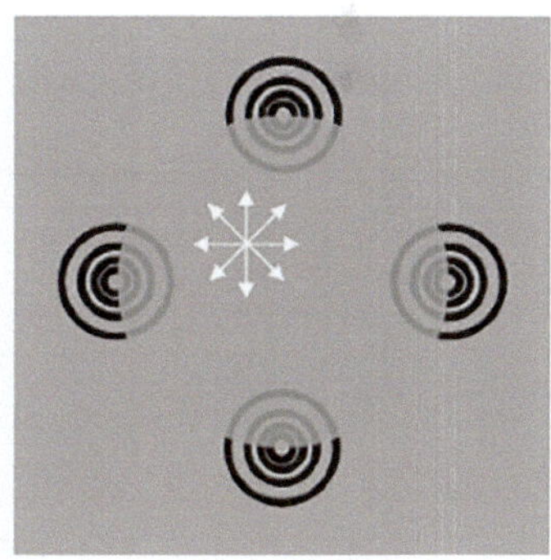

Complementary Properties of Boundaries and Surfaces

Boundary Completion	Surface Filling-In
Inward	Outward
Oriented	Unoriented
Insensitive to direction-of-contrast	Sensitive to direction-of-contrast

FIGURE 2.16 The properties of boundary completion and surface filling-in are computationally complementary. Reprinted with the author's permission from Grossberg, S. (2014). How visual illusions illuminate complementary brain processes: Illusory depth from brightness and apparent motion of illusory contours. *Frontiers in Human Neuroscience*, doi: 10.3389/ fnhum.2014.00854.

The complementary properties of boundary and surface processing are needed for each process to succeed. For example, filling-in needs to be *unoriented* so that it can cover an entire surface. This unoriented flow of brightness can only be effectively contained by an *oriented* boundary. In other words, a seeing process needs to fill-in entire surfaces, so it cannot build boundaries around them. Both types of processes are needed for either process to work well. They interact to overcome each other's complementary deficiencies.

Figure 2.17 illustrates the main brain areas where boundaries and surfaces are formed. First, the photosensitive retina activates the lateral geniculate nucleus, or LGN. The LGN then activates, in parallel, both the *interblob* stream and the *blob* stream of the visual cortex. Boundaries are formed within the interblob stream, whereas surfaces are formed within the blob stream. Although both boundaries and surfaces are activated in parallel, they respond in different ways to the same external stimulus because their properties are computationally complementary (Figure 2.16).

Because boundaries and surfaces are co-activated by the same visual images and scenes, it is often hard to separate an invisible boundary from the surface filling-in process that may render it visible, as in the percept of the Kanizsa square in Figure 2.4. The filling-in of feature contours within boundary contours provides a clear explanation of how this happens in Figure 2.12.

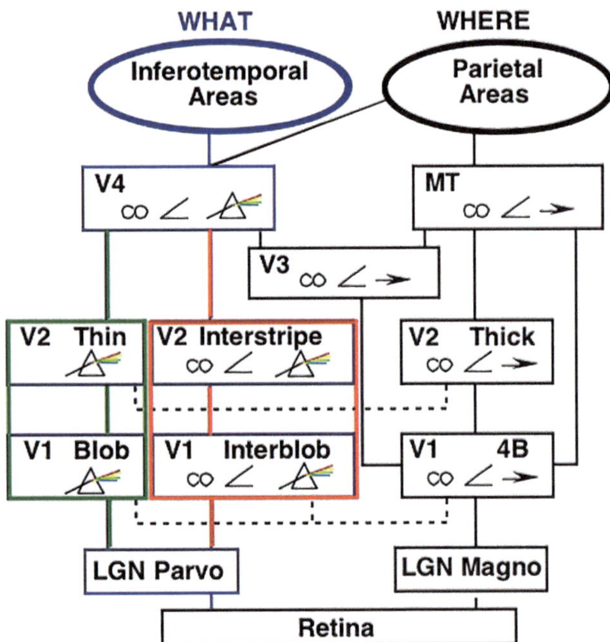

FIGURE 2.17 A macrocircuit of the main processing stages of the visual cortex. Three main visual cortical streams respond in parallel to visual inputs that reach the retina. The two parvocellular streams process visual surfaces (blob stream) and visual boundaries (interblob stream). The magnocellular stream processes visual motion. Adapted with permission from Figure 1 in DeYoe, E. A., and van Essen, D. C. (1988). Concurrent processing streams in monkey visual cortex. *Trends in Neurosciences*, 11, 219–226.

2.6 From complementarity to complementary consistency

Once I understood the complementary properties of visual boundaries and surfaces and the brain mechanisms that created them, I was faced with the daunting question: How does our visual system combine these complementary properties into a higher synthesis, which we experience as unified conscious visual percepts? Said in another way, the challenge is to reconcile the *complementarity* of boundary and surface properties with the *consistency* of conscious percepts. I call this resolution the property of *complementary consistency*.

I explained above why the boundary stream needs to send completed boundaries to the surface stream in order to contain surface filling-in. I will explain below how complementary consistency is achieved by feedback signals from completed surfaces to the boundaries that surround them. These feedback signals strengthen and confirm the closed boundaries that successfully contain filling-in, while inhibiting irrelevant boundary fragments, so that only consistent boundaries and surfaces survive.

This analysis also shows, remarkably, that the process that achieves complementary consistency also triggers the process of *figure-ground separation*, whereby object percepts, including percepts of partially occluded objects, are separated in depth from each other and from their backgrounds, so that we can begin to learn to recognize them, both in natural scenes and in paintings and sculpture.

2.7 Discounting the illuminant

There are multiple reasons why filling-in is needed. One of them, as noted above, is to compensate for the holes in our retinal representations that are caused by the blind spot and retinal veins. Another basic reason why filling-in is needed is that we see the world when it is illuminated by many different light sources that change their color and intensity across scenes and throughout the day. Somehow the brain compensates for this variability in lighting to avoid confusing variable illumination with unvarying properties of object shape and color. This process is often called "discounting the illuminant."

I will explain how this process works using a kind of neural network that occurs throughout our brains. This network did not have to be invented by evolution just to discount the illuminant!

Although great scientists like Helmholtz were aware that a process like discounting the illuminant occurs, he did not have an adequate model to explain how and why it happens. A neural network model of how filling-in works in concert with boundary completion in response to 2D pictures was accomplished in my articles about brightness perception in variable illumination conditions with Michael Cohen (Cohen and Grossberg, 1984), Ennio Mingolla (Grossberg and Mingolla, 1985a, 1985b), and Dejan Todorovic (Grossberg and Todorovic, 1988). These insights were generalized to include both 2D pictures and 3D scenes by the FACADE and 3D LAMINART models (e.g., Bhatt et al., 2007; Cao and Grossberg, 2005; Fang and Grossberg, 2009; Grossberg, 1987a, 1987b, 1994, 1997; Grossberg, Kuhlmann, and Mingolla, 2007; Grossberg and McLoughlin, 1997; Grossberg and Swaminathan, 2004).

In particular, in order to eliminate (most of) the variability of illumination across a scene that could cause wildly unstable shape and color percepts, feature contours are computed at positions where luminance or color contrasts change rapidly enough across space; e.g., Figures 2.12B, 2.13B, and 2.18 (top row). Such positions often occur along a surface's boundaries, which I call boundary contours to distinguish them from feature contours (Figures 2.12C, 2.13C, and 2.18 (top row)).

After discounting of the illuminant takes place, the surviving feature contours compute brightness and color signals that are relatively uncontaminated by varying illumination levels. They have this property because the contrast changes where they are computed are due primarily to changes in the material properties, called *reflectances*, of the underlying objects, whereas the illumination level changes little, if at all, across such a contrast change. The reflectance of a surface is the *relative* amount of light that it reflects at a given wavelength. The reflectances of a surface are constant, even though the absolute amounts of light reflected from the surface increase in proportion to its incident intensity.

After the reflectance-selective feature contour signals are computed, they can, at a subsequent processing stage, trigger filling-in of the corresponding brightnesses or colors across the surface until they hit the boundary contours that enclose the surface (Figures 2.12D, 2.13D, and 2.18 (bottom left)).

Boundaries are computed at multiple depths and trigger filling-in of colors and brightnesses only at those depths where the feature contours and boundary contours are parallel and adjacent to one another, and the boundaries are closed (Figures 2.18 (top) and 2.19), so that they can contain the filling-in process (Figure 2.18, bottom left). If a boundary is not closed, then brightness and color can spill out of gaps in the boundary (Figure 2.18, bottom right).

Filling-in occurs in networks within the surface stream that I call Filling-In-DOmains, or FIDOs. Multiple FIDOs exist to enable selective filling-in of multiple opponent colors (red–green, blue–yellow) and achromatic brightnesses (light–dark) at multiple depths. The three pairs of opponent FIDOs at every depth in Figure 2.19 schematize this property.

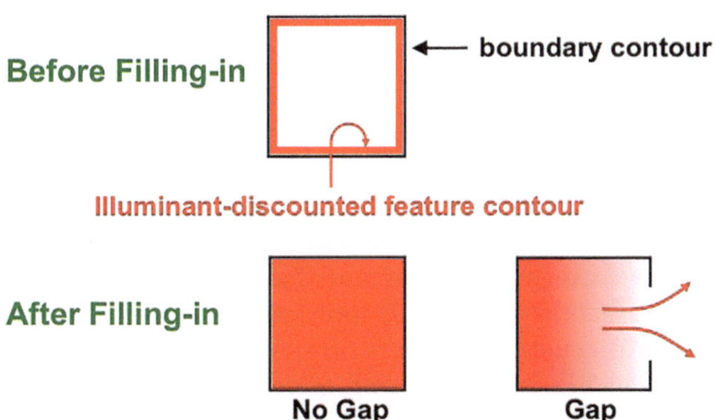

FIGURE 2.18 How closed boundaries contain filling-in of feature contour signals about boundary contours (top row and bottom-left image), whereas open boundaries allow color to spread to both sides of the boundary (bottom-right image). Author created.

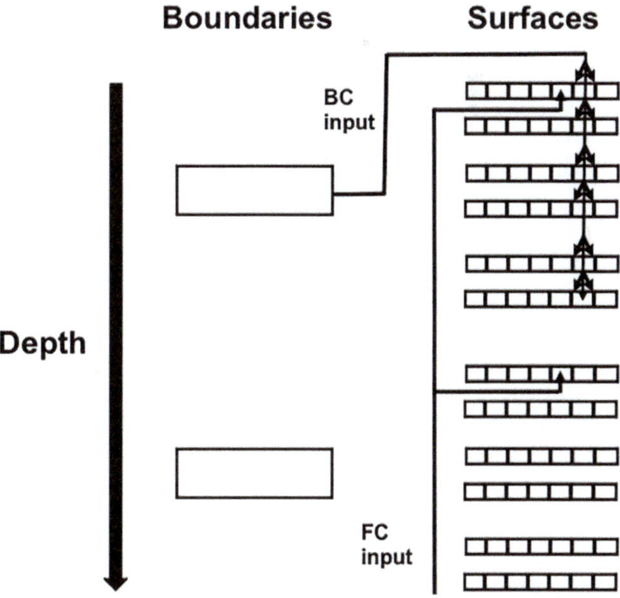

FIGURE 2.19 Depth-selective boundaries, or boundary contours (BC), interact with depth-selective, illuminant-discounted brightness and color contrasts, or feature contours (FC), to activate depth-selective surface filling-in. Whether or not the BC and FC interaction leads to filling-in is explained in Figure 2.18. Author created.

Many painters have struggled to represent the difference between the real colors of objects and the illumination that is reflected from them and onto other objects. Their success in doing so is all the more remarkable, considering that they see the world only after the illuminant has been discounted, which is also true of us as we look at their paintings.

2.8 Gist as a coarse texture category

Given that boundaries and surfaces are the functional units of visual perception, how is gist perceived? This is an active area of research in both biological vision and computer vision. I proposed how gist is computed with my PhD student Hung-Chen (Tren) Huang using the ARTSCENE model that we developed together (Grossberg and Huang, 2009). Our article also reviews other scene classification models and summarizes comparative benchmark simulations of these models.

The ARTSCENE model classifies the gist of a scene as a *texture category* that is derived from coarse boundary and surface representations that are activated by a single glimpse of the scene. After being activated by the scene, this texture representation activates inputs to a subsequent processing stage that can learn to categorize, or classify, textures into different scene types (e.g., coast, forest, mountain, countryside). The same kind of categorizing ART network can learn to classify textures, objects, and many other kinds of events that are experienced through vision, audition, or other sensory modalities. The ART model will be reviewed in the next section.

Grossberg and Huang (2009) further proposed how, after a glimpse of a scene enables its gist to be classified, the eyes may wander around the scene to accumulate additional information about it. ARTSCENE assumes that the eyes look at bigger regions of a scene with higher probability, other things being equal. For example, the largest region in a coastal scene may be the ocean, so the eyes may look at the ocean with a higher probability. ARTSCENE explains how *spatial attention* highlights such a region, at the same time that it down-regulates processing of the rest of the scene. Spatial attention can do this because it is realized by a top-down, modulatory on-center, off-surround network, just like other ART attentional circuits. The highlighted texture can then trigger learning of its own *recognition category*.

Before going on, I will say more about what recognition categories do. A recognition category may, for example, learn to respond *selectively* to the spatial pattern of distributed features in a particular scene or painting, and not to very different scenes or paintings. Typically, such a category is realized by a single population of cells, or a small number of cell populations, that is activated by the scene or painting. A recognition category is represented by a much smaller number of cells than the *feature detector cells* that are activated by the scene or painting that they represent. In this sense, a category *compresses* the scene or painting into a selective recognition event. It is a learned *symbol* of the scene or painting. Whereas individual pixels of a scene or painting have no meaning, the selective activation of a category by a spatially distributed input like a painting provides a context-sensitive internal representation of the painting.

Coming back to the ARTSCENE model: After the gist of a scene is categorized by a texture-selective category, eye movements shift the model's attentional focus a few times, thereby allowing spatial attention to focus upon the biggest textures of the scene; e.g., ocean, sky, grass, trees, etc. These finer textures are also categorized, and output signals from all of these recognition categories then *vote* for the scenic category that best represents their co-occurrence. This simple voting procedure generates a more accurate prediction of scene identity than gist alone, indeed a prediction that achieves human levels of gist recognition, and that did better than other, more complicated, models for computing gist.

In summary, one of the first things that a viewer of a painting may notice is the gist of the painting. This may include information that it is a painting of a certain kind of scene or that it is a painting by a particular artist. The subsequent text will explain some of the processes that enable our brains to make such distinctions.

2.9 Learning, recognition, and prediction by adaptive resonance

In order to recognize the contents in a painting, one needs to be able to recognize the textures and the objects of which it is composed, even if these objects are just simple lines and curves. Learning about any texture or object needs to solve a basic problem that confronts every learner who experiences the world moment-by-moment in real time. I call this problem the *stability-plasticity dilemma* (Grossberg, 1980, 2021).

The stability-plasticity dilemma asks how the brain is able to learn quickly about new objects and events without just as quickly forgetting previously learned, but still useful, memories. In other words, how do our brains learn quickly without experiencing *catastrophic forgetting*?

ART proposes a rigorous neural network model and computational theory of how the stability-plasticity dilemma is solved (Asfour et al., 1993; Bhatt, Carpenter, and Grossberg, 2007; Bradski and Grossberg, 1995; Carpenter and Grossberg, 1987a, 1987b, 1991, 1993; Carpenter et al., 1992; Carpenter, Grossberg, and Reynolds, 1995; Carpenter, Grossberg, and Rosen, 1991a, 1991b; Grossberg, 1976b, 1980. 2013, 2017, 2018, 2021). ART is currently the most advanced cognitive and neural theory that can learn to attend, recognize, and predict objects and events in a changing world that is filled with unexpected events. This claim is supported by several kinds of evidence, which I mentioned in Chapter 1, but will repeat here for completeness and emphasis:

First, ART has explained and predicted much more psychological and neurobiological data than other neural models.

Second, all of the foundational hypotheses from which ART was derived have been supported by such data.

Third, and perhaps most important, ART was derived in 1980 in an oft-cited article that I published in *Psychological Review* (Grossberg, 1980) from a *thought experiment* about how *any system* can *autonomously learn to recognize and predict* objects and events in a changing world that is filled with unexpected events. The small number of hypotheses used to derive ART in the thought experiment are facts of life that are familiar to us all because they are ubiquitous environmental pressures that, working together, drove the evolution of our brains.

Moreover, these hypotheses do not mention mind or brain.

ART is thus the *unique* and *universal* class of systems that solves this autonomous learning, classification, and prediction problem.

To solve the stability-plasticity dilemma, ART uses *attentive matching* between bottom-up input patterns (upward-facing green excitatory pathways in Figures 1.31 and 2.20) and learned top-down expectations (downward-facing green excitatory arrow in Figure 2.20) at feature-selective cells that get activated at the Feature Patterns level in Figure 2.20. A learned top-down expectation is released by a currently active recognition category in the Categories level of Figure 2.20. Such expectations

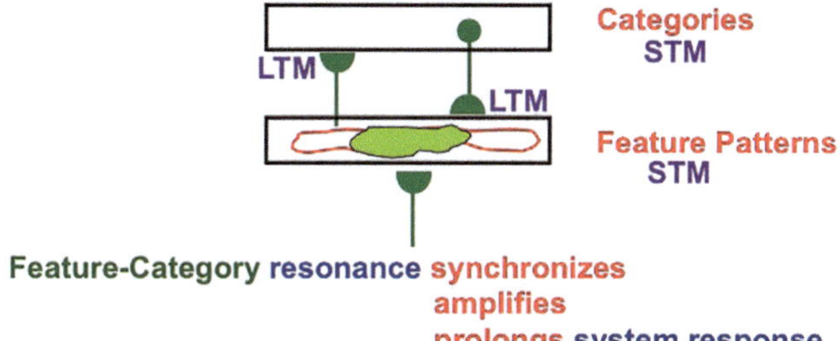

FIGURE 2.20 A two-level neural network as it experiences an adaptive resonance. See the text for details. Author created.

may be thought of as *predictions* that the category emits about input patterns it expects to find active at the feature-selective cells based on past experiences. These predictions are also called *prototypes*, although the critical feature patterns in an ART prototype are not the same as the ones postulated in earlier cognitive science. These cognitive science prototypes just averaged across previously experienced input patterns. Among other problems with them, they are incompatible with neurobiological data and do not help to solve the stability-plasticity dilemma.

In contrast, ART prototypes are defined by the *critical feature patterns* that are learned from past experience to determine predictive success. These critical feature patterns are thus incorporated during learning into the adaptive weights, or LTM traces, that are learned in ART, and are the ones to which we pay attention. The critical feature pattern is denoted by the light green region in Figure 2.20.

2.10 Expectations, learning, priming, and matching of critical features

In the absence of bottom-up inputs, top-down expectations only prime, sensitize, or modulate their target cells with the critical feature patterns that they have learned. These cells hereby only reach subthreshold activation levels that are insufficient, by themselves, to drive their activities to suprathreshold levels.

A prime can, for example, be activated by asking a viewer of a painting to look for a particular object in it. If the viewer activates an expectation of seeing that object, the expectation, by itself, does not usually generate a conscious percept of the object. That would be akin to hallucinating the object. Rather, when the viewer's eyes happen to look at or near the primed object, the object will be more quickly recognized, and with a higher probability, than if the expectation were not active. This happens because, when a top-down expectation is active, a bottom-up input pattern that *matches* the expectation well enough at feature-selective cells of the Feature Patterns level can more rapidly and vigorously activate these cells to suprathreshold values than in the absence of the expectation.

In Figure 2.20, I have drawn, for simplicity, only a single category receiving a single bottom-up input pathway and emitting only a single top-down pathway. In both the brain and the ART network, bottom-up output pathways from feature-selective cells are *one-to-many*; that is, each feature-selective cell can send pathways to many category cells. As a result, bottom-up pathways to the Category level converge in a *many-to-one* way at each category cell. The same is true for every category that has been learned.

Figure 2.20 labels activations, or *short-term memory* (STM) traces, and adaptive weights, or learned memories, or *long-term memory* (LTM) traces. The STM traces are properties of cells. The LTM traces occur at the ends, or *synapses*, of each adaptive pathway. Such a location for an LTM trace enables it to learn when its pathway is carrying a big enough signal at the same time when the post-synaptic cell that it abuts is also active. This kind of correlational learning is often called *associative learning*.

Humans can learn many types of information and knowledge using associative learning, including sequences of events occurring through time, such as the alphabet A, B, C,.... In an adaptive resonance network, the spatial pattern of LTM traces in the bottom-up, many-to-one adaptive pathways from feature-selective cells to a category cell acts like an *adaptive filter*. The spatial pattern of LTM traces in the top-down, one-to-many adaptive pathways from a category cell to feature-selective cells is a *learned expectation*.

How is a top-down expectation learned? In particular, how does learning even get started? In particular, when a novel input pattern activates a spatial pattern of activity at feature-selective cells, which then activates a new category cell, why does that category cell not read out a top-down expectation that mismatches the spatial pattern at the feature-selective cells? If it did, the mismatch could suppress activity across the feature-selective cells before any learning could begin.

This problem does not occur for the following reason: The *initial* values of *all* top-down LTM traces are large within the one-to-many pathways from each cell in the Category level to cells in the Feature Pattern level. These large, broadly distributed LTM traces can *match* whatever spatial pattern of feature detector activations occurs at the Feature Pattern level. This kind of match is called a *superset* match. A superset match is a good enough "match" not to reset ART because of the way that ART matching works, which I will explain below. I will also explain why a good enough match is needed to allow learning to begin.

Top-down learning on subsequent learning trials *prunes* the LTM traces until they match the critical feature pattern (marked by the light green region at the feature level in Figure 2.20) that is learned from the input patterns that activate the category. The critical features are thus the ones that are learned by the LTM traces in the bottom-up adaptive filters and the top-down expectations. Critical features also control predictions about what is expected to happen next.

All other LTM traces than those that learn critical features become very small, or zero, by being associated with very small, or zero, STM activities at their target cells.

The above example illustrates the general fact that learning occurs within an ART system when a good enough *match* occurs between a bottom-up input pattern and a top-down expectation on a given learning trial. A good enough match triggers a context-sensitive *resonance* between the currently active cells in the feature and category levels. Such a resonance is supported by the mutual exchange of bottom-up excitatory signals within the adaptive pathways from active feature cells to the active category and of top-down excitatory signals within the adaptive pathways from the active category to the active feature cells (Figure 2.20).

The resonance prolongs, synchronizes, and amplifies the activities of both the attended critical feature pattern and the active category; hence the name *feature-category resonance* for this dynamical state. A feature-category resonance triggers fast learning by the adaptive weights in both the bottom-up and top-down pathways that join these representations; hence the name *Adaptive* Resonance Theory.

ART hereby solves the stability-plasticity dilemma using a combination of top-down matching, convergence over learning trials of LTM traces to a critical feature pattern, and inhibition of irrelevant outlier feature cells that do not resonate during the match-based learning process.

2.11 ART Matching Rule: Attention, biased competition, and predictive coding

Attention to a pattern of critical features within an ART circuit obeys the ART Matching Rule. This Rule is embodied by top-down, modulatory on-center, off-surround networks that read out the learned expectations to be matched against bottom-up input patterns (Figure 1.21). This matching process has the following properties:

When a bottom-up input pattern is received at a stage that processes Feature Patterns (Figure 2.20), it can activate its target cells, if nothing else is happening (Figure 2.22a). The yellow pattern in Figure 2.22a signifies that the cells which are activated most have the highest peaks.

Likewise, if nothing else is happening when a top-down expectation pattern is received at this stage, it can provide excitatory modulatory, or priming, signals to cells in its on-center, and inhibitory signals to cells in its off-surround. The on-center is modulatory because the off-surround also inhibits the on-center cells (Figures 2.21 and 2.22b), and these excitatory and inhibitory inputs are approximately balanced ("one-against-one").

When a bottom-up input pattern and a top-down expectation are both active, cells that receive both bottom-up excitatory inputs and top-down excitatory priming signals can fire ("two-against-one") with amplified and synchronized activities, while other cells are inhibited. The cell activities that survive top-down matching are schematized in Figure 2.22b by the yellow spatial pattern. The inhibited cells are schematized by the white spatial pattern. In this way, only those cells can fire whose features are "expected" by the top-down expectation, and an attentional focus starts to form at these cells. This attentional focus eventually converges upon the critical feature pattern as learning trials proceed.

Psychological, anatomical, and neurophysiological data have provided support for the ART Matching Rule prediction of how attention works. These data support all the main predicted properties, including the modulatory on-center, off-surround network; excitatory priming of features in the on-center; suppression of features in the off-surround; and amplification of matched data (e.g., Bullier, Hupé, James, and Girard, 1996; Caputo and Guerra, 1998; Downing, 1988; Hupé *et al.*, 1997; Mounts, 2000; Reynolds, Chelazzi, and Desimone, 1999; Sillito, Jones, Gerstein, and West, 1994; Somers, Dale, Seiffert, and Tootell, 1999; Steinman, Steinman, and Lehmkuhle, 1995; Vanduffell, Tootell, and Orban, 2000). The way in which the off-surround competition biases attention using the ART Matching Rule has led to the term "biased competition" by various experimental neurophysiologists (Desimone, 1998; Kastner and Ungerleider, 2001). The property of the ART Matching Rule that bottom-up signals may be enhanced when matched by top-down signals is supported by neurophysiological experiments that have reported the facilitatory effect of attentional feedback (Luck et al., 1997; Roelfsema et al., 1998; Sillito et al., 1994).

Various other data have supported ART predictions about how what I call the CLEARS processes of Consciousness, Learning, Expectation, Attention, Resonance, and Synchrony are related in the brain. For example, attention and learning are linked in a manner consistent with ART predictions about visual perceptual learning (e.g., Ahissar and Hochstein (1993, 1997), Ito, Westheimer, and Gilbert (1998), and Lu and Dosher (2004)), auditory learning (e.g., Gao and Suga, 1998), and somatosensory learning (e.g., Krupa, Ghazanfar, and Nicolelis, (1999) and Parker and Dostrovsky (1999)). ART has also predicted links between attention and synchronous oscillations that were subsequently experimentally reported (e.g., Buschman and Miller (2007), Engel, Fries, and Singer (2001), Gregoriou et al. (2009), Grossberg (2009), and Pollen (1999)), as well as links between synchronous oscillations and consciousness (e.g., Lamme (2006), Llinas et al. (1998), and Singer (1998)).

The above experimental confirmations are a subset of a large experimental literature. All the predictions of ART have also received subsequent experimental support. Illustrative experiments will be cited throughout this book, as appropriate.

2.12 A brain without Bayes

Before leaving the topic of how expectations embody predictions about what may or may not happen next in the world, it is important to realize that not all theories of predictive coding are supported by the experiments that support the ART Matching Rule. In particular, these data do not support Bayesian "explaining away" models, in which bottom-up matches with top-down feedback cause only suppression, leaving only the unmatched data, or "prediction error," for further processing (Mumford, 1992; Rao and Ballard, 1999).

Such inhibitory matching does occur in the brain, but it typically controls spatial and motor representations (see Figure 1.34). There are fundamental reasons for this difference in matching schemes between perceptual-cognitive ART-like excitatory matching and spatial-motor inhibitory matching, also called VAM-like matching. These differences between how perceptual-cognitive and spatial-motor processes are computed are one illustration of the paradigm of Complementary Computing, which I introduced to clarify how our brains are globally organized into parallel systems that obey computationally complementary properties (see Figures 1.34, 1.35, and 1.41). Thus, whereas the ART Matching Rule in perceptual-cognitive brain networks solves the stability-plasticity dilemma (see Carpenter and Grossberg (1987) for a mathematical proof), matching in the spatial-motor brain networks does not, and thereby allows our spatial maps and motor commands to adapt throughout life to our changing bodies.

2.13 Symbol grounding by complementary computing

Several examples of complementary computing occur in an ART neural network. For one, the resonant state itself combines computationally complementary properties of the ART network's processing levels into a context-sensitive representation of an object or event. In particular, individual pixels or features of a painting or scene are meaningless. Pixels and features become meaningful only as part of the *spatial context* of the pixels and features that surround them. This spatial context is implicitly represented by the spatial pattern of activation across feature detectors at level F_1 in Figures 2.20 and 2.21. Activating the recognition category of a painting, or scene, at level F_2 in these figures provides context-selectivity because the category responds selectively to specific spatial patterns of activation across the features that define a painting, but not to spatial patterns that characterize different paintings. The category itself has no knowledge, however, of what feature patterns cause it to fire, or indeed any information about the object or painting that activates it.

In brief, individual F_1 cells process features without knowing the spatial patterns to which they belong, and individual F_2 cells respond selectively to these spatial patterns without knowing what features they represent.

A resonant state binds together the distributed feature pattern of the painting with its category. This *bound state* represents both the identity of the painting and the spatial pattern of features that define it. In this way, a *feature-category resonance* (Figure 2.20) provides a solution to the symbol grounding problem (Harnad, 1990).

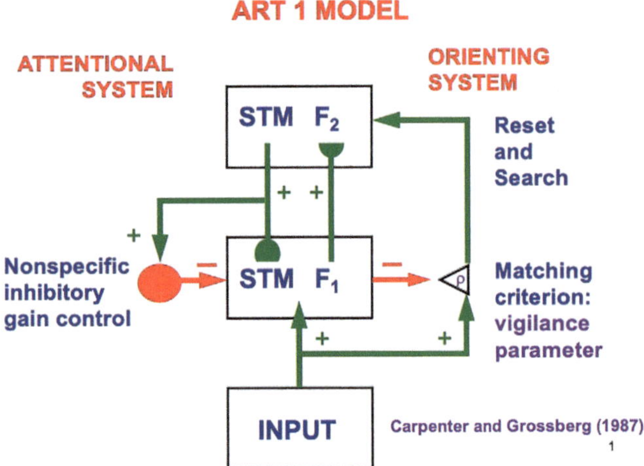

FIGURE 2.21 The ART 1 model circuit includes an Attentional System where attention, learning, and resonance occur in response to expected events, interacting with an Orienting System where mismatches and novel events are processed in response to unexpected events. Attention is realized by a top-down, modulatory on-center, off-surround network that is said to obey the ART Matching Rule. These systems obey computationally complementary laws and interact to discover and learn new recognition categories. See the text for details. Author created.

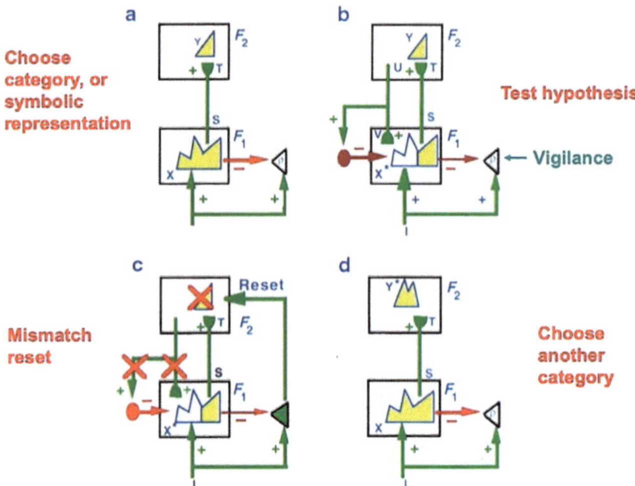

FIGURE 2.22 In response to input patterns, ART triggers a hypothesis testing and learning cycle to discover and learn recognition categories. See the text for details. Author created.

2.14 Attentive learning and memory search are complementary

Another example of complementary computing in ART occurs at the system level. Category learning in ART is controlled by interactions between an attentional system and an orienting system (Figure 2.21). These two systems embody yet another type of computationally complementary laws (Grossberg, 1980, 2000). The attentional system (levels F_1 and F_2 in Figures 2.21 and 2.22) carries out processes like *attention, category learning, expectation, and resonance.* The orienting system (defined by the triangular region enclosing the *vigilance parameter* ρ in Figure 2.22) interacts with the attentional system to enable the attentional system to learn about novel information using processes like *reset, memory search, and hypothesis testing.* The attentional system includes brain regions like temporal cortex and prefrontal cortex (PFC). The orienting system includes brain regions like the nonspecific thalamus and hippocampus.

If an input pattern activates a category (Figure 2.22a) that causes a sufficiently bad mismatch to occur within the attentional system (Figure 2.22b), it will activate the orienting system (Figure 2.22c). The orienting system, in turn, inhibits, or resets, the currently active category. Inhibition of the active category triggers a bout of hypothesis testing, or memory search, for a better-matching category, possibly an entirely new one (Figure 2.22d). The four panels in Figure 2.22 summarize one cycle of the ART hypothesis-testing and learning cycle.

This cycle continues until a match state occurs, either by matching the current input pattern with an already known category, that was not active when the hypothesis-testing cycle began, or by choosing previously unused category cells and initiating learning of a new category with them.

Within the orienting system, a *vigilance parameter* ρ (Figure 2.22) determines the generality of the categories that will be learned. If vigilance is high, then learning of a concrete or specific category occurs, such as a category that recognizes a frontal view of a friend's face, or of the painting called *Impression, Sunrise* by Monet. If vigilance is low, then learning of an abstract or general category occurs, such as a category which recognizes that everyone has a face, or that one is viewing an Impressionist painting.

In general, vigilance is chosen as low as possible to conserve memory resources, without causing too great a reduction in predictive success. Because the baseline level of vigilance is initially set at the lowest level that led to predictive success in the past, ART models try to learn the most general category that is consistent with the data that they have processed. This tendency can lead to the type of overgeneralization that is seen in young children (Brooks et al., 1999) until subsequent learning leads to category refinement (Tomasello and Herron, 1988).

When a given task requires a finer categorization, vigilance is raised. Vigilance can be automatically adjusted to learn either specific or general information in response to predictive failures, or disconfirmations, within each environment. Such a predictive failure could occur, for example, if a viewer of a painting believes that the painting is by Claude Monet before being informed that it is actually by Alfred Sisley. Within ART, such a predictive disconfirmation causes a shift in attention to focus on a different combination of features that can be learned and used to recognize that this particular painting is by Sisley, and perhaps to recognize other paintings by him as well.

2.15 Match tracking achieves minimax learning

One type of automatic adjustment of vigilance in response to a predictive disconfirmation is *called match tracking*, for the following reason: Suppose that on every learning trial, a predictive failure causes vigilance to increase by the smallest amount that can trigger a memory search for a new category that can correct the error. As a result of such a memory search, a category will be learned that is just general enough to eliminate the error.

When this happens, match tracking is said to occur, because vigilance tracks the degree of match between the input pattern and the expected prototype (Figure 2.23). Match tracking leads to a type of *minimax learning*, or learning that can *minimize* predictive errors while it conjointly *maximizes* category generality. This is because match tracking uses the minimum additional memory resources that are needed to correct the predictive error. Because vigilance can vary during match tracking in a manner that reflects current predictive success, recognition categories capable of encoding widely differing degrees of generalization or abstraction can be learned by a single ART system (Carpenter and Grossberg, 1987, 1991).

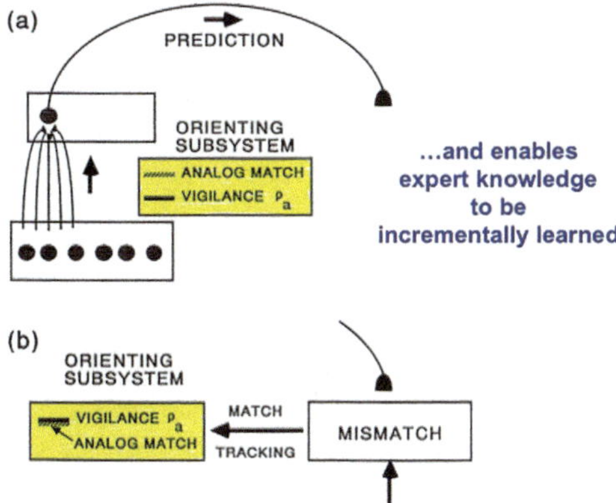

FIGURE 2.23 In a supervised ARTMAP system, like fuzzy ARTMAP, a predictive mismatch causes a memory search for a better matching category. A memory search that carries out *match tracking* realizes a *minimax learning principle* that sacrifices the minimum amount of learned category generality to correct a predictive error. Using the notation of Figure 2.24, the predictive mismatch takes place in ARTb, whereas the memory search and discovery of a better matching recognition category takes place in ARTa. Author created.

Since the concept of vigilance control was first published by Carpenter and Grossberg (1987), quite a bit of new data has been published that supports ART predictions about how vigilance may be regulated in the brain. In particular, brain regions like the nucleus basalis of Meynert are activated by big enough predictive mismatches and thereby release neurotransmitters such as acetylcholine throughout layer 5 of neocortex, thereby increasing vigilance (Grossberg and Versace, 2008; Palma, Grossberg, and Versace, 2012a, 2012b).

It has also been predicted that vigilance cannot dynamically adjust itself in some clinical patients, leading to problems with attention, learning, and recognition. Grossberg and Seidman (2006) have, for example, proposed that various individuals with autism have their vigilance stuck at an abnormally high value, leading to the learning of abnormally concrete and hyperspecific recognition categories, as well as to a correspondingly narrow focus of attention.

2.16 Unsupervised and supervised learning: ARTMAP systems

Much of the recognition learning that we do is *unsupervised*. In other words, it goes on without an explicit teacher telling us the answers. Unsupervised learning *clusters* different input patterns into the same category based on the *similarity* of their feature patterns using the world itself as a teacher. An unsupervised ART system can do this. However, there are many instances where *supervised* learning is needed. During supervised learning, a teacher provides the answers. Here, the recognition category learns to activate the answer that the teacher provides. A supervised ARTMAP system can do this (Figure 2.24).

Supervised learning may be needed, for example, to help a young learner distinguish between the letters E and F: Suppose that the learner first learns the letter F. When E is then presented, it would be natural to process E as a variant of F, especially if the bottom horizontal limb of the E is short. Supervised learning makes it possible to quickly learn to distinguish the two letters. When an ARTMAP system (Figure 2.24) predicts erroneously that letter E is classified as a letter F, and it is given the correct answer, the mismatch between the two letters E and F in ARTb can drive a bout of hypothesis testing, or memory search in ARTa that shifts to a new focus of attention whose combination of critical features can be learned in a new category that recognizes the letter E.

Supervised learning enables us to learn what type of scene one is viewing and classifying in ARTa by answering with the *name* of the scene type in ARTb; e.g., E, F, beach, forest, mountain, etc. Supervised learning is also needed to learn the name of the artist who has created a particular painting, or what artistic movement that painting

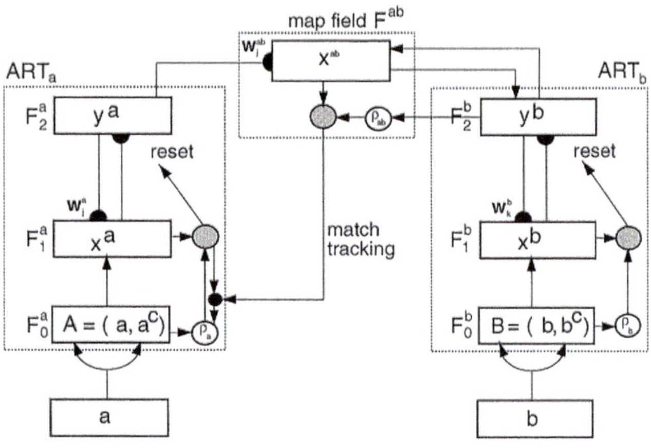

FIGURE 2.24 The fuzzy ARTMAP neural architecture for category learning in ARTa and in ARTb, joined by a learned associative map from ARTa to ARTb via a map field Fab. For example, learning to recognize visual objects can take place in ARTa, and learning the names of the objects can take place in ARTb. After learning, a familiar object can be recognized and named by the model because both ARTa and ARTb have adaptive bottom-up and top-down pathways. As a result of these connections, a familiar object can activate bottom-up pathways in ARTa, the associative map in the map field Fab, and then top-down pathways in ARTb. Author created.

illustrates, or the name of a painting itself. In the Fuzzy ARTMAP neural network in Figure 2.24, a prediction about a painting, such as its name, is coded in the ARTb network, just as the names of the letters E and F are, whereas categories that recognize the painting itself are learned in the ARTa network, just as the recognition categories of the letters E and F are. When such an ARTMAP system predicts erroneously that a painting by Monet is actually by Sisley, and it is given the correct answer, the mismatch between the two names, Monet and Sisley, in ARTb can drive a bout of hypothesis testing, or memory search in ARTa that shifts to a new focus of attention whose learned combination of critical features can recognize the painting as a Sisley in ARTa.

ARTMAP systems hereby use predictive mismatches to drive bouts of mismatch-mediated hypothesis testing to discover and learn new categories with which to correct predictive errors. Much experimental evidence supports the hypothesis that a similar process goes on in human observers, as reviewed in the above archival articles and Grossberg (2021).

2.17 Invariant category learning, recognition, and search

The discussion in Section 2.7 mentioned the utility of scanning a scene or picture with eye movements in order to accumulate more information about it, as the ARTSCENE model does to recognize scene type. However, such scanning raises a basic question. Much of our learning about objects takes place by classifying multiple views of a novel object into a view-invariant object category, without the benefit of a teacher, as our eyes freely scan a scene or picture. Invariant category learning binds together different appearances of an object, not only as it is seen from the perspective of multiple views, but also at multiple positions, orientations, and sizes on our retinas. In this way, we can recognize a family member or friend at a glance anywhere in our visual field. Invariant category learning avoids the combinatorial explosion that might otherwise have occurred if our brains could only learn and store view-, position-, orientation-, and size-dependent representations of objects. Invariant categories provide a greatly compressed representation of the objects to which our brains need to attend.

The invariant binding process raises new questions: What prevents the promiscuous binding together of features from multiple objects in a scene? Why do not views of multiple paintings in a museum erroneously get fused into the learning of a single category that is supposed to represent just one of these paintings?

And even if promiscuous binding can be avoided during the learning of invariant object categories, so that we can learn an invariant category for individual paintings in the museum, there remains the problem of how we use invariant categories to search the museum to discover and view any particular painting? This is a problem because an *invariant* object category, being positionally invariant, does not represent the location of the object that it represents. But if we could not learn information about the location of the painting in the museum, then how could we ever learn where to go to view it?

How do we get the benefits of positionally invariant recognition without suffering the cost of never knowing the position of the object whose category we have learned?

To overcome this problem, I thus need to explain: How does activation of a positionally invariant category representation of a painting trigger search for the painting in a particular place in the museum? As I will discuss in Chapter 4 in the context of explaining how we learn language meanings, this problem is often called the Where's Waldo Problem. Our brains use their solution of the Where's Waldo Problem whenever we efficiently search for a familiar painting in a familiar museum, or indeed for any valued object in any scene, familiar or unfamiliar.

The 3D ARTSCAN Search model (Figures 2.25 and 2.26) explains and simulates on the computer how valued objects are learned, recognized, and searched as our eyes freely scan a 3D scene or 2D painting (Cao, Grossberg, and Markowitz, 2011; Chang, Grossberg, and Cao, 2014; Fazl, Grossberg, and Mingolla, 2009; Foley, Grossberg, and Mingolla, 2012; Grossberg, 2009; Grossberg, Markowitz, and Cao, 2011; Grossberg, Srinivasan, and Yazdanbkahsh,

2014). The model shows how our brains avoid the promiscuous grouping of features from multiple objects from occurring during invariant category learning. It accomplishes this by using interactions between the dorsal, or Where, cortical stream and the ventral, or What cortical stream (Figure 1.35). The Where cortical stream restricts learning to views of the same object by the What cortical stream. The model also solves the problem of finding Waldo using interactions between the two cortical streams, but in this case from an invariant category for Waldo in the What stream to the position of Waldo in the Where cortical stream.

To solve both problems, the model defines and simulates interactions between multiple brain regions whose dynamics embody processes of spatial and object attention, invariant object category learning, predictive remapping, reinforcement learning and motivation, and attentive visual search.

I will summarize how 3D ARTSCAN Search models the way that spatial attention modulates invariant category learning. This analysis also clarifies how spatial attention can focus on the different textures that make up a scene, as was discussed for the ARTSCENE model in Section 2.7. Within 3D ARTSCAN Search, the ART model circuitry that was reviewed in Sections 2.8–2.11 and Figure 2.21 is used to learn view-specific and position-specific categories that are bound together into invariant categories by the regulatory machinery of the more comprehensive 3D ARTSCAN Search circuitry.

In so doing, 3D ARTSCAN Search provides functional explanations and predictions of neurobiological data about interactions between multiple brain regions, including cortical areas V1, V2, V3a, V4, PPC, LIP, ITp, ITa, and PFC, as illustrated in Figures 2.25 and 2.26. 3D ARTSCAN Search is thus not just a neural model. Rather, it is a neural *architecture* that proposes an explanation of how a significant fraction of the visual brain works together to attend, learn, recognize, and search for valued objects in complicated scenes. With this neural architecture in hand, it is possible to explain how a person who sees a painting from any viewpoint can understand not only its properties as a texture, as during recognition of its gist but also the objects that are represented within it, as well as their scenic context and position in the surrounding space.

2.18 Attentional shrouds modulate invariant category learning

Many visual scientists have observed that spatial attention tends to fit itself to the shape of an attended object. Tyler and Kontsevich (1995) called such a form-fitting distribution of spatial attention an *attentional shroud*. The ARTSCAN model, on which the 3D ARTSCAN Search model was built, proposed how an attentional shroud maintains its activity within the posterior parietal cortex, or PPC, of the Where cortical stream, and thereby restricts active scanning to that attended object's surface. ARTSCAN also predicts how such an active shroud regulates learning of an invariant object category in the What cortical stream of the scanned views of the attended object's surface (Fazl, Grossberg, and Mingolla, 2009; Grossberg, 2007, 2009). This Where-to-What interaction can also sustain spatial attention upon a painting as the eyes scan it and learn how to combine multiple views of the painting into an invariant recognition category for it. This process is proposed to work as follows:

When the eyes fixate on a particular view of an attended object, the brain can rapidly learn a view-specific category of that object using an ART circuit. To fix ideas, suppose that such a view-specific category is learned in the posterior inferotemporal cortex, or ITp (Figure 2.25). The first view-specific category to be learned in ITp activates some cells in the anterior inferotemporal cortex, or ITa. As the observer scans the object, this ITa cell population will become an invariant object category as it is associated with multiple view-specific categories in ITp that represent different views of the object. An attentional shroud keeps the ITa cell population active while multiple view-specific categories get associated with it through time.

This invariant category learning process unfolds through time in the following way: After the first view-specific category is learned in ITp, the eyes will continue to scan the object's surface. As they do so, they may fixate on different views of the object than the initial one. When this happens, the first view-specific category is inhibited by ART mismatch (Figures 2.22b and 2.22c), thereby enabling a view-specific category of the new object view to be learned in ITp (Figure 2.22d).

When the first view-specific category in ITp is inhibited, the cell population that it activated in ITa is *not* inhibited. This ITa cell population remains active so that the second view-specific category activated in ITp can be associated with it. As the eyes continue to scan the object, *all* the newly learned view-specific categories in ITp can continue to be associated with the active ITa cell population. In this way, the ITa cell population learns to fire in response to all the ITp categories that have been associated with it. The ITa cell population thereby becomes a view-invariant object category.

For this process to work, the ITa cell population must remain active while all the view-specific categories are associated with it. And for that to happen, the ITa population cannot be inhibited when the first view-specific category in ITp is inhibited, even though this ITp population activated the ITa population in the first place, and the ITp inputs to ITa are eliminated when the first view-specific category is inhibited.

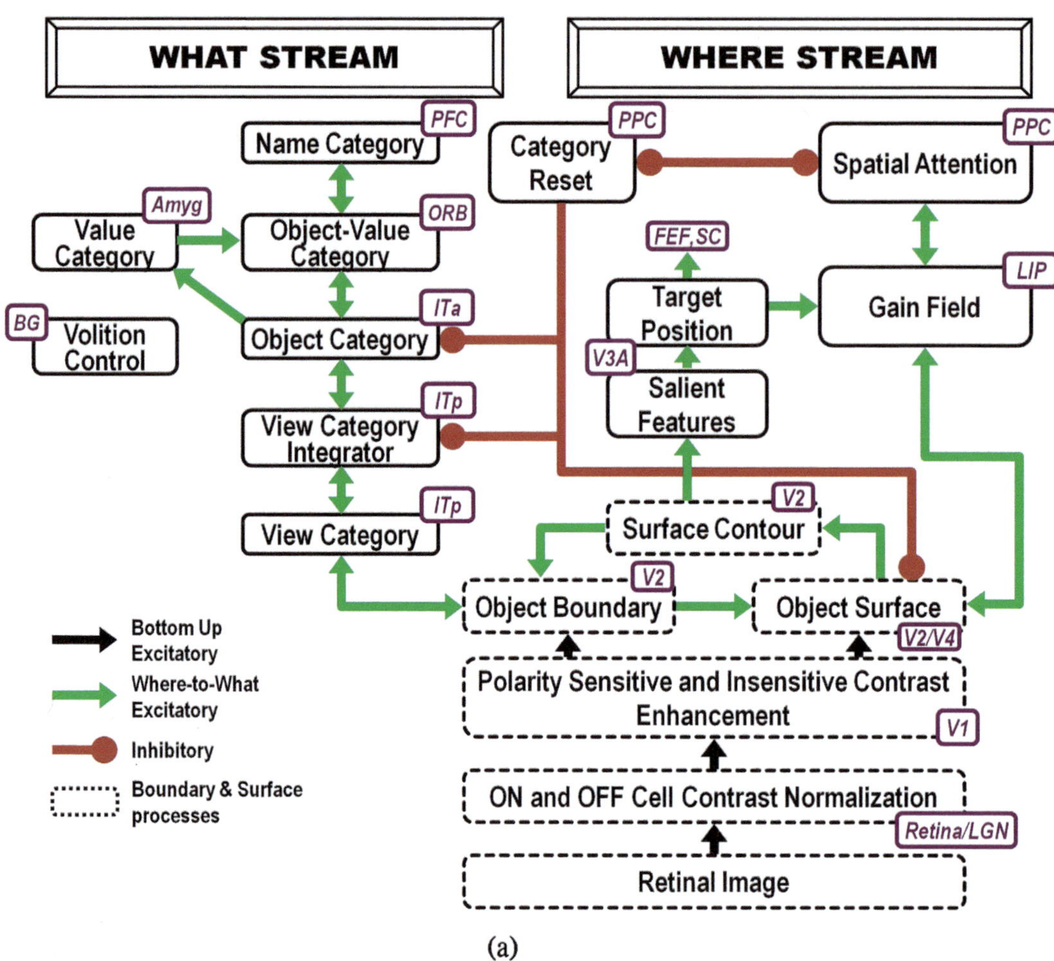

FIGURE 2.25 The 3D ARTSCAN Search model explains and simulates on the computer how valued objects are learned and recognized as our eyes freely scan a 3D scene or 2D painting. In particular, ARTSCAN Search model processes enable it to learn to recognize and name invariant object categories. Interactions between spatial attention in the Where cortical stream, via surface-shroud resonances (via the interaction between Spatial Attention and the Object Surface processes, mediated by the coordinate-changing Gain Field), and object attention in the What cortical stream, that obeys the ART Matching Rule, coordinate these learning, recognition, and naming processes (see top-down interactions from Object-Value Category to Object Category and then to View Category Integrator processes). Reprinted with the author's permission from Chang, H.-C., Grossberg, S., and Cao, Y. (2014). Where's Waldo? How perceptual cognitive, and emotional brain processes cooperate during learning to categorize and find desired objects in a cluttered scene. *Frontiers in Integrative Neuroscience*, doi: 10.3389/fnint.2014.0043.

Thus, the main problem that we need to solve is: Why is the ITa population not inhibited along with the ITp population that initially activated it?

The ARTSCAN model, and its 3D ARTSCAN Search generalization, predicts that *the attentional shroud protects the ITa cells from getting reset*, even while view-specific categories in ITp are reset, as the eyes explore multiple views of an object. The shroud (labeled Spatial Attention in Figure 2.27) does this by inhibiting the ITa *category reset* mechanism in the parietal cortex (dashed horizontal red connection from Spatial Attention to Category Reset) that would otherwise have inhibited ITa (dashed red connection from Category Reset to Object Category in Figure 2.27). When the eyes move to attend to a new object, the shroud collapses, thereby removing inhibition from the Category Reset mechanism. The Category Reset mechanism is thus disinhibited and can now inhibit the ITa population, which is labeled Object Category in Figure 2.27. The neural circuitry is then ready to learn an invariant category for another object. In particular, the collapse of the shroud enables the eyes to move to another surface and attend to it using a new shroud, whereupon new view-specific and view-invariant object categories can be learned. The cycle can then repeat itself ad infinitum.

At around the same time that the ARTSCAN model was published, Chiu and Yantis (2009) published data that provided strong support for this search cycle, notably the prediction that spatial attention in PPC regulates the inhibitory reset of invariant object categories in ITa.

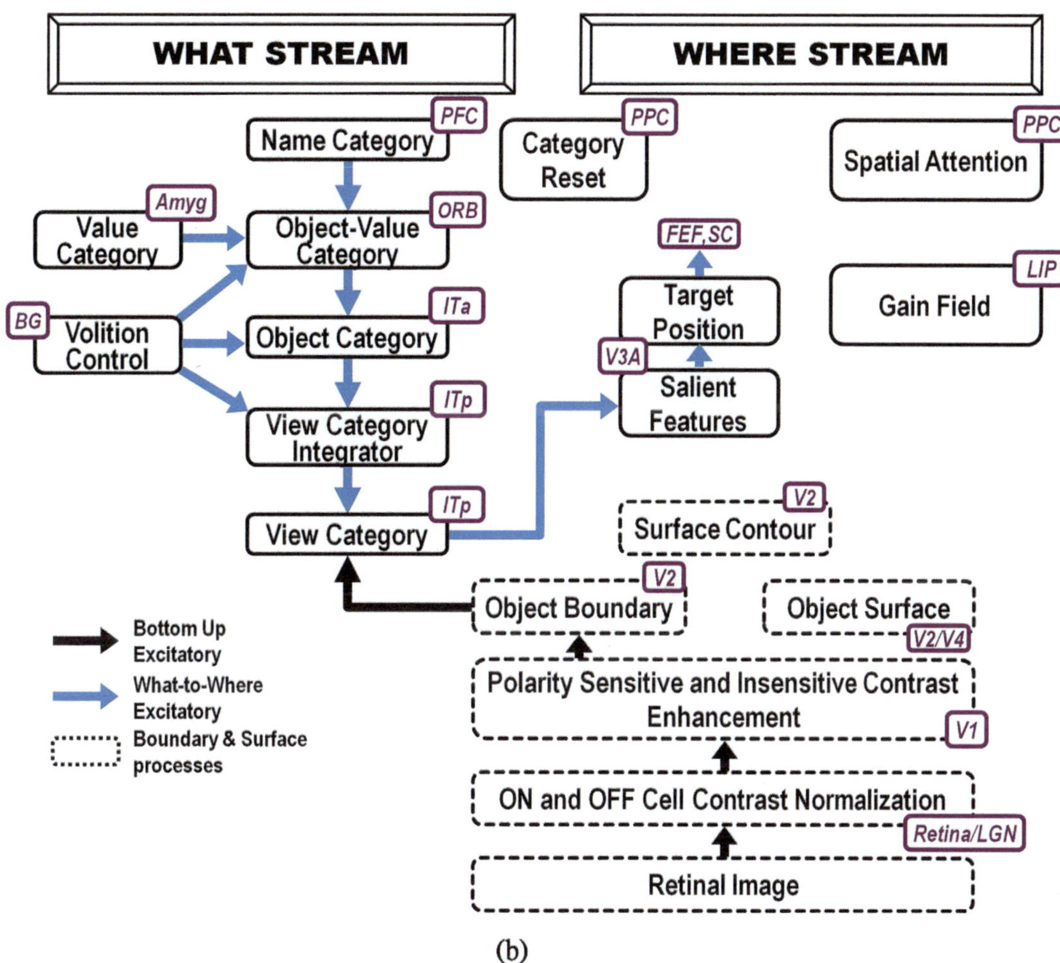

FIGURE 2.26 The 3D ARTSCAN SEARCH model enables the name of a learned category to trigger a search for it in a cluttered scene, thereby providing a solution of the Where's Waldo problem. Reprinted with the author's permission from Chang, H.-C., Grossberg, S., and Cao, Y. (2014). Where's Waldo? How perceptual cognitive, and emotional brain processes cooperate during learning to categorize and find desired objects in a cluttered scene. *Frontiers in Integrative Neuroscience*, doi: 10.3389/fnint.2014.0043.

Using rapid event-related MRI to record brain activity in humans, they found that a shift of spatial attention evokes a transient, domain-independent signal in the medial superior parietal lobule that corresponds to a shift in categorization rules. In ARTSCAN, these data properties translate into the following brain mechanisms: Collapse of an attentional shroud (spatial attention shift) disinhibits the parietal reset mechanism (transient signal) that inhibits the previous view-invariant object category and allows activation of a new one (shift in categorization rules).

The transient signal is *domain-independent* because *all* objects inhibit the same population of parietal category reset cells while their shrouds are active, and these reset cells can inhibit *all* invariant object category representations in ITa (dotted red inhibition connections in Figure 2.27). To the best of my knowledge, this anatomical prediction about a *many-to-one convergence* of inputs to the parietal reset cells, and a *one-to-many divergence* of outputs from them, has not yet been tested.

Grossberg (2013, 2017) provides additional data explanations and cites other articles that contributed to the development of the 3D ARTSCAN Search architecture.

2.19 Seeing and knowing: Surface-shroud and feature-category resonances

In order for the above brain mechanisms to work, a viewer needs to *sustain spatial attention* long enough on an object, scene, or picture to scan its interesting features, and thereby learn an invariant category with which to recognize it. This observation raises the mechanistic question: How is shroud activity chosen to focus on an object of interest and sustained during active scanning of the object?

Looking at an object automatically, or pre-attentively, generates a surface representation of the object in the prestriate visual cortex, including cortical area V4. This

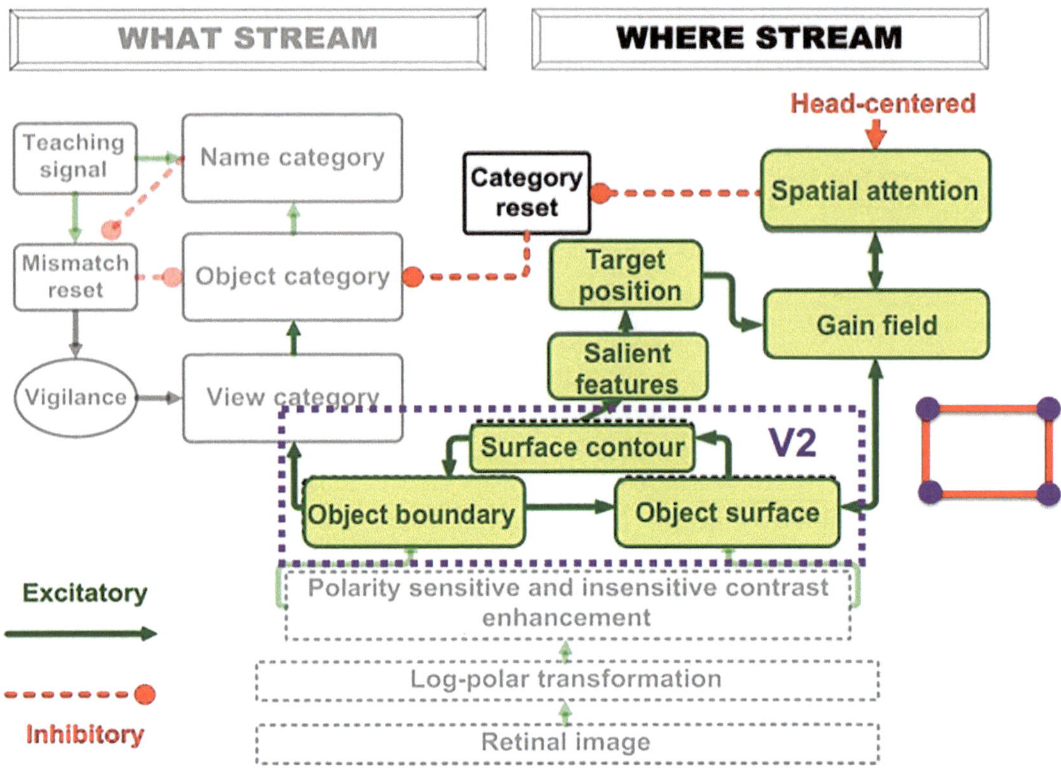

FIGURE 2.27 The 3D ARTSCAN Search model explains how our brains learn both view-specific categories (View category) and view-*invariant* object categories (Object category), and uses them to search for desired objects in a cluttered scene, such as favorite paintings in a museum, thereby solving the classical Where's Waldo Problem. Spatial attention is sustained on an object via a surface-shroud resonance (between Spatial attention and Object surface) as an invariant object category of it is being learned. Adapted with the author's permission from Grossberg, S. (2017). Towards solving the Hard Problem of Consciousness: The varieties of brain resonances and the conscious experiences that they support. *Neural Networks*, 87, 38–95. See the text for details. Author created.

surface representation, in turn, activates a shroud in the PPC. But how is a shroud selected for just one of several different objects in a scene that may simultaneously be competing for spatial attention in PPC? All of these spatial attentional foci also send positive feedback signals back to their generative surfaces. Taken together, all these excitatory and inhibitory interactions form a recurrent on-center off-surround network (Figure 2.28): The on-center consists of the positive feedback signals between the surface in V4 and its shroud in PPC. The off-surround consists of the recurrent inhibitory signals within PPC, whereby the shrouds compete with each other.

It has been known for many years how such a recurrent on-center off-surround network can contrast-enhance, or render more active, the representation that starts with the most activity (e.g., Grossberg, 1973, 1980). When this happens, the winning shroud focuses spatial attention on its surface (Figure 2.28). This positive feedback loop *sustains* spatial attention on the object surface while it amplifies and synchronizes the activities of the attended cells, thereby leading to what I call a *surface-shroud resonance* between a surface representation (e.g., in cortical area V4) and spatial attention (e.g., in PPC).

The concept of a surface-shroud resonance also provides new insights into basic issues such as: How do we consciously see? How is conscious seeing related to conscious recognition, or knowing? How does conscious seeing enable us to look at and reach objects of interest? In 2009, I proposed that we consciously see using surface-shroud resonances (Grossberg, 2009). That is, we see the visual qualia of a surface when they are synchronized and amplified within a surface-shroud resonance. Such a resonance can propagate both top-down from V4 to lower cortical levels, such as V2 and V1, as well as bottom-up to higher cortical areas, such as the PFC, thereby synchronizing an entire cortical hierarchy.

The hypothesis that a surface-shroud resonance supports conscious seeing of visual qualia provides an answer to a basic question that was raised by two of my previous predictions, but which I could not answer for many years:

The first prediction was made when ART was first being introduced (e.g., in Grossberg, 1980); namely, that "all conscious events are resonant events."

The second prediction was made when the properties of boundaries and surfaces were articulated (e.g., in Grossberg, 1987a, 1987b, 1994); namely, that "all

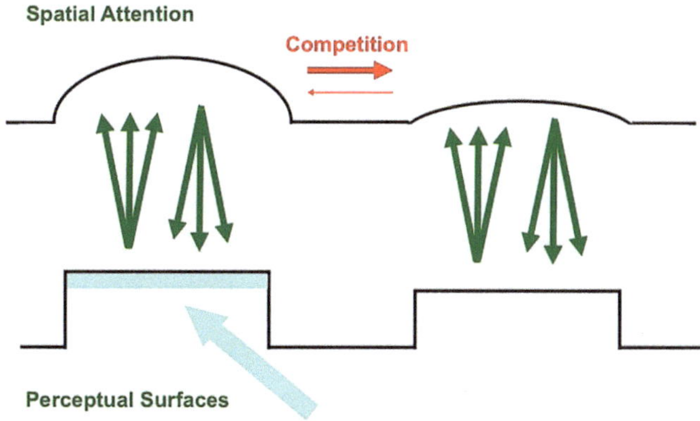

FIGURE 2.28 A cross-section of a simple filled-in surface (e.g., in cortical area V4) is shown in which a more contrastive bar is to the left of a less contrastive bar. Each position in the surface sends topographic bottom-up excitatory signals to the spatial attention region (e.g., the Posterior Parietal Cortex, or PPC). Each activated spatial attention cell sends topographic top-down excitatory signals to the surface, as well as broad off-surround inhibitory signals to other spatial attention cells. These synchronous bottom-up and top-down interactions generate a *surface-shroud resonance* that contrast-enhances the more active spatial attention cells, while inhibiting the less active ones. A form-sensitive distribution of spatial attention, or *attentional shroud*, forms as a result that focuses spatial attention upon the more contrastive surface, while also increasing its effective contrast. Adapted with the author's permission from Grossberg, S. (2017). Towards solving the Hard Problem of Consciousness: The varieties of brain resonances and the conscious experiences that they support. *Neural Networks*, 87, 38–95. See the text for details. Author created.

conscious percepts of visual qualia are surface percepts"; see Section 2.

Combining these two predictions leads to the question: What kind of resonance supports conscious percepts of visual qualia? The answer proposed in Grossberg (2009) is: a surface-shroud resonance. Grossberg (2017) went on to explain data about conscious and unconscious visual perception in normal individuals and clinical patients using surface-shroud resonances.

Surface-shroud resonances support conscious *seeing*. Conscious seeing is not, however, the same process as conscious *recognition*. Seeing, as well as other kinds of perception, is a *general-purpose* process because we can see both unfamiliar and familiar objects and scenes. In contrast, we can recognize, or know about, only objects and scenes that we have previously learned.

Given that seeing and recognition are different processes, we need to ask: What kind of resonance supports conscious recognition? How do we consciously recognize familiar objects and scenes as we consciously see them?

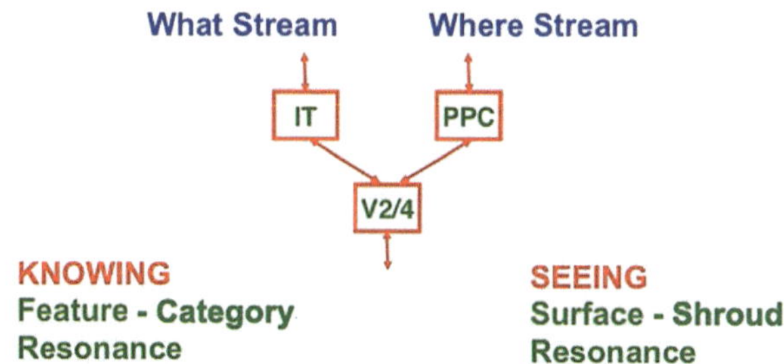

FIGURE 2.29 When we observe a familiar object and consciously see and recognize it, a surface-shroud resonance for seeing, and a feature-category resonance for recognition, is synchronously activated. The category occurs in the What stream, while the shroud occurs in the Where stream. Adapted with the author's permission from Grossberg, S. (2017). Towards solving the Hard Problem of Consciousness: The varieties of brain resonances and the conscious experiences that they support. *Neural Networks*, 87, 38–95. See the text for details. Author created.

While a surface-shroud resonance is being activated between V4 and PPC, the brain is also activating known recognition categories, or learning new ones, via interactions between brain regions such as V2, V4, ITp, and ITa. These latter interactions cause a *feature-category resonance* (Figure 2.20) to occur. Surface-shroud resonances for seeing, and feature-category resonances for recognizing, are linked via prestriate cortical areas like V4, with the surface-shroud resonances including Where cortical stream regions like PPC, and the feature-category resonances including What cortical stream regions like IT (Figure 2.29). When the two kinds of resonances get synchronized to each other via their shared representations in visual cortex, we can simultaneously see *and* know things about an attended familiar object.

The above concepts clarify how an observer may consciously see multiple views of a painting or sculpture, and simultaneously know things about them, as the observer's spatial attention shifts from one part of the work of art to the next.

2.20 Scene understanding, search, and episodic learning and memory

The ARTSCENE model (Section 2.7) clarifies how the gist of a scene can be learned, and how gist may be refined by spatial attention shifts across space that learn finer textures of a scene (Grossberg and Huang, 2009). Indeed, each attention shift enables a newly focused region to be attended to using its own attentional shroud, which then restricts learning to that region's texture category. The 3D

ARTSCAN Search model clarifies, in addition, how we can learn to recognize and search for valued objects in such a scene.

Neither of these models proposes how, as the eyes scan a scene, they learn to accumulate contextual information about the kind of scene that it is. For example, when a refrigerator and a sink are viewed in sequence, an expectation of other kitchen appliances, like stoves and microwaves, may be primed, rather than of beds or beaches. If the refrigerator and sink are familiar, and in familiar positions relative to one another, then more definite expectations of particular kitchen appliances and their positions may be primed. The same is true for any scene, including a painting, leading to expectations whose confirmation or violation after a shift of attention may influence both our recognition and our aesthetic appreciation, of it.

The ARTSCENE Search model (Figure 2.30; Huang and Grossberg, 2010) goes beyond the ARTSCENE model to propose how object and spatial contexts about a scene, or painting, can be created as our eyes sequentially scan it, and then be used to more efficiently search for particular objects within the painting based upon this stored information. ARTSCENE Search has been used to quantitatively simulate many challenging data about what is called *contextual cueing* within the cognitive science literature; e.g., data of Brockmole, Castelhano, and Henderson (2006), Chun (2000), Chun and Jiang (1998), Jiang and Chun (2001), Jiang and Wagner (2004), Lleras and von Mühlenen (2004), and Olson and Chun (2002).

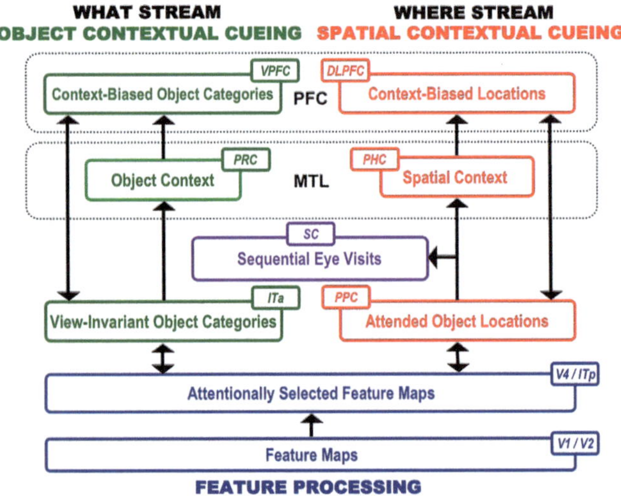

FIGURE 2.30 The ARTSCENE Search model simulates how the gist of a scene is learned, and how gist is refined by spatial attention shifts across space that learn finer texture categories of the scene. All the learned categories then vote together to determine scene type. Both What and Where cortical streams interact to accomplish this goal. See the text for details. Reprinted with the author's permission from Huang, T.-R., and Grossberg, S. (2010). Cortical dynamics of contextually cued attentive visual learning and search: Spatial and object evidence accumulation. *Psychological Review*, 117, 1080–1112.

An *object context* (left side of Figure 2.30) accumulates information about the sequence of objects that the eyes foveate as they scan a scene, like a painting or sculpture, or a series of such works of art. Such sequential information is stored in an *object working memory*. The contents of this working memory can then be used to learn a special kind of recognition category that is called an *object list chunk*, or plan, that responds selectively to that particular sequence of experienced objects. Learning and activation of such a plan represent that particular object context.

In much the same way, a *spatial context* (right side of Figure 2.30) accumulates information about the sequence of positions, or locations, that the eyes foveate as they scan a scene. Such sequential information is stored in a *spatial working memory*. The contents of this working memory can then be used to learn a special kind of recognition category that is called a *spatial list chunk*, or plan, which responds selectively to that particular sequence of experienced locations.

Object and spatial plans can be used to direct a search for additional objects that are expected after familiar object and spatial contexts have been experienced, as well as to read out sensory expectations of these objects that prime us to expect to see them. In this way, if we are looking at a familiar painting or series of paintings in the Louvre or the Museum of Modern Art (MoMA), we are also primed to find other paintings that we like in their permanent collections.

The output signals from active object and spatial plans may be combined to "vote" for the object that a viewer most probably expects to see, as well as the position where this object is expected to be seen. This kind of knowledge begins to achieve context-sensitive *scene understanding*. It helps us to understand a painting as an entity within which certain combinations of objects, textures, and colors occur where we expect them to be. We can also learn in this way to understand a room filled with favorite Impressionist paintings in MoMA.

ARTSCENE Search clarifies how the brain achieves scene understanding using interactions between multiple brain regions. In addition to the brain regions that are modeled by the 3D ARTSCAN Search model, the ARTSCENE Search model also includes interactions of the perirhinal and parahippocampal cortices with the temporal, parietal, and prefrontal cortices (Figure 2.30). Object contexts are computed in the perirhinal cortex, and spatial contexts are computed in the parahippocampal cortex. These object and spatial contextual representations interact with object and spatial plans in the prefrontal cortex to read out visual expectations of what and where new objects, textures, and colors are expected to be seen.

With all these mechanisms in play, a viewer of a painting can more intelligently explore different parts of a painting based upon other regions of the painting that have already been viewed.

3 How boundaries are completed and surfaces filled-in

3.1 Complementary boundaries and surfaces and complementary consistency

As summarized in Section 2, all visual percepts are built up from interactions between boundary and surface representations. These representations are processed in parallel cortical processing streams (Figure 2.17) that individually compute computationally complementary properties (Figure 2.16). Interactions between these streams, across multiple processing stages, overcome their complementary deficiencies to compute unified representations of the world. Said in another way, these interactions convert computations that obey complementary laws into a consistent percept, thereby achieving the property of *complementary consistency*. In order to better understand important properties of many different kinds of paintings, additional information is needed about how boundaries are completed and surfaces filled-in to generate a conscious percept of a work of art.

3.2 Oriented filtering, spatial and orientational competition, and bipole grouping

The boundary signals that activate the brain may be incomplete either because they receive incomplete inputs from the world, as illustrated by the famous example of a Dalmatian in Snow (Figure 2.3), or because they are processed by the retina across the blind spot or retinal veins (Figure 2.15). Oriented filtering by simple cells and complex cells (Figures 2.5–2.7) cannot complete these incomplete boundary fragments. Additional processing by subsequent brain regions is needed to do that. To this end, output signals from complex cells input to a subsequent processing stage where boundary completion is achieved by *bipole grouping* (Figures 2.11 and 2.31).

Bipole grouping cells can cooperate together to complete boundaries in response to pairs of (almost) collinear and (almost) like-oriented contours (Figure 2.4). Bipole grouping thereby completes boundaries *inwardly*

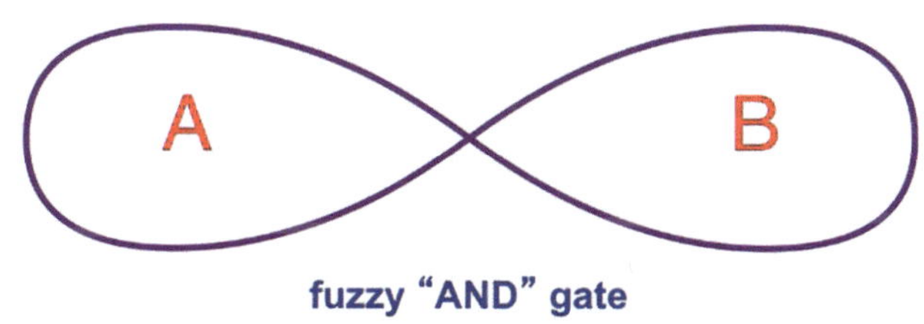

FIGURE 2.31 Interactions among a network of bipole cell receptive fields is a late stage in perceptual grouping. Author created.

between pairs, or greater numbers, of (almost)-collinear and (almost)-like-oriented cells that are activated by these contours. Because bipole cells are activated by complex cells, bipole cells can group inducers with like contrast polarities, as in Figure 2.4 (top row, left image), or with opposite contrast polarities, as in Figure 2.4 (top row, right image). In other words, bipole grouping is *insensitive to contrast polarity*.

There is a large experimental literature about properties of the perceptual grouping process of which bipole grouping forms a part. In 1984, I predicted and began to explain the grouping properties of bipole cells with two of my postdoctoral fellows, Michael Cohen and Ennio Mingolla, who later became faculty in our department (Grossberg, 1984; Cohen and Grossberg, 1984; Grossberg and Mingolla, 1985a, 1985b). At around the same time, von der Heydt, Peterhans, and Baumgartner (1984) reported their discovery, using neurophysiological methods, of neurons with the predicted bipole properties in cortical area V2; see Figure 2.32. Our laminar cortical models of how the brain sees (e.g., Grossberg (1999) and Grossberg and Raizada (2000)) proposed how the bipole grouping property can be achieved by neurons in a laminar neocortical circuit. We predicted how the bipole property could arise from recurrent long-range excitatory interactions between neurons in layer 2/3 of V2, balanced by short-range, recurrent inhibitory interneurons (Figure 2.33). This prediction functionally explained how the bipole grouping property could arise from interactions between known anatomical cell types.

A comparison of Figure 2.10 with Figure 2.33 shows a remarkable similarity, and thus parsimony, of the circuits that generate binocular simple cell properties with those that create bipole grouping cell properties. In both cases, pairs (or greater numbers) of oriented excitatory

BIPOLES: FIRST NEUROPHYSIOLOGICAL EVIDENCE (V2)

Stimulus: / Probe location: •	Cells in V2 Response?	
		Grossberg, 1984, prediction
•	YES	von der Heydt, Peterhans, and Baumgartner, 1984
•	NO	
•	NO	
•	YES	Peterhans and von der Heydt, 1988
• (more contrast)	NO	
•	YES	

Evidence for receptive field:

FIGURE 2.32 Neurophysiological properties of bipole cells in cortical area V2 were discovered by von der Heydt, Peterhans, and Baumgartner (1984). See the text for details. Author created.

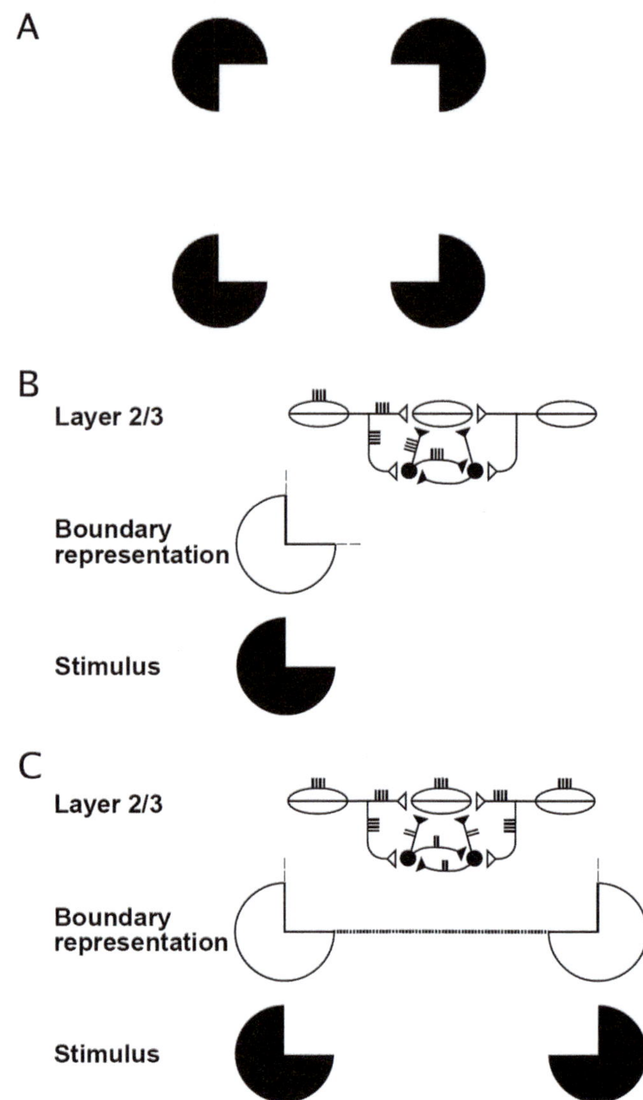

FIGURE 2.33 Anatomical circuitry that can carry out boundary completion of a Kanizsa square using bipole cell interactions to form long-range perceptual groupings, including illusory contours. See the text for details. Author created.

input signals converge on a target cell and are balanced by inhibitory signals whose total input strength is self-normalized by recurrent inhibitory interneurons.

3.2.1 HOW DO INITIALLY FUZZY GROUPINGS END UP SHARP?

Figures 2.4, 2.12, and 2.13 illustrate that completed boundaries are typically both oriented and spatially *sharp*. We should not take this combination of properties for granted. It requires considerable processing of visual inputs to achieve for the following reasons:

To initiate boundary processing, the brain uses simple cells and complex cells, which are oriented local contrast detectors in cortical area V1, as noted in Figures 2.5–2.7. The property of orientationally selective tuning by these cells balances between advantages and limitations, however. By averaging input contrasts across their receptive fields, simple cells and complex cells enable the brain to avoid the need to use a large number of different detectors that are individually designed to respond selectively to edges, or shading, or texture, or some other highly specific visual property. Instead, a single simple cell and complex cell can respond to edges, shading, *and* texture, among other visual properties, that roughly share the orientation and spatial scale of their receptive fields. For example, in Figure 2.5, a vertically oriented simple cell with a dark-light polarity can respond at the right edge of a vertical black bar on a white background. Such a simple cell can also respond to a contrast difference that is caused by two vertically oriented black dots on a white background, as well as other types of contrast gradients. However, such a simple cell cannot respond to the white background itself because the background has uniform luminance. Nor can a horizontally oriented simple cell respond to a vertical black bar.

This type of response selectivity also leads to computational uncertainties. For example, the oriented receptive fields of simple cells and complex cells do not respond at the ends of thin lines, or other high curvature contours, as illustrated by the computer simulation of cell responses at a line end in Figure 2.34 (left image). I call such a hole at the end of a boundary an *end gap*. Without further processing, the end gaps in the boundaries at line ends would allow brightness and color to flow out of every line end due to surface filling-in, leading to perceptually disastrous consequences.

To prevent this perceptual catastrophe, additional processing stages are needed subsequent to complex cells.

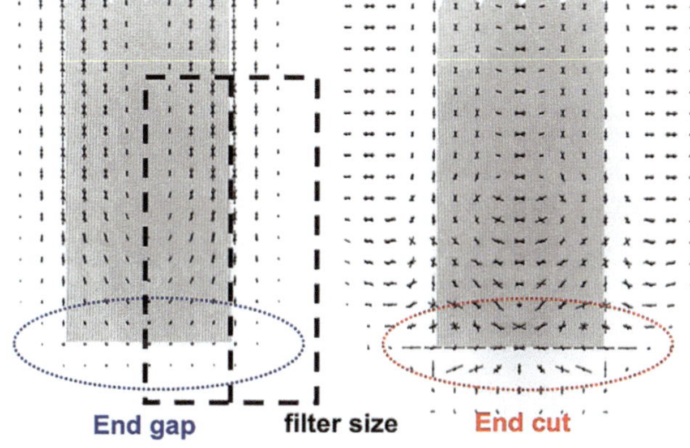

FIGURE 2.34 Computer simulation by Grossberg and Mingolla (1985a, b) of two stages in boundary completion at line ends. See the text for details. Author created.

My articles Grossberg (1984) and Grossberg and Mingolla (1985a) predicted and modeled how this additional processing can be accomplished in two stages:

As illustrated in Figure 2.11, complex cells are proposed to input to a subsequent processing stage via a *spatial competition*. Cells at this next stage are often called *hypercomplex* cells (Hubel and Wiesel, 1968). Hypercomplex cells receive inputs from complex cells via an on-center off-surround network: Each complex cell excites the hypercomplex cell at its position with the same orientational preference, while inhibiting nearby hypercomplex cells with the same, or similar, orientational preferences. Due to the off-surround, the responses of hypercomplex cells are sensitive to the length of input lines, since line inputs that fall within a strong region of inhibition within the off-surround are suppressed, a property that is used to help classify cells as hypercomplex cells in neurophysiological experiments. The cells that are part of this on-center off-surround spatial competition are said to constitute the *first competitive stage*.

Many boundaries would still remain incomplete if boundary processing stopped with the first competitive stage. For example, line-end boundaries would still have a hole in them, and the two Kanizsa squares in Figure 2.4 would continue to be seen and recognized as four spatially disjoint pac-man figures.

This does not happen because of subsequent processing stages. In particular, the cells of the first competitive stage activate cells at a *second competitive stage*. These latter cells compete with other cells at the same position that are tuned to respond to different orientations (Figure 2.11). The maximal mutual inhibition occurs between cells that prefer *perpendicular* orientations.

Cells at the second competitive stage are tonically active, which means that they are continuously activated by an internal energy source. Their balanced mutual inhibition suppresses their tonic activity when none of these cells receives an external input. Suppose, however, that a cell in the second competitive stage lies just beyond the end of a line. Then, in addition to its tonic inhibition, it will be inhibited via cells at the first competitive stage that are located near the line end.

When these cells in the second competitive stage are inhibited, their inhibition of other cells at the same position in the second competitive stage will also be reduced, notably inhibition of cells that code the perpendicular orientation, and orientations close to it. These latter cells are hereby *dis*inhibited, or freed from inhibition, and can create a boundary at the line end. Such a boundary is called an *end cut* (Figure 2.34, right image). The spread of brightness and color signals out of every line end is prevented by its end cuts. Taken together, the first and second competitive stages form a Double Filter (Figure 2.11).

In summary, the two competitive stages compensate for the missing boundaries at the end of a vertical thin line as follows: The competition across the hypercomplex cells at the first competitive stage is *across position* and *within orientation*, whereas the competition across the hypercomplex cells at the second competitive stage is *within position* and *across orientation*.

At the first competitive stage, active vertically oriented complex cells near the line end inhibit vertically oriented hypercomplex cells just beyond the line end. When these cells are inhibited, their inhibition of cells at their position that are tuned to other orientations is removed. The most inhibition is removed from cells that are preferentially tuned to the horizontal orientation. These horizontally oriented cells are hereby disinhibited, and create line ends, or end cuts, that complete the boundary at the line end, and are capable of containing the line's brightness and color within its borders.

3.2.2 SERIFS DOUBLY COMPLETE END CUTS AT LINE ENDS

Designers of letter fonts often attach a small line, or *serif*, to the end of a stroke in a letter. Many of the letters printed in this book possess such serifs. There are many possible historical explanations of serifs, or "Roman" typefaces, but one psychological benefit of them is to strengthen the boundary at a line end, just as an end cut does. In this sense, *serifs doubly complete line ends* to make them more visible.

Although cells at the second competitive stage can complete the boundaries at line ends, they cannot complete boundaries that can be created in response to textures with spatially discrete features, as occurs in the Dalmatian in Snow image (Figure 2.3), or across the blind spot and retinal veins (Figure 2.15). Cells at the second

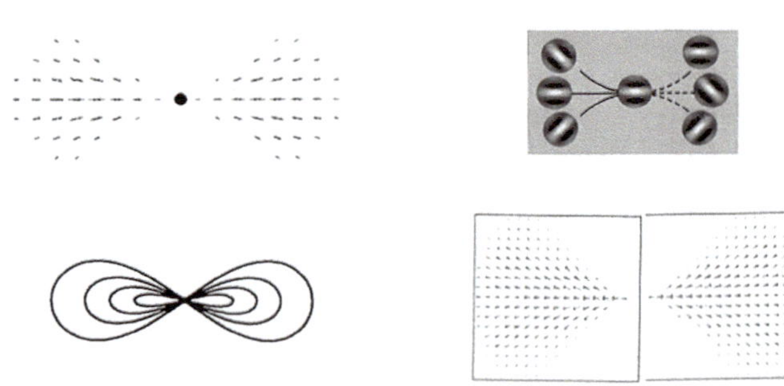

FIGURE 2.35 Some bipole cell receptive fields. The predicted bipole cell receptive field (upper-left corner) has been supported by both neurophysiological data and psychophysical data, and used in various forms by many modelers. This figure was adapted from lecture notes on neural models of visual perception created by S. Grossberg and E. Mingolla: https://sites.bu.edu/steveg/files/2016/06/VSS-2005-Grossberg-Mingolla.pdf

competitive stage input to bipole cells that can carry out boundary completion (Figure 2.11).

Because objects in the real world often have curved boundaries, when these boundaries are occluded at unpredictable locations, bipole groupings need to link object features that may have similar, but unequal, orientations, and that are not collinear across space. Our brains can cope with these uncertainties because individual cells at the second competitive stage respond to a line end by activating a band of similar orientations (Figure 2.34, right image), all of which input to bipole cells that can also respond to a band of similar orientations (Figure 2.35). The band of orientations at the second competitive stage is created by the bands of orientations that are activated near the edges of the line end (Figure 2.34, left image). As a result of these fuzzy inducers, bipole cells can group across space in any of several orientations, depending upon which of these orientations line up best across space and have the most support from inputs in the scene or picture.

3.3 Hierarchical resolution of uncertainty: From fuzzy to sharp

These spatial and orientational competitive interactions are part of a process that I call *hierarchical resolution of uncertainty* (Grossberg, 1984, 1994; Grossberg and Mingolla, 1985a), whereby uncertainties caused by lower levels of cortical processing are compensated and overcome by processes at higher cortical levels.

Why is hierarchical resolution of uncertainty needed? Why does not the brain just define perfect edge detectors that do not fail to detect line ends? A basic reason for this has already been briefly noted; namely, simple cells are not just edge detectors. They are oriented local contrast detectors that can respond to edges, textures, shading, and depth differences. If the brain defined edge detectors instead, then it would also have to define many other types of specialized detectors, perhaps leading to a combinatorial explosion of detectors.

Our brains would then also be faced with the challenging problem of figuring out how to put together the specialized information that is computed by all these different kinds of detectors. Even if this problem could be solved in some situations, such a fusion of information could be rendered impossible in situations that often occur in scenes and paintings, where edges, textures, shading, and depth are all overlaid in the same locations. Specialized detectors would be defeated by such an overlay of properties.

Our brains have evolved to deal effectively with such environmental uncertainties. Instead of using innumerable specialized detectors, cortical visual processes start to build boundaries with simple cells, or oriented local contrast detectors, which can respond with coarse oriented and positional estimates to all of these different types of properties, albeit imperfectly, and then can use hierarchical resolution of uncertainty to contextually complete orientationally sharp and positionally hyperacute boundary representations.

Several hierarchical resolutions of uncertainty are needed to generate the visual percepts that we consciously see. Three more of them are summarized below: one having to do with the process of boundary grouping, another with surface filling-in, and the third with figure-ground separation. These hierarchical resolutions of uncertainty clarify why multiple processing levels occur in the anatomy of visual cortex (Figures 1.40, 2.11, and 2.17).

Our brains thus do not eliminate uncertainty too soon. They take advantage of uncertainty until processing stages are reached at which uncertainty can effectively be drastically reduced or eliminated by context-sensitive interactions to support a final visual percept. Many artists have understood this point, at least implicitly, with the Impressionists, such as Claude Monet (see Section 4.1), and other plein air painters being notable examples.

The next section explains in greater detail how, despite the imperfect alignment of brush strokes in paintings, they can be grouped into emergent objects in scenes, much like the black splotches in the Dalmatian in Snow image are (Figure 2.3).

3.4 Neon color spreading: End gaps, end cuts, and fuzzy-to-sharp groupings

Many experimental data and percepts support the idea of hierarchical resolution of uncertainty and how it works. For example, visible evidence of how end cuts usually prevent spurious filling-in from occurring can be seen in some percepts due to the fact that, although the two competitive stages work most of the time, they do not prevent spurious color spreading from occurring in response to all scenes and pictures.

A famous example of color spreading from line ends is shown in Figure 2.36. This percept is called *neon color spreading*. Neon color spreading illustrates key properties of color filling-in, in addition to demonstrating how end gaps and end cuts work. The image in Figure 2.36 consists of circular annuli, part of which are black and part blue. When viewing this figure, an illusory square can be seen that is filled with blue "neon" color, even though the only blue in the image is in the blue circular arcs that form part of the four sets of concentric circles that induce the percept.

Neon color spreading was reported by Dario Varin in Varin (1971). Varin there described a "chromatic spreading" effect that was induced when viewing an image like the one in Figure 2.36. Harrie F. J. M. van Tuijl soon after introduced images that gave rise to percepts that he called "neon-like color spreading" (van Tuijl, 1975).

To explain the percept in Figure 2.36, first note that the black and blue arcs in Figure 2.36 both create boundaries in our brains. The contrast of the black arcs with respect to the white background is chosen larger than the contrast of the blue arcs with respect to the white background. The boundaries formed by these contrasts are contrast-sensitive. As a result, the boundaries formed by the black-white contrasts are stronger than those formed by the blue-white contrasts. The black-white boundaries

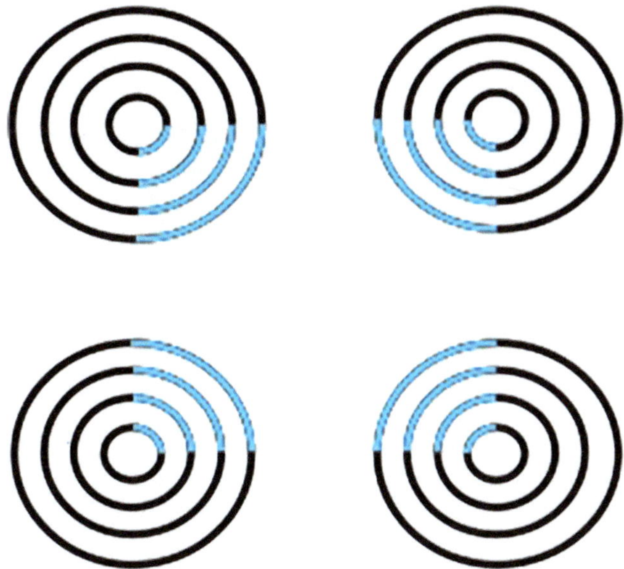

FIGURE 2.36 Neon color spreading. Adapted with the author's permission from Grossberg, S., and Yazdanbakhsh, A. (2005). Laminar cortical dynamics of 3D surface perception: Stratification, transparency, and neon color spreading. *Vision Research*, 45, 1725–1743.

can thus inhibit the weaker, spatially abutting blue-white boundaries, more than conversely, using the like-oriented spatial competition of the first competitive stage (Figure 2.11). At the positions where the differently colored boundaries touch, the boundaries caused by the black arcs hereby cause end gaps to occur at the ends of the abutting boundaries caused by the blue arcs.

Although the heuristics of these operations are simple to describe, actually making them work requires that one uses the correct neuronal laws and network interactions. This fact is illustrated by the answer to the following question: If spatial competition occurs along an entire line, then why does not the balanced inhibition along the entire interior of the line suppress its boundary entirely? The line *could* have been inhibited if spatial competition just *subtracts* activity from its target cells.

Instead, a form of competition is used that is called *shunting* competition, which fortunately is the kind of competition that is obeyed by the membrane equations that define neuronal dynamics (see Figure 1.21). Shunting competition does not just subtract activity from inhibited cells. Instead, it also *divides* the activity of these cells. This property can easily be seen by solving the equation in Figure 1.21 at equilibrium. The dividing terms are due to the interaction terms—$C_i x_i$ and $E_i x_i$ in Figure 1.21 that multiply, or shunt, the terms that are added up within the square brackets [ ].

Shunting competition can be designed so that, when the strength of competition between nearby cells is the same, such as along the interior of a black arc, then the activities of these cells are reduced, but not eliminated, by an amount that is sensitive to the *relative* activities of all the neighboring cells. This property is called *contrast normalization*.

At positions in the neon color display in Figure 2.36 where the weaker blue arc inputs receive shunting competition from nearby black arc inputs, these competitive signals are not balanced. Instead, holes are caused in the blue arc boundaries at positions that lie just beyond the ends of the black arc boundaries. These holes are the *end gaps*.

The spatial competition that causes end gaps in response to a neon color spreading image can then cause *end cuts* when the first competitive stage inhibits cell activities at the second competitive stage, as shown in Figures 2.11 and 2.34. Because the second competitive stage carries out orientational competition of tonically active cells at each position, inhibiting cells at the second competitive stage with a given orientational preference *dis*inhibits cells that are tuned to other orientations, notably the perpendicular orientation. These disinhibited cell activations are the end cuts.

In summary, both end gaps and end cuts can occur where the black arcs touch the less-contrastive blue arcs in a neon color spreading image.

These end cuts occur in bands of almost-perpendicular activations at multiple line ends in response to the neon color spreading image in Figure 2.36. These active cells can activate like-oriented bipole cells that can cooperate across space to complete a boundary with the shape of an illusory square. It is this illusory square that bounds the filling-in of blue color in response to the image in Figure 2.36.

As I noted above, bipole grouping can occur even when adjacent end cuts do not have the same orientational preference. Grouping can happen when these orientations are similar enough to be grouped by the orientationally and positionally fuzzy bipole cell receptive fields. These receptive fields prefer to group collinear orientations, but they can also group non-collinear orientations with a strength that decreases with the orientational difference from the preferred orientation and the physical distance from the bipole cell body (Grossberg and Mingolla, 1985b), as noted in Figure 2.35. These receptive field properties arise during development of the bipole cells as they are exposed to the statistics of many scenes and images. See Grossberg and Raizada (2000), Grossberg and Seitz (2003), Grossberg and Swaminathan (2004), Grossberg and Williamson (2001), Leveille, Versace, and Grossberg (2010), and Raizada and Grossberg (2001) for computer simulations of how bipole receptive fields can develop.

Due to the spatial and orientational uncertainty in the bipole cell receptive fields, bipole cells can interact together to complete boundaries in response to collinear, perpendicular, or oblique inducers (Figures 2.4 and 2.37). Each of these final boundary groupings typically appears to be spatially sharp, despite the uncertainty, or fuzziness, in the number of possible inducing orientations before a final grouping is chosen. This property of transforming an initially fuzzy orientational and positional grouping choice into a sharp final grouping is a second example of hierarchical resolution of uncertainty (Figure 2.38).

This particular hierarchical resolution solves the following problem: The probability that pairs of nearby visual inducers are perfectly aligned across space is vanishingly small, especially when one considers that the design and placement of neurons are not perfect. Despite this limitation, it is essential to be able to initiate grouping between them. The fuzzy bipole cell receptive fields enable this to occur (Figure 2.35), and a band of grouping possibilities is initially generated (Figure 2.38, left image).

Once grouping begins, however, cooperative and competitive bipole cell interactions choose the strongest grouping and inhibit the weaker ones, thereby converting a fuzzy initial grouping into a sharp final grouping (Figure 2.38, right image). When the strongest inducers differ in orientation and position, the final boundary groupings may go through the locally preferred end-cut orientations; namely, the orientations that are perpendicular to their line ends (Figure 2.37, top row). However, if there is no sufficiently simple grouping that can go through the perpendicular end cuts, then a grouping may form that goes through weaker diagonal end-cut orientations (Figure 2.37, bottom row). In all these cases, the boundary completion process chooses the grouping that has the most evidence in support of it, using the same cooperative and competitive interactions that induce end gaps and end cuts. Boundary completion hereby realizes a kind of real-time *probabilistic hypothesis testing* that chooses the best-supported grouping from uncertain data.

The simulations in Figures 2.34 and 2.39 illustrate another important property of boundary completion. Note in Figure 2.39 that, in addition to the end cuts that enable line ends to group, there are also small horizontal boundaries that are created on both sides of the vertical boundary inducers in each input. However, these small horizontal boundaries do not create emergent horizontal boundaries between pairs of vertical inputs along the entirety of their

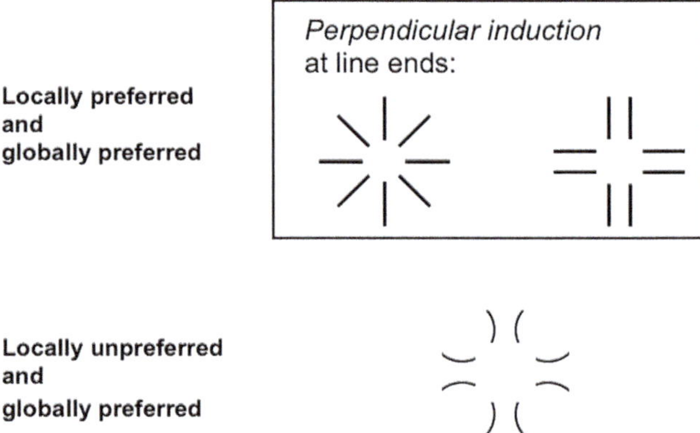

FIGURE 2.37 How boundaries can be completed perpendicular to line ends (top row) or diagonal to them (bottom row). Boundaries are completed with the orientations that receive the largest total amount of evidence, or support. Some can form in the locally preferred orientations that are perpendicular to an inducing line, while others can form through orientations that are not locally preferred. These examples illustrate that there is initially a fuzzy band of almost perpendicular initial grouping orientations at the end of each line (see Figure 2.38). The orientation with the greatest inducer support typically wins the competition to contribute to the final percept. Author created.

FIGURE 2.38 Boundary completion begins with fuzzy bands of possible boundaries (left image) and ends with a sharp final boundary that has the most support from all inducers (right image). Author created.

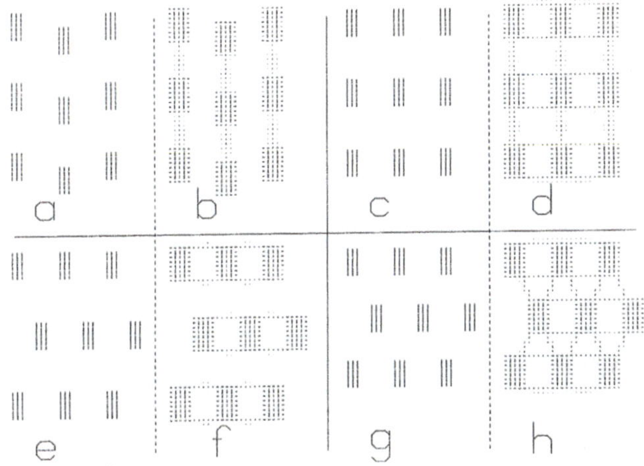

FIGURE 2.39 Computer simulations of boundary groupings in response to multiple oriented inducers at different relative positions in space. Parallel, perpendicular, and oblique groupings can all occur. Computer simulations in b, c, f, and h of groupings in response to different spatial arrangements in a, c, e, and g of inducers that are composed of short vertical boundaries. Note the emergent horizontal groupings in d, f, and h, and the diagonal groupings in h, despite the fact that all its inducers have vertical orientations. Reprinted with the author's permission from Grossberg, S. and Mingolla, E. (1987). Neural dynamics of surface perception: Boundary webs, illuminants, and shape-from- shading. *Computer Vision, Graphics, and Image Processing*, 37,116–165.

sides. If this did happen, then many spurious boundary groupings would occur.

These spurious boundaries do not occur because of a property that I call *spatial impenetrability* (Grossberg and Mingolla, 1985b). Spatial impenetrability is achieved by the fact that, for example, horizontally oriented bipole cells are inhibited by vertically oriented hypercomplex cells at the same positions (Figure 2.11). In the middle of a vertical line input, a combination of horizontal and vertical responses prevents horizontally oriented bipole cells from "penetrating" these vertical inducers. Only at line ends are there enough like-oriented cell responses to be able to induce boundary groupings, whether collinear (Figure 2.39b and 2.39d), perpendicular (Figures 2.39d, 2.39f, and 2.39h), or obliquely oriented (Figure 2.39h) with respect to the orientations of the inducing lines.

In summary, the illusory contours that are generated in our brains by the images above, and many paintings, can be both orientationally and positionally sharp, despite the fuzziness of end cut orientations and bipole cell receptive fields. Boundary sharpening happens when bipole cells interact via feedback with spatial and orientational competition to inhibit weaker cell responses (Figure 2.11).

Sharpening of end cuts during illusory contour formation is another example of hierarchical resolution of uncertainty: The initial fuzziness is needed to initiate boundary grouping (Figure 2.38), but risks a loss of acuity. The ensuing sharpening by cooperative-competitive feedback interactions permits a final sharp grouping to form without a loss of acuity.

When considering properties of this boundary sharpening, or choice, process, another question arises whose answer also requires that cellular and network computations be done in the correct way? This question asks how strong the chosen boundary will be. Many examples illustrate the need for boundary strength to covary with the number, spatial separation, relative orientation, and contrast of the boundary inducers. These constraints are often called Gestalt grouping laws. One can summarize them heuristically by saying that the *degree of commitment—in this case, boundary strength—should covary with the amount of evidence*. The computational challenge is that cooperative-competitive feedback interactions that make a boundary choice could easily cause the winning cells to attain their *maximal* activities, and thus become insensitive to the amount of evidence. This could prevent the amount of evidence from determining adaptive choices at later processing stages. My colleagues and I have shown that one computational advantage of the *laminar circuits* of the visual cortex is that they can make choices robustly, without losing analog sensitivity to the Gestalt grouping laws (e.g., Grossberg, Mingolla, and Ross, 1997; Grossberg and Raizada, 2000). I call this property *analog coherence*.

3.5 Neon color spreading: Discounting illuminant before surface filling-in

In response to the neon color spreading image in Figure 2.36, blue color can flow out of the end gaps in the broken boundaries. This spreading, or *surface filling-in*, of blue color across space continues until the color hits the square that is made up of both pac-man and illusory boundaries, which together prevent its further spread. Thus, both real and illusory boundaries act as barriers to the filling-in of brightness or color.

In summary, the process of end cutting that completes line ends, and thus illustrates the first hierarchical resolution of uncertainty, also creates end gaps through which brightness or color can flow when a stronger boundary abuts a collinear weaker boundary. The end cuts also cooperate via bipole grouping to complete an illusory boundary that can contain the flow of brightness or color, thereby illustrating a second hierarchical resolution of uncertainty.

Neon color spreading also provides a third example of hierarchical resolution of uncertainty. Recall from Section 2.7 that the process of discounting the illuminant computes feature contour signals that significantly eliminate, or discount, effects of the illuminant, in order to compute surface properties of the object itself, such as its reflectances. Feature contour signals tend to occur along boundary contours (see Figures 2.12, 2.13, and 2.18) because those are the positions where image contrasts may occur rapidly across space. Intuitively, one can think of feature contours as strips of color or brightness contrast that are localized in space.

In order to recover surface properties of an object, such as its color and brightness, in a way that is relatively insensitive to illumination changes, filling-in of feature contours occurs at a processing stage that is subsequent to discounting the illuminant (Figure 2.19). Thus, the filling-in that is observed during neon color spreading illustrates a hierarchical resolution of uncertainty, whereby feature contours are computed to discount the illuminant, and then, at a later processing stage, the feature contours induce filling-in within their boundary contours to recover surface color and brightness.

In summary, the charming but seemingly unimportant percept of neon color spreading may be understood as the combined effect of three hierarchical resolutions of uncertainty, each of which is essential for the visual system to work well. These mechanistic insights about how neon color spreading works will be used below to clarify aesthetic effects that various artists have achieved in their paintings.

3.6 Seeing a 2D image of a shaded ellipse as a 3D surface: Boundary webs

The examples to the present have focused on 2D properties of visual percepts. However, even in response to the Kanizsa square in Figure 2.4, the completed square can consciously be seen "in front of" four partially occluded discs whose unoccluded parts are consciously seen as pac-man figures.

This percept leads to the general question: How does a 2D picture give rise to a 3D percept? Were it not for this property, 2D paintings could never have been invented to represent aspects of the 3D world that humans experience every day. FACADE theory (e.g., Grossberg, 1994, 1997; Grossberg and McLoughlin, 1997), and its refinement and generalization by the 3D LAMINART model of how the laminar circuits of visual cortex see (e.g., Cao and Grossberg, 2005; Grossberg, 1999; Fang and Grossberg, 2009; Grossberg and Raizada, 2000; Grossberg and Swaminathan, 2004; Grossberg and Yazdanbakhsh, 2005), explain how the brain mechanisms that have evolved to see the 3D visual world can automatically generate 3D representations of 2D pictures. Aspects of how this is proposed to happen will now be reviewed.

The 3D percept that is generated by a 2D picture of a shaded ellipse is a good example to start this explanation (Figure 2.40, left panel). This simple picture illustrates

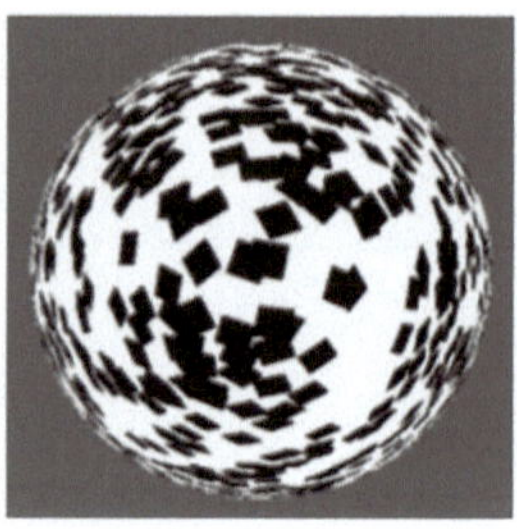

FIGURE 2.40 Two-dimensional images of a shaded ellipse (left image) and a textured disk (right image) can both give rise to vivid conscious percepts of curved surfaces in depth. Left image author created. Right image adapted from Todd, J., and Akerstrom, R. (1987). Perception of three- dimensional form from patterns of optical texture. *Journal of Experimental Psychology: Human Perception and Performance*, 13, 242–255.

fundamental issues about how vision works and has influenced the art of many painters. One of the first issues concerns the difference between an *edge* detector and a *boundary* detector. Unlike an edge detector, a boundary detector is sensitive to edges, shading, texture, and depth. Recall that the brain does not use edge detectors if only to avoid a combinatorial explosion of detectors, which would require having to process and fuse information from all of them. Instead, oriented simple cells are used (Figures 2.5 and 2.6), whose uncertain computations are compensated by hierarchical resolutions of uncertainty.

If the brain did use edge detectors, then such a detector could compute only the bounding edge of the ellipse. If this were the only boundary that formed, however, then the gray colors within the ellipse could all spread throughout its interior via surface filling-in to generate a flat percept of a uniformly gray ellipse, just as occurs during filling-in of a Kanizsa square (Figures 2.12 and 2.13). This is not, however, what is seen.

Instead, the brain computes a dense array of boundaries that configure themselves along the *isophotes*, or equal luminance positions, of the gradients of shading and texture in an image or scene. I call such an oriented array a *boundary web* (Grossberg, 1987; Grossberg, Kuhlmann, and Mingolla, 2007; Grossberg and Mingolla, 1987). Oriented simple cells of a given size respond to the elliptical form of the shaded gradient along these isophotes to generate a boundary web. The web's boundaries parallel one another, all the way from the bounding contour of the ellipse inward toward its interior. Each of the small compartments in the web can trap its local shade of gray and prevent it from spreading via filling-in to a larger region of the ellipse.

If this is indeed the case, then why is this boundary web not visible? There are two answers to this question. The first answer is that "all boundaries are invisible." The second answer is that the invisible boundary web reveals itself through the smooth gradient of gray color that it

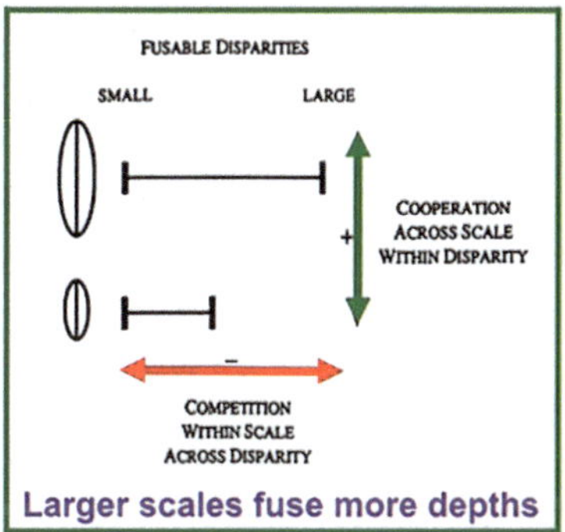

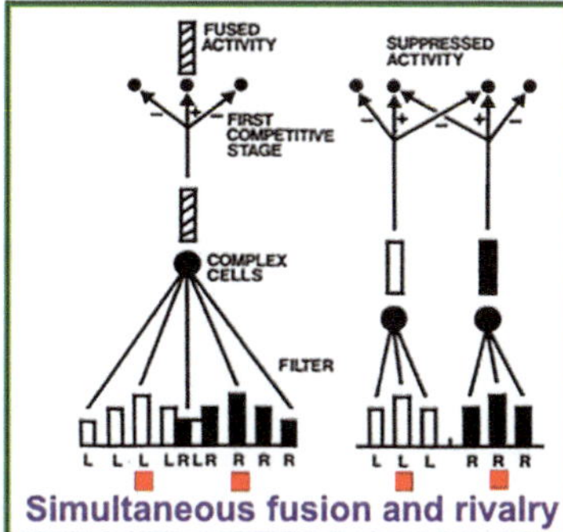

FIGURE 2.41 Multiple receptive field scales, followed by spatial and orientational competitive interactions, set the stage for generating 3D percepts from 2D images, as well as from 3D objects in the natural world. See the text for details. Reprinted from Figure 11.27 in Grossberg (2021) with the author's permission.

traps within its compartments to form a consciously seen, shaded *surface* percept.

Using simple cells with a single receptive field size, or spatial scale, is not sufficient to generate a percept of a depthful, or 3D, shaded ellipse. For this to happen, multiple simple cell spatial scales need to simultaneously respond to the image. Other things being equal (which is not always the case), large boundary scales code "near" and small scales code "far," if only because objects of a given size cast larger images on the retinas as they are viewed at nearer distances. This property is often called the *size-disparity correlation* (Julesz and Schumer, 1981; Kulikowski, 1978; Richards and Kaye, 1974; Schor and Tyler, 1981; Schor and Wood, 1983; Schor, Wood, and Ogawa, 1984; Tyler, 1975, 1983). In particular, complex cells with larger receptive fields can binocularly fuse a broader range of binocular disparities than can cells with smaller receptive fields and thus can represent a larger range of depths, with the largest disparities often coding the nearest depths (Figure 2.41, left panel).

In response to a picture of a 2D shaded ellipse, each scale of simple cells activates its own network of complex, hypercomplex, and bipole cells (Figure 2.41, right panel), and each network of bipole cells responds differently to the gradient of gray shading in the ellipse (Figure 2.42). The smallest scale generates the narrowest band of boundaries in its web. This boundary web lies adjacent to the bounding contour of the ellipse. The largest scales generate the broadest bands of boundaries. Larger boundary webs can hereby penetrate more deeply into the interior of the ellipse. Said in another way, other things being equal, smaller-scale simple cells can fire more easily nearer to the bounding edge of the ellipse. As the spatial gradient of shading becomes more gradual with distance from the bounding edge, it becomes harder for smaller scales to respond to this gradient. Thus, other things being equal, larger scales tend to respond more as the distance from the bounding edge increases. As a result of the size-disparity correlation, the boundary webs nearer to the middle of the ellipse code a nearer depth than the ones near the bounding edge of the ellipse.

Larger scales do not always code for nearer depths due to the way in which multiple spatial scales interact with grouping properties (Figure 2.43). These multiple-scale interactions will not be needed to make our main points here. They are explained in Grossberg (1994, 2021), along with a more detailed explanation of when and how larger spatial scales do code for nearer depths.

Each boundary scale maximally activates a Filling-In DOmain, or FIDO, within which filling-in of surface brightness and color occurs at the corresponding depth (Figure 2.19). During the filling-in of the percept generated by a shaded ellipse, each boundary web traps gray shading within itself at the corresponding

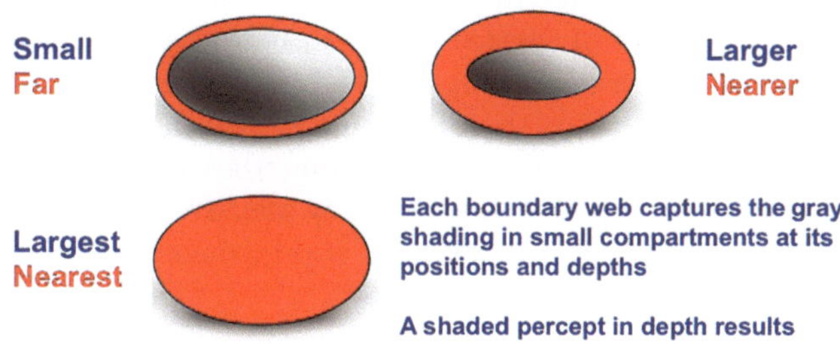

FIGURE 2.42 Multiple-scale boundary webs trap brightness and color signals at different positions. The ensemble of all these filled-in surfaces generates a percept of 3D form. In particular, webs of boundaries of the smallest scales are closer to the bounding edge of the ellipse. Increasingly large-scale boundary webs penetrate ever deeper into the ellipse image, because there is enough shaded evidence for them to fire. The final result is a multiple-scale depth-selective boundary web that can capture depth-selective surface filling-in, and thereby generate a depthful shaded percept of an ellipsoid. Reprinted from Figure 11.29 in Grossberg (2021) with the author's permission.

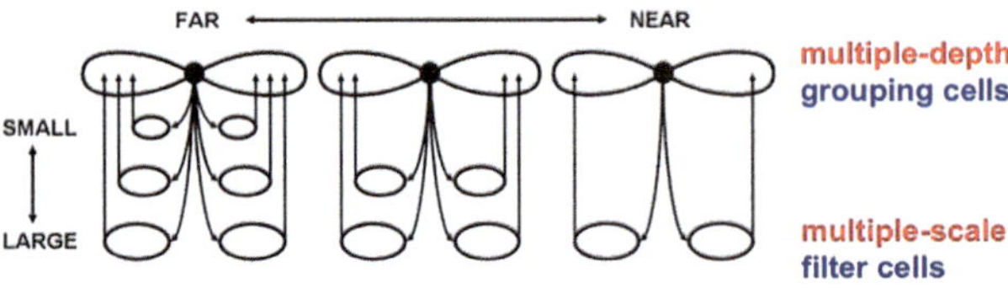

FIGURE 2.43 Multiple scales can be activated by a single depth. Many-to-one scale-to-depth maps and one-to-many depth-to-scale maps interact to disambiguate a final perceived depth from the ambiguous depth signals from multiple scales. Reprinted from Figure 11.30 in Grossberg (2021) with the author's permission.

FIDO. The spatial distribution of filled-in contrasts across all these depth-selective FIDOs represents the perceived 3D shape of the elliptical surface (Figure 2.42). Many details need to be carefully developed for this simple scheme to work properly, but the main idea should be clear.

These details are mathematically modeled in a model whose acronym, LIGHTSHAFT, abbreviates its ability to simulate LIGHTness-and-SHApe-From-Texture (Grossberg, Kuhlmann, and Mingolla, 2007). Percepts such as the smooth 3D shape that is generated by the spatially discrete 2D texture in Figure 2.40 (right panel) are also explained by LIGHTSHAFT, whose explanation is supported by quantitative simulations of parametric psychophysical data about such percepts. I elaborate more about these simulations below and show how well they fit the judgments of human observers.

In both the shading and the texture examples, one needs to ask: If different boundary web scales generate a depthful percept of surface form and lightness by differentially filling-in of shading and texture contrasts within them, then why are these boundary webs not visible? Again the answer is that invisible boundary webs reveal themselves by the way in which they selectively organize the filling-in of surface contrasts within their respective Filling-In Domains.

3.7 Depthful percepts from chiaroscuro, trompe l'oeil, and perspective

How multiple-scale boundary webs can create 3D percepts from a 2D image, as in response to the images in Figure 2.40, also explains the effects achieved by the painterly technique of *chiaroscuro*. Leonardo da Vinci was one of the first painters to use the chiaroscuro technique to create bulging 3D percepts from 2D paintings. Figure 2.44 (left panel) illustrates how the technique, as used by Rembrandt in an early self-portrait, generates a vivid percept of depth. This portrait was painted in 1629 when Rembrandt was twenty-three years old. It now hangs in the Mauritshuis, The Hague. Vivid percepts of depth can also be seen when viewing the *trompe l'oeil* paintings of the contemporary British painter, Graham Rust (Rust, 1988; Figure 2.44, right panel).

The same sorts of multiple-scale boundary webs can be used to explain 3D percepts that are caused by *perspective* in a 2D picture or the systematic reduction of spatial scale as a vanishing point on the horizon is approached (Figure 2.45). Perspective has been regularly used in paintings since the time of the Renaissance. The bigger scales,

FIGURE 2.44 Painterly techniques like chiaroscuro (left image) and trompe l'oeil (right image) can create a vivid percept of depth in a painting. (left image) Rembrandt van Rijn, Self-portrait as a Young Man, 1628, Rijksmuseum, Amsterdam. (right image). Graham Rust (1988). http://grahamrust.squarespace.com/home/. By permission of Graham Rust.

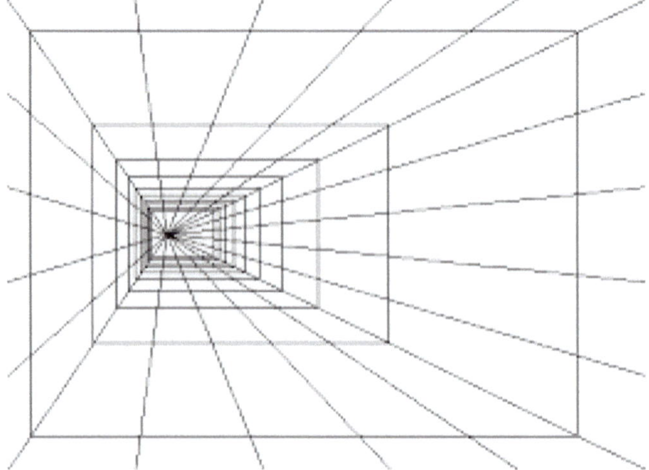

FIGURE 2.45 Perspective creates a percept of receding depths partly due to the size-disparity correlation, with larger-size scales appearing closer, other things being equal. Author created.

other things being equal, create a percept of a nearer surface, while the smaller scales create a percept of a surface that is farther away, as in the size-disparity correlation.

The ambiguous depths that multiple-scale filters code are disambiguated to create percepts of relative depth by a combination of scale-to-depth and depth-to-scale maps (Figure 2.43), cooperative-competitive boundary interactions, and the depth-selective filling-in of surface representations under the control of the resultant boundaries, in much the same way as the LIGHTSHAFT model explains the percept of Figure 2.40 (right panel). Let me emphasize that scale-to-depth and depth-to-scale interactions are needed because multiple scales can contribute to the percept of a single depth (Figure 2.43).

3.8 Recognition without seeing, and conscious seeing to reach

We are almost ready to explain how various painters have exploited for their aesthetic goals different combinations of brain processes that control how we consciously see. Before doing so, it is relevant to the goals of this book to ask: Why do we consciously see at all?

Many people, if they have thought about this issue at all, may conclude that "we see things to recognize them." These people include artists who have thought deeply about how they see the world. For example, Mann (2016, p. 6) has written: "The main purpose of this conscious macular vision is to enable us to recognize," where *macular vision* is the kind of high acuity vision that is achieved using the fovea (Figure 2.15). However, the fact that "all boundaries are invisible" contradicts this claim (see Section 2.1), at least as a general explanation of why we consciously see, because these invisible boundaries enable us to consciously recognize many emergent structures without seeing them; e.g., Figure 2.2 (top row, left panel), Figure 2.3, Figure 2.4 (top row, right panel), and (see below) Figure 2.46 (left panel).

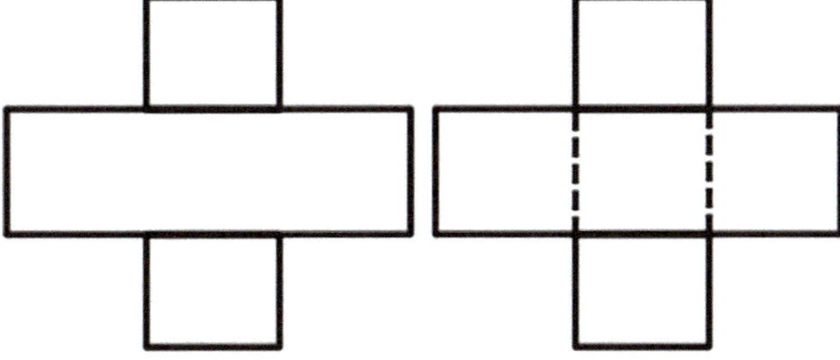

FIGURE 2.46 Figure-ground perception: The abutting three rectangles in the left image generate a vivid percept of a horizontal bar in front of a partially occluded vertical bar, as schematized by the right image, where the dashed vertical lines complete a vertical rectangle whose occluded section is recognized but not consciously seen. Author created.

I have proposed how a surface-shroud resonance (Section 2.19) may support conscious seeing when it is triggered between prestriate visual cortical area V4 and the PPC while propagating top-down to cortical areas such as V2 and V1, and bottom-up to cortical areas such as the prefrontal cortex, or PFC. This analysis also explains how a surface-shroud resonance enables a consciously seen surface percept in V4 to control effective looking and reaching movements toward the unoccluded surfaces of overlapping objects, as shown in Figure 2.46 (left panel). I have also predicted that a similar link between conscious perception and action occurs in other modalities, including how conscious hearing enables effective vocalizing and speaking, and conscious feeling enables effective actions toward valued goal objects in the world (Grossberg, 2017, 2021). These conclusions about how conscious percepts are intimately linked to the actions that they control are supported by explanations of psychological and neurobiological data from normal subjects and clinical patients that have no other explanations at present.

4 Brain processes that enable us to consciously see paintings

4.1 Claude Monet, coarse boundary webs, and uncertainty

Claude Monet, who lived from 1840 to 1926, was one of the founders of the French Impressionist movement, which derived its name from his painting called *Impression, Soleil Levant (Impression, Sunrise)* (Figure 2.47). This painting depicts a scene in the port of Le Havre, France. The scene is enveloped in a mist that provides a hazy background for the French harbor. An orange sun and its reflections in the water contrast vividly with the silhouetted dark vessels carried along by a sea current that is represented by spatially discrete brushstrokes.

This painting was seen by the public in 1874 in an exhibition that was organized by Monet and his associates as an alternative to the Salon de Paris. A critic who attended the exhibition, M. Louis Leroy, wrote an article in *Le Charivari* where he used the term "Impressionist" based on the title of Monet's painting. Painters such as Degas, Monet, and Renoir thereafter called themselves Impressionists.

Monet's use of color significantly departed from the Academic painting style that dominated the Salon de Paris. The surfaces of his paintings of natural scenes are built up from small brushstrokes. These discrete, but carefully coordinated, color elements induce percepts

FIGURE 2.47 The painting called *Impression, Soleil Levant (Impression, Sunrise)* by Claude Monet, first seen in public in 1874, gave the Impressionist artistic movement its name. From a free printable pdf on the web: https://www.claude-monet.com/impression-sunrise.jsp#google_vignette

of amodal boundary webs in viewers' brains, which organize the strokes into recognizable object forms. These emergent forms can be seen and recognized. The individual compartments from which they are composed are also visible when attentional shrouds focus spatial attention upon them.

Without a certain degree of orientational uncertainty embedded in the human visual system (e.g., Figures 2.34, 2.35, and 2.38), long-range cooperative boundary completion would be nearly impossible in response to Monet's paintings. His paintings exemplify the paradoxical prediction that orientationally "fuzzy" computations often lead to sharp boundary groupings (Figure 2.38) because there are hardly any sharp edges in Monet's paintings. Edges, forms, and the light that falls on them are perceived despite the fact that the paintings are irregular, noisy surfaces composed of varying densities of luminance and color patches.

Monet's brushstrokes induce boundary webs in which many of the compartments *within* a form are nearly equiluminant, while there may be stronger boundaries *between* forms. This combination facilitates color spreading within forms and separation of brightness and color differences between forms. Figure 2.48 (upper row) shows four paintings from Monet's Rouen Cathedral series. Their grayscale versions are below them (Figure 2.48, lower row). This famous series was created in the 1890s and includes over thirty paintings of similar views of the Rouen Cathedral at different times of day.

The grayscale versions of these paintings reveal the near-equiluminance of the brushstrokes within forms, as well as the places where brightness and color differences support groupings that differentiate between forms, notably between the cathedral and the sky. Monet's paintings hereby exploit the statistical uncertainty that our visual system is designed to overcome using both color and luminance differences, rather than luminance alone, as in the first and fourth images from left to right.

The grayscale versions of Monet's paintings also eliminate some critical closed boundaries that allow us to segregate forms. This is because the boundary system in the visual cortex pools *both* chromatic color and brightness signals at complex cells that boundaries can be computed entirely from color differences, without a corresponding achromatic brightness difference, as in the fourth image to the right in Figure 2.48 (upper row).

Figure 2.49 contains images of two paintings from Monet's Rouen cathedral series that illustrate how different lighting of the same object can lead to dramatically different percepts, in part by causing different boundary webs to form. The left painting depicts the lighting on the cathedral facade while the sun is setting, whereas the right painting depicts it in full sunlight. The right painting generates a more depthful percept than does the left one. This is due to multiple factors interacting together. One factor is that the colors in the left painting are more equiluminant, and the ensuing boundary webs are coarser and more uniform. The strong gradients of light that are evident in the right painting, including effects of sharp architectural features and shadows, are not seen in the left painting. Note in addition that, due to the greater range of contrasts in the right painting, many nearly horizontal boundaries are clearly seen to be occluded by vertical boundaries.

Such occlusions often occur at visual features that are called T-junctions. T-junctions can generate percepts of depth in the absence of any other visual cues. An explanation of how this may happen is given in Section 4.8 when *figure-ground separation* is discussed. As I will explain there, T-junctions can help to trigger figure-ground separation using basic properties of the boundary completion and surface filling-in processes, without the need for specialized T-junction detectors.

4.2 Boundary webs "hang like a cloud" in color field paintings of Jules Olitski

Nearly a century after the Impressionists, Color Field painting emerged in the United States. The term "Color Field painting" is generally used to describe painters such as Mark Rothko and Barnett Newman, who were part of the larger Abstract Expressionist movement, and their successors, like Jules Olitski and Helen Frankenthaler, who painted in the 1960s and 1970s.

The Color Field painters were primarily concerned with exploring the pure emotional power of color, particularly color that fills large surfaces. Jules Olitski created

FIGURE 2.48 Four paintings by Monet of the Rouen cathedral under different lighting conditions (top row) and their monochromatic versions (bottom row). See the text for details. Claude Monet, Rouen Cathedral, Musée d'Orsay, Paris, France.

FIGURE 2.49 The Rouen cathedral at sunset (left image) generates very different boundary webs than it does in full sunlight (right image). Left image: Claude Monet, Rouen Cathedral, Setting Sun (Symphony in Pink and Grey), National Museum, Cardiff, Wales. Right image: Claude Monet, Rouen Cathedral, Full Sunlight, Musée d'Orsay, Paris, France.

several such paintings in New York City. By 1965, Olitski developed a technique in which he sprayed paint onto unprimed canvases and created some of his most famous paintings, which are accordingly called "spray paintings" (Figure 2.50).

Spraying paint onto a canvas in fine mists gave Olitski control over the color density on the canvas. In contrast to Monet's paintings, in which a viewer can see lots of individual brushstrokes within a coarse boundary web, it is not possible to see discrete colored units within the fine boundary webs in Olitski's spray paintings, much as in the example of a shaded ellipse (Figure 2.40, left panel). However, unlike the boundary webs of the shaded ellipse, the boundary webs that cover the surface of Olitski's spray paintings cause a percept of ambiguous depth, as if the viewer is staring into a space filled with colored fog, or at the sky during a sunset with no clouds.

Olitski intentionally created this effect. He wrote: "When the conception of internal form is governed by edge, color...appears to remain on or above the surface. I think...of color as being seen in and throughout,

FIGURE 2.50 Four color-field spray paintings of Jules Olitski. See the text for why they generate surface percepts with ambiguous depth. (Top-left image) This painting is in the Museum of Fine Arts, Boston, Massachusetts. © 2020 Estate of Jules Olitski/Licensed by VAGA at Artists Rights Society (ARS), NY. (Bottom-left image) This painting is in the Tate Museum, London, Great Britain. © 2020 Estate of Jules Olitski/Licensed by VAGA at Artists Rights Society (ARS), NY. (Top-right image) This painting is in the Guggenheim Museum, New York. © 2020 Estate of Jules Olitski/Licensed by VAGA at Artists Rights Society (ARS), NY. (Bottom-right image) Comprehensive Dream, 1965. © 2020 Estate of Jules Olitski/Licensed by VAGA at Artists Rights Society (ARS), NY. Images reprinted with permission © Estate of Jules Olitski/ VAGA at ARS, NY and DACS, London 2025.

not solely on, the surface" (Riggs, 1997). Art critics also recognized this quality. As Rosalind Krauss wrote: "...the very seeing of the painting in all its literalness poses a question about where the surface is. To see Olitski's color means to see the surface itself as elusive and unaligned" (Riggs, 1994). Occasional sharp spatial discontinuities in some of these paintings (e.g., Figure 2.50, upper left image) sharpen the viewer's awareness of the fact that drawing plays little role in creating the surface percept that is perceived across the painting.

The principles of FACADE theory and the 3D LAMINART model (Section 1.1) explain why Olitski's paintings create the sense that one is looking into a colored mist within a 3D space. These models highlight the importance of closed boundary contours to create percepts of surfaces that are restricted to specific depth planes (Figures 2.18, 2.19, and 2.27). Within the colored mists in Olitski's paintings, sharp, closed boundary contours do not occur. Instead, gradual chromatic and luminance gradients coexist, thereby enabling multiple scales to form boundary webs and thereby capture color shading on multiple depth-selective surfaces whose gradients often coexist at similar positions. Unlike the example of the shaded ellipse (Figure 2.40 (left panel) and Figure 2.42), these gradients and their boundary webs do not vary in either orientation or scale in a systematic way across space. Thus, the shaded surfaces of these paintings are not perceived within one depth plane, thereby achieving Olitski's goal to paint color that is "seen in and throughout."

Another factor that contributes to the percept of amorphous depth is gradual brightness and color gradients on the surface. The property of *proximity-luminance covariance* (Dosher, Sperling, and Wurst, 1986; Egusa, 1983) describes the fact that increasing brightness can make a surface look closer.

A similar property is obtained when brighter Kanizsa squares look closer. The percepts that we see in response to images in Figure 2.4 illustrate this property. In particular, the Kanizsa square (upper left panel) looks brighter than its background, as well as closer to the viewer. The reverse-contrast Kanizsa square (upper right panel) looks neither brighter nor closer than its background. The Kanizsa square with four additional straight line inducers (lower left panel) looks even brighter than, and closer than, the Kanizsa square (upper left panel) under carefully controlled experimental conditions.

In Grossberg (2014a), I provided the explanation that I here review of this property using mechanisms that help to ensure complementary consistency (Section 2.5) and figure-ground separation (see Section 4.8).

Many paintings, including the Olitski paintings, are painted on rectangular canvases that have a sharp rectangular boundary with respect to the wall on which they are hung. Why is the "colored mist" of an Olitski painting seen separate from its rectangular boundary? Why does not the mist seem to be "attached" to the rectangular canvas, unlike the boundary web of the shaded ellipse in Figure 2.40 (left panel) which does appear to be attached to its bounding contour?

One factor is that the orientations of the boundary webs induced by the interior of an Olitski painting are parallel to the painting's isophotes. These isophotes are incongruent with the orientation and scale of the rectangular boundary of the canvas. This is unlike the case of the shaded ellipse, whose boundary webs are consistent with the bounding contour of the ellipse. These incongruities in Olitski's paintings prevent the attachment of the boundary webs within the painting to the bounding contour of the rectangular frame of the painting and help to explain why the color in Olitski's spray paintings "hangs like a cloud, but does not lose its shape" ("Lysander-1", 2015).

The only visible boundary contours within Olitski's spray paintings are the edges of the canvas itself and the framing lines he introduced along the edges of some of the paintings, which the famous American critic, Clement Greenberg, named *selvage* (Olitski, 1994) (Figure 2.50, upper left image). The word "selvage" is used in many contexts, such as textiles and manufacturing, but it generally refers to an edge added to a material to finish it, prevent it from fraying, or allow it to be handled. In contrast to the ambiguous depth of the color mists, the depth plane of the selvage is perceived as being on the surface of the canvas due to the sharp boundaries that contain it, and thus is perceived in the depth plane of the picture's frame.

4.3 Watercolor illusion, Jo Baer, and Mach bands

Before turning to a discussion of figure-ground separation due to partially occluded boundaries, let us consider other examples of how boundary webs with multiple spatial scales can generate depthful percepts, even if there are no occluded boundaries.

I will first illustrate this phenomenon with a display that induces the *watercolor illusion* (Figure 2.51). Variants of this illusion have been studied by Baingio Pinna and his colleagues since the 1980s (Pinna and Grossberg, 2005). For an overview, see http://www.scholarpedia.org/article/Watercolor_illusion. As I will further discuss below, the watercolor illusion has also been included in paintings of visual artists, notably those of Jo Baer, since at least 1963. After explaining how depth can be perceived from a watercolor illusion display, I will then discuss Jo Baer's paintings that exhibit watercolor illusion properties.

Figure 2.51 induces the watercolor illusion percept because a more contrastive and undulating purple band abuts a less contrastive yellow band. Their interaction causes the yellow color of the inner band to fill-in the interior of the region that is surrounded by the inner yellow boundary. Pinna and Grossberg (2005) explained this percept, and its many variations, using properties of spatial competition, or the first competitive stage (Figure 2.11), of the boundary contour system, in which a complex cell at a given position

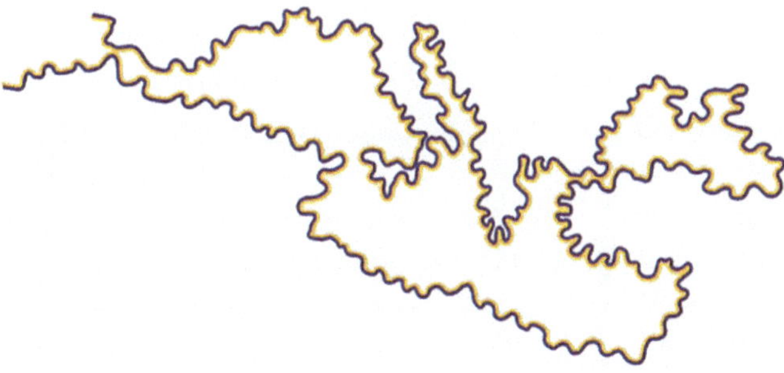

FIGURE 2.51 The watercolor illusion of Baingio Pinna can be explained using spatial competition at the first competitive stage (the first stage of the double filter in Figure 2.11) between like-oriented boundary signals of different strength. Why the percept seems to bulge in depth may be explained using multiple-scale, depth-selective boundary webs, much as in the explanations of Figure 2.40. Reprinted with permission from Pinna, B. (1987). Un effetto di colorazione. In Il laboratorio e la città. XXI Congresso 9780190070557_Book.indb 643 31-Mar-21 19:14:33 degli Psicologi Italiani, V. Majer, M. Maeran, and M. Santinello (Eds.). Edizioni SIPs, Societá Italiana di Psicologia, Milano, 1987, p. 158.

excites a population of hypercomplex cells with the same orientational preference at that position, while inhibiting like-oriented hypercomplex cells at nearby positions. In other words, the first competitive stage is an on-center, off-surround network *within* orientation and *across* position.

The yellow color spreads because the boundary that is formed between the purple band and the white background is stronger than the boundary that is formed between the yellow band and the white background. The boundary that is formed at the purple-yellow contour is also weaker than that at the purple-white contour, but stronger than that at the yellow-white contour. Due to spatial competition, the stronger boundaries inhibit the yellow-white boundary more than conversely, thereby weakening the yellow-white boundary and allowing yellow color to spill out beyond the yellow band to fill-in the entire surface that it surrounds. The undulating shape of the contours creates longer boundaries per unit area through which color can spread, thereby intensifying the effect. Basic processes of perceptual grouping, that are part of a hierarchical resolution of uncertainty, hereby help to explain the watercolor illusion and its variations.

As in our percept of the shaded ellipse (Figure 2.40), the watercolor illusion can also generate a 3D percept. In response to the image in Figure 2.51, a spatial array of successively weaker boundaries forms, from the purple-white to purple-yellow to yellow-white boundaries, as the distance increases from the most contrastive purple-white edge into the display's interior. These successively weaker boundaries together form a boundary web, albeit a more spatially discrete one than the one generated by a shaded ellipse. When this multiple-scale boundary web traps the corresponding colors, it can generate a rounded appearance of the percept, with the filled-in yellow surface looking a little closer, as a result of the same size-disparity correlation mechanisms that also play a role in creating chiaroscuro percepts (Figure 2.44, left panel).

Jo Baer incorporated the watercolor illusion in some of her paintings from the 1960s in New York City. She was then interested in working with black, white, and color in a minimalist style, without direct reference to anything outside the painting (Boersma, 1995). Her series of three paintings called *Primary Light Group: Red, Green, Blue* (Figure 2.52) exhibits watercolor properties. This triptych is now part of the Museum of Modern Art collection in New York.

Baer studied biology at the University of Washington in Seattle and physiological psychology at the New School in New York. Her studies equipped her to understand why the colors that were used to create her paintings are different from the colors that are perceived when viewing them. In an interview published in 1995 in *BOMB*

Magazine (Boersma, 1995), Baer said: "I was always curious why the color on the palette was different than the color on the painting. I knew what I wanted something to look like, and I found that the means to do it were so different than the end result. And then I was very pleased to discover the reasons why."

Baer wrote about some of these reasons in an article for *Aspen Magazine* (1970–1971): "Tucked in between the white and black, the narrow color band gains a vast brightness due to...distinct edge effects working at the color interfaces...[these]...brightness contrast effects push the color band...higher into luminosity through a physiological neural phenomenon called *Mach bands*" (Baer, 1970-1971). Baer wrote that she "intuitively fashioned" the paintings in the style of those shown in Figure 2.52 with her knowledge of Mach bands and retinal physiology data "somewhere in mind" (Baer, 1970-1971). Baer's interest in how humans see is reflected in the title of her triptych, which includes the word "light," and her description of these paintings as containing "primary colors of light" ("Jo Baer, primary light group: red, green, blue", n.d.).

The Mach bands to which Baer refers are an optical illusion that enhances the perceived contrasts at positions adjacent to abutting regions that have slightly different luminances (Figure 2.53). The same thing happens when viewing Baer's paintings in Figure 2.52. Here, the black band can enhance the contrast of the less luminous color band, and the color band can influence the perceived contrast of the white interior.

Although viewing Baer's paintings does create percepts of enhanced contrasts, their watercolor illusion properties, and thus the main percept that makes them interesting, are not local contrast effects like Mach Bands. Rather, a combination of spatial competition within the boundary stream, followed by color filling-in within the surface stream, makes them interesting (Figures 2.11 and 2.16). Indeed, when viewing the paintings in Figure 2.52, the white central area of the painting on the left appears to have a red/pink tint, the middle painting appears a green tint, and the right painting a blue/purple tint.

The same spatial competition that helps to explain the watercolor illusion also helps to explain the percepts in the Baer paintings because the luminance contrast between the black band and the white wall is larger than that between the colored band and the white interior of the painting. As a result, the boundary between the black band and the white wall is stronger than the boundary between the colored band and the white interior. Spatial competition between the stronger, white-black boundary and the weaker color-white boundary weakens the color-white boundary, leading to color spreading, or surface filling-in, within the interior white surface of each painting. When viewing several of these paintings side-by-side in the same room, as in Figure 2.52, these filling-in percepts become more vivid.

JO BAER

Primary Light Group: Red, Green, and Blue

1964–1965

FIGURE 2.52 Each of the paintings in the triptych of Joe Baer, called *Primary Light Group: Red, Green, and Blue*, generates a watercolor illusion percept. When they are displayed side by side in a museum, the differences between the filled-in colors are more striking. See the text for details. Digital Image © The Museum of Modern Art/ Licensed by SCALA/Art Resource, NY. By permission of Jo Baer.

FIGURE 2.53 Mach bands: Along the boundaries when adjacent shades of gray abut, lateral inhibition makes the darker area appear even darker, and the lighter area appear even lighter. This image was taken from the web, where it is freely available. See http://opticals.joachim-wedekind.de/mach-bands/

4.4 Lightness anchoring and self-luminosity

Some paintings, such as those of Ross Bleckner that are discussed below in Section 4.6, generate percepts of self-luminosity. Before explaining these painterly percepts, I will explain how matte paint can ever generate a self-luminous percept. In fact, two different brain processes can induce self-luminous percepts. Both processes are exploited in Bleckner's paintings.

As discussed in Sections 2.7, 3.4, and 3.5, reliable object colors may be perceived under widely different lighting conditions because the visual system can "discount the illuminant" to compute surface properties based on changes in reflectance across boundaries, where reflectances are the *relative* amounts of light reflected from a surface to the eyes within each wavelength of light. Discounting the illuminant is an important example of contrast normalization in on-center off-surround networks with shunting dynamics. Contrast normalization can also occur when these networks process luminances and colors within a scene or picture.

Contrast normalization does not, however, compute the *absolute* perceived brightnesses of a scene. This is because discounting the illuminant does not lead to an image representation that uses the *full dynamic range* of neuronal cells. An additional process is used by our brains to compute the full dynamic range of neurons, and thereby to compute estimates of absolute brightnesses of objects and scenes.

The process whereby our brains compute absolute brightnesses also leads to examples of self-luminous percepts. Other famous effects include the Gelb effect, whereby a black surface can look white when it is intensely illuminated; and the Area effect, which is often demonstrated by placing a viewing subject's head within a dome that is divided into two regions. When the highest luminance area occupies more than half of the visual field, it appears white, while the darker part looks gray. When the darker luminance area occupies more than half of the visual field, it tends to look increasingly white, while the lighter area appears to be self-luminous.

Alan Gilchrist and his colleagues (e.g., Gilchrist et al., 1999) have been leaders in demonstrating such effects, which they attribute to a process of *anchoring*; namely, the process whereby the brain uses its full dynamic range to compute brightnesses and colors. Great predecessors in the study of how anchoring may work include Hermann von Helmholtz (1866) and Hans Wallach (1948, 1976). Indeed, Wallach proposed the highest-luminance-as-white (HLAW) rules whereby the perceptual quality "white" is assigned to the highest luminance in a scene, and gray values of less luminous surface regions are assigned relative to the white standard. Although this rule works in some situations, there are other situations in which it leads to the wrong answer.

I developed the anchored Filling-In Lightness Model (aFILM) with my PhD student, Simon Hong, to provide a neural model of how the full dynamic range of neurons is used to compute the absolute lightness of scenes (Grossberg and Hong, 2006; Hong and Grossberg, 2004). aFILM explains and simulates all of the experimental anchoring effects that are reviewed in Gilchrist et al. (1999), including how percepts of self-luminosity occur.

aFILM achieves this expanded explanatory range by revising the Wallach HLAW rule by using a *blurred-highest-contrast-as-white* (BHCAW) rule that overcomes problems with the Wallach rule and, when embedded in suitable model processing stages (Figure 2.54), can explain and simulate many anchoring data.

Figure 2.55A illustrates how the BHCAW rule uses the full dynamical range of neurons by anchoring whiteness and, with it, the gray shades in a scene. Figure 2.55B proposes how a percept of self-luminosity may be generated in response to different scenic conditions. In particular, the aFILM model simulates how a scene may be anchored to the highest luminance that is computed after it is processed through a blurring kernel. If the area of highest luminance in the scene is the same size or larger than the blurring kernel, as occurs in Figure 2.55A, then this region is perceived as white. However, if the area of highest luminance in the scene is smaller than the blurring kernel, as occurs in Figure 2.55B, then the scene is again anchored to the blurred highest luminance as white, but "white" now corresponds to a luminance that is lower than the highest luminance in the scene. This allows the highest luminance in the scene to be perceived as being "brighter than white," or self-luminous.

Self-luminosity is one of many percepts that the Wallach HLAW model cannot explain. This is because self-luminous regions are typically perceived as having a lightness value that is higher than white, but the HLAW model sets the highest luminance in a stimulus to equal white.

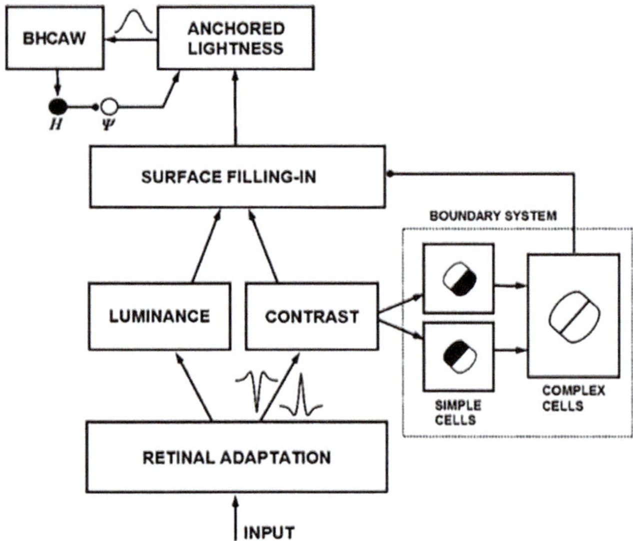

FIGURE 2.54 The anchored Filling-In Lightness Model, or aFILM, explains and simulates how lightness anchoring occurs, including self-luminous percepts in pictures. The model's Blurred Highest Luminance As White rule sometimes normalizes the highest luminance to white, but at other times normalizes it to be self-luminous. Reprinted with the author's permission from Grossberg, S. and Hong, S. (2006). A neural model of surface perception: Lightness, anchoring, and filling-in. *Spatial Vision*, 19, 263–321.

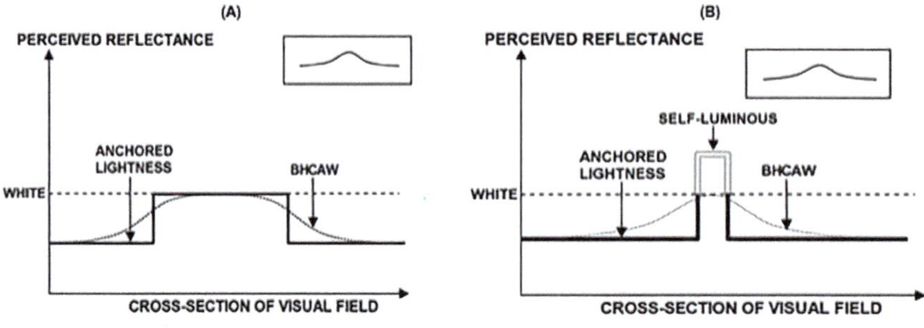

FIGURE 2.55 This figure shows how Blurred Highest Luminance As White (BHLAW) rule sometimes normalizes the highest luminance to white (left panel) and sometimes normalizes it to be self-luminous (right panel). Reprinted with the author's permission from Grossberg, S. and Hong, S. (2006). A neural model of surface perception: Lightness, anchoring, and filling-in. *Spatial Vision*, 19, 263–321.

4.5 Boundary webs, glare, double brilliant illusion, and gloss

There is a second way that the spatial context in which a luminance occurs within a scene can create a percept of self-luminosity, without the need for the BHCAW rule. This second way arises when a region of prescribed luminance is surrounded by a suitable form-selective luminance gradient. This can be seen in examples shown in Figure 2.56, notably the glare effect (Zavagno, 1999; Zavagno et al., 2004), the double brilliant illusion (Bressan, 2001), and percepts of gloss (Beck and Prazdny, 1981).

In the glare effect (Figure 2.56, top left), the interior white region of all four figures is the same, but an increasingly strong luminance gradient surrounds this region as one proceeds from the left figure in the first row to the right figure in the second row. The percept in response to the strongest gradient in the lower right image is one of a glowing self-luminosity, with lesser effects occurring in response to the weaker luminance gradients.

This effect may be explained as the result of the boundary webs that are generated in response to the luminance gradients and how they control the filling-in of lightness within themselves and abutting regions. In particular, such a boundary web is like a continuous version of the boundary web that enables color spreading to occur out of the weakest boundary in the watercolor illusion (Figure 2.51) or Jo Baer's paintings (Figure 2.52). Due to these boundary web gradients, more lightness can spread into the central square as the steepness of the boundary gradient increases.

Similar concepts help to explain the double brilliant illusion that is illustrated in Figure 2.56 (top right), along with a schematic in Figure 2.56 under the stimulus images of how the aFILM model explains and simulates this percept.

The gloss effect that is perceived in response to the image in Figure 2.56 (bottom left) where two almost identical vases are shown. A highlight was added to a local surface region of the left vase. Remarkably, this local change makes the entire vase look glossy compared with the matte vase to the right from which it was constructed. I explained this glossy percept with Ennio Mingolla (Grossberg and Mingolla, 1987) using the way in which the boundary web that is induced by the highlight is assimilated into the boundary web that is generated by the rest of the vase's form. The lightness within the highlight can hereby spread to abutting areas of the vase, rendering its appearance glossy.

As with the case of the shaded ellipse image in Figure 2.40, this boundary web is invisible and is seen only through the gradients of surface lightness that it traps within its compartments. In support of this explanation, Beck and Prazdny (1981) showed that changing the gradual luminance gradient of the highlight to a spatially abrupt one, or changing its relative orientation with respect to the rest of the vase, can eliminate the glossy percept.

4.6 Ross Bleckner's self-luminous paintings

Ross Bleckner has worked as a painter in New York for the past several decades. He has described his paintings as attempts to "eke out of a formal code a maximum

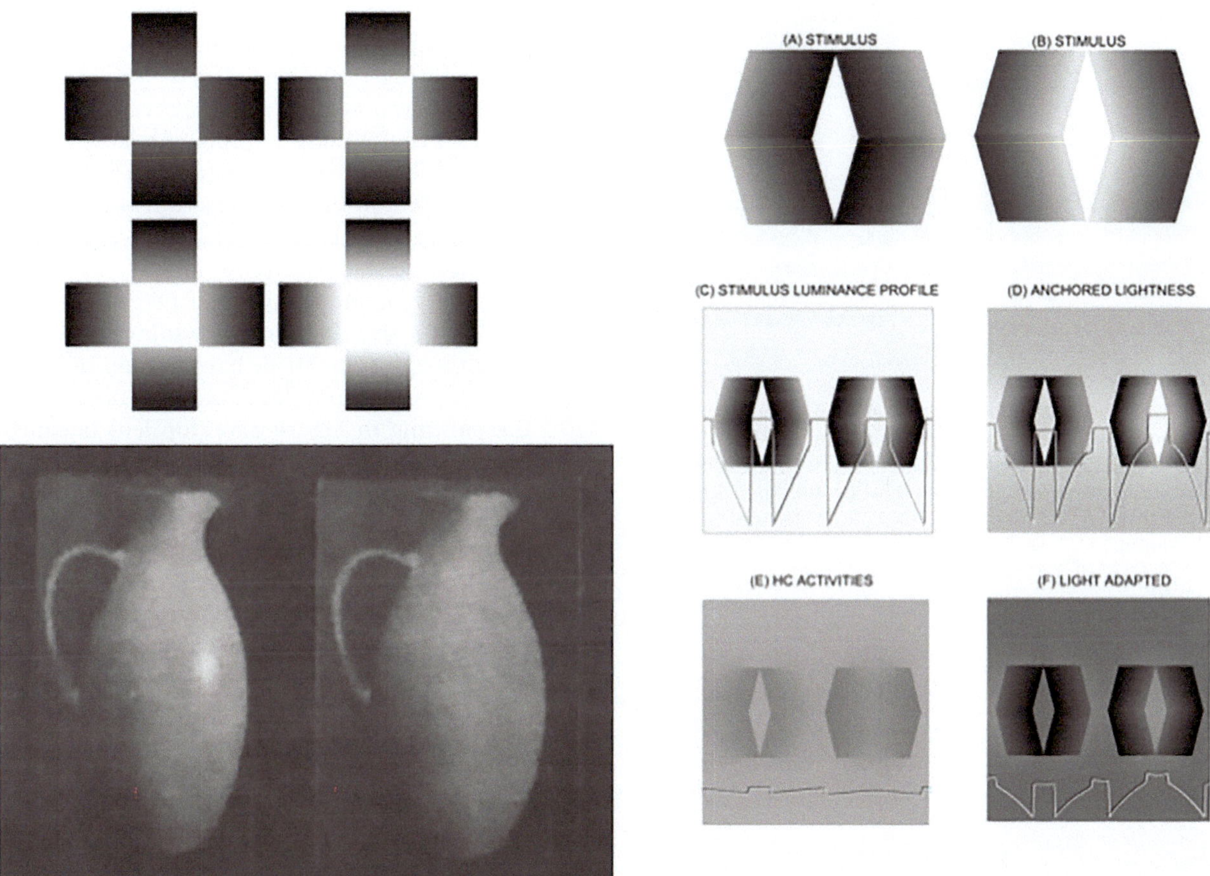

FIGURE 2.56 Different ways that spatial gradients in boundary webs can cause self-luminous percepts. Upper-left figure found in Zavagno, D. (1999). Some new luminance-gradient effects. *Perception*, 28, 835–838. Lower-left figure adapted from Beck, J. and Prazdny, S. (1981). Highlights and the perception of glossiness. *Perception and Psychophysics*, 30, 407–410. Upper-right figure adapted from Bressan, P. (2001). Explaining lightness illusions. *Perception*, 30, 1031–1046. Lower-left figures adapted from Grossberg, S. and Hong, S. (2006). A neural model of surface perception: Lightness, anchoring, and filling-in. *Spatial Vision*, 19, 263–321.

amount of light" (Rankin, 1987). Bleckner does this, not by painting large surface areas with high reflectances or bright colors, but rather by creating compositions of small, star-like, circular regions that are perceived as self-luminous (Figure 2.57).

At least three interacting properties of the paintings in Figure 2.57 contribute to their self-luminosity: high luminance areas relative to a dark background, their small size, and the smooth luminance gradient that surrounds many of them. The large luminance difference enhances the effects of brightness contrast, whereas the small size can create self-luminosity via the BHCAW rule (Figure 2.55). The luminance gradient can enhance self-luminosity using boundary webs, much as in percepts like the glare and double-brilliant percepts that are induced by images in Figure 2.56. When the points of light surrounded by a boundary web are magnified, as in the center of Figure 2.57, they still appear self-luminous, despite their larger size, thereby illustrating the independence of the BHCAW rule and the self-luminosity effects of boundary webs in Figure 2.56, since the boundary web is present even when the figure is too large for the BHCAW rule to cause self-luminosity.

The two images by Bleckner in Figure 2.58, a painting (right) and a photo (left), further illustrate how self-luminosity can be generated by surrounding a high-luminance point with a dense boundary web. In the image on the left, the large yellow circular regions appear to be self-luminous. In contrast, the yellow circular regions in the painting on the right are surrounded by sharp, black contours that prevent them from being enhanced by spreading color from surrounding regions, since sharp boundaries act as filling-in barriers, as in the percept of neon color spreading (Figure 2.36). Even the smallest high-luminance points in the painting in Figure 2.58 are thus not as self-luminous as the points within Figure 2.57, which receive the added benefits of self-luminosity due to boundary webs.

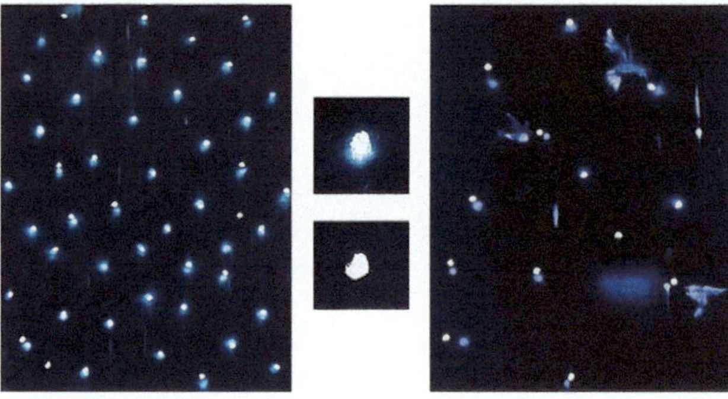

FIGURE 2.57 Examples of Ross Bleckner's self-luminous paintings. (Left image) Ross Bleckner, *Galaxy* painting, 1993. Linda Pace Foundation Collection, San Antonio, Texas. Courtesy of the artist and Petzel, New York. (Right image) Ross Bleckner, *Galaxy with Birds* painting, 1993. Rankin, A. (1987). Ross Bleckner. BOMB Mag. 19, Spring.

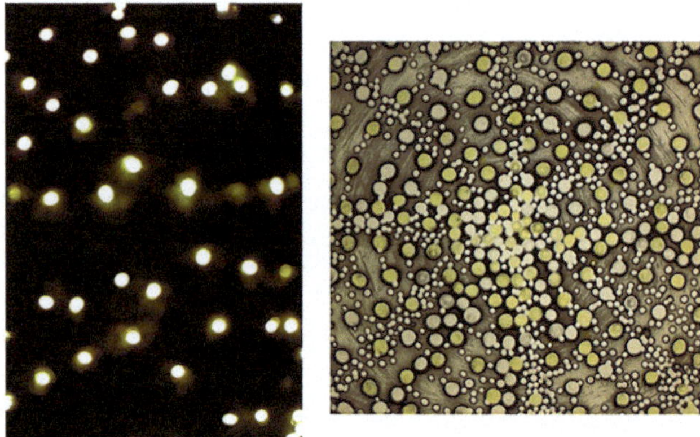

FIGURE 2.58 A painting (right, Insertion Sequence, 2002) and a photograph (left, Untitled) by Ross Bleckner that illustrate how luminance gradients, or the absence thereof, may or may not support self-luminous percepts. Reprinted from Rankin, A. (1987). Ross Bleckner. BOMB Mag. 19, Spring.

4.7 Gene Davis: Proximity-luminance covariance and color assimilation

Depth can be conveyed through several different means: *Binocular disparity* between left and right eye views of an object in depth is one important cue. Size is another important factor, if only because of the *size-disparity correlation* (see Section 3.6), which summarizes the property that larger objects tend to appear closer, while smaller objects tend to appear farther away, other things being equal. *Shading* can also create a 3D percept by generating multiple-scale form-sensitive boundary webs, with examples in the works of Monet (Figures 2.48 and 2.49), Olitski (Figure 2.50), and Bleckner (Figures 2.57 and 2.58) of how different painterly percepts can be generated by different boundary web configurations and how they organize the filling-in of surface brightness and color.

The paintings of Gene Davis provide additional examples of how boundary groupings can influence painterly percepts that induce percepts of relative depth. Gene Davis was a journalist who began to paint seriously in Washington D.C. in 1950. He is best known for his vertical stripe compositions (Figure 2.59), which comprise bulk of his life's work. These paintings illustrate how *scene stratification* can generate a variety of percepts, even if they do not contain size differences, shading, or recognizable objects, although some stripe paintings do contain stripes with shading and/or physical size differences.

The painting in Figure 2.59 (top left) is called *Black Popcorn*. Despite the lack of explicit depth cues, the "brightest" stripes—namely, the four intermingled pale yellow and pale blue stripes located approximately one-third of the length of the canvas from the left—appear to be in the nearest depth plane. In contrast, in *Flamingo* (Figure 2.59, top right), both individual stripes and stripe groupings can be perceived as nearer or further in depth, again with brighter stripes appearing closer.

Some of the neural mechanisms that influence these depth percepts are the following: Color similarities and/or near-equiluminance between stripes in these paintings can influence whether spatial attention is drawn to individual stripes or groups of stripes. Achromatic versions of the two paintings more clearly show regions where color assimilation is facilitated, notably the much smaller differences in luminance of the stripes within *Black Popcorn* (Figure 2.59, bottom left) than within *Flamingo* (Figure 2.59, bottom right). Boundaries between stripes

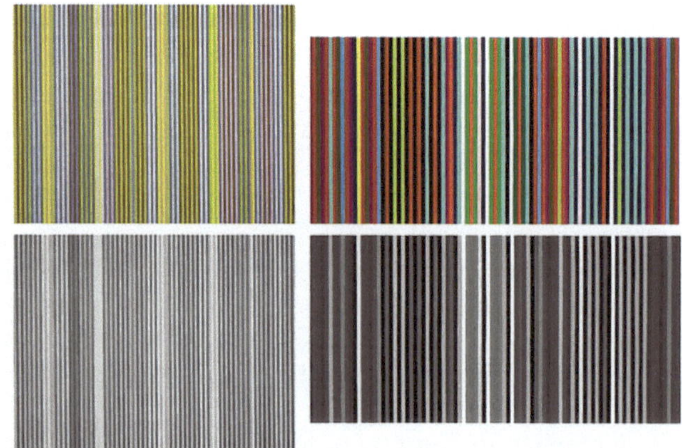

FIGURE 2.59 Two of Gene Davis's paintings are shown in full color (top row) and in monochromatic versions (bottom row). See the text for an explanation of how they achieve their different percepts of grouping and relative depth. (Left image) *Black Popcorn* is in the Smithsonian American Art Museum, Washington, D.C. © 2020 Estate of Gene Davis/Artists Rights Society (ARS), New York. (Right image) *Flamingo*, 1965. Photo by Joshua Nefsky. © ARS, NY and DACS, London 2025.

that have low luminance contrast are weak, thereby facilitating increased surface filling-in across several stripes. Such color assimilation calls to mind the watercolor illusion (Figure 2.51) and the paintings of Jo Baer (Figure 2.52). As a result of weak boundaries and color assimilation, an attentional shroud (Figure 2.28) that is larger than one stripe is more easily formed during viewing of *Black Popcorn*, whereby viewers can perceive multiple stripes as one object under the same shroud.

In *Flamingo*, individual bright stripes tend to be isolated between individuals or groups of dimmer stripes and the luminance and color of adjacent stripes vary more across space compared to those in *Black Popcorn*. As discussed in the context of multiple-scale boundary webs and proximity-luminance covariance in Sections 3.6 and 4.2, respectively, both relative size and relative brightness influence depth percepts: *Larger* objects tend to appear closer to the viewer because they more strongly activate larger spatial scales in the visual system. Because of this, the grouping of bright stripes into larger units results in larger filters being sensitive to them, which can cause them to appear closer to the viewer because of the size-disparity correlation. Their *relative brightness* can, in addition, cooperate with this effect to make them appear closer because of proximity-luminance covariance.

These properties help to explain why the four-stripe unit in the left part of *Black Popcorn* appears to be nearer than the individual bright white stripes in *Flamingo*, despite the fact that the latter stripes are brighter than any individual stripe in *Black Popcorn*. These *Flamingo* stripes can benefit from proximity-luminance covariance, but not from the size-disparity correlation. Thus, despite the apparent simplicity of the iterated vertical stripes in the paintings of Gene Davis, the interactive effects of relative boundary strength, color assimilation, and spatial attention influence their percepts of relative depth using interactions between the neural processes that create both the size-disparity correlation and proximity-luminance covariance.

4.8 Figure-ground separation: Seeing and recognizing partially occluded objects

Many percepts that I have already discussed depend upon boundaries that are completed directly from their image contrasts. When boundaries overlap in 2D pictures of partially occluded objects, as they do vividly in the Monet painting in Figure 2.49 (right panel), additional cues to depth become available and influence what is consciously seen and recognized. Figure 2.46 (left panel) illustrates this fact in a simple example. This figure is composed of three abutting rectangles in a 2D picture. It is hard, however, not to perceive it as a large horizontal rectangle that partially occludes a large vertical rectangle that lies "behind" the horizontal rectangle (Figure 2.46, right panel). Examples such as these clarify how brain designs for 3D vision in the natural world have enabled 2D pictures to generate representations of the 3D world, thereby enabling pictorial art, movies, TV, and all other cultural advances that use 2D canvases and screens to represent pictorial information.

This process plays a major role in how we consciously see and recognize paintings. It is therefore important to understand how the simple 2D picture in Figure 2.46 (left panel) gives rise to a 3D percept that includes a partially occluded object, even though nothing can be "behind" something else in a 2D image.

One might at first guess that this is due to what we have learned from many experiences with partially occluded objects. However, Kanizsa (1979) devised many ingenious examples (e.g., Figure 2.2 (right panel) and Figure 2.4 (bottom row)) showing that similar perceptual properties of figure-ground perception occur even when we view unfamiliar pictures. In fact, the percept in Figure 2.46 (left panel) follows directly from basic properties of boundary grouping that I have already reviewed. This fact helps to answer the question: How did such a seemingly sophisticated process like figure-ground perception arise during evolution? FACADE theory and the 3D LAMINART model clarify how the key properties of figure-ground perception arise from basic properties of perceptual grouping and the complementary consistency of boundary and surface representations (Section 3.1).

To understand why this is so, consider one of the T-junctions in Figure 2.46 (left panel); that is, a place where a horizontal boundary (the top of the T) is intersected by a vertical boundary (its stem), as shown in the top left image of Figure 2.60. The top middle image of Figure 2.60 depicts how the excitatory and inhibitory inputs to a bipole cell are spatially distributed. In particular, the excitatory receptive field consists of two long-range oriented receptive fields throughout which inputs can cooperate to activate the bipole cell body. The inhibitory receptive field is defined by a shorter-range competition whereby nearby bipole cells inhibit each other, including bipole cells that are sensitive to different orientations at the same position. I will first consider bipole cells that have the horizontal orientational preference of the top of the T, and then bipole cells that have the vertical orientational preference of the stem of the T.

As in my explanation of neon color spreading (Sections 3.4 and 3.5), a bipole cell that has the same horizontal orientation as the top receives excitatory inputs from both sides of its receptive field, and can thus strongly inhibit nearby bipole cells that respond preferentially to different orientations. A bipole cell that has the same vertical orientation as the stem gets excitatory inputs from only one

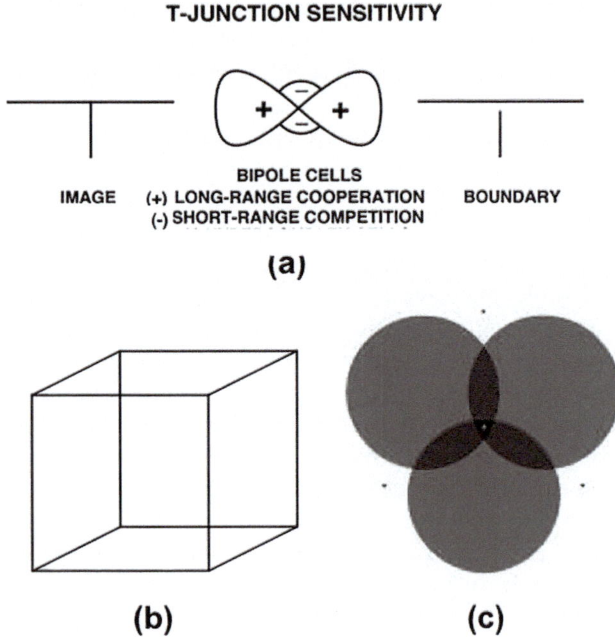

FIGURE 2.60 (a) How bipole cells cause an end cut between the horizontal and vertical lines in the T. (b) The Necker cube generates a bistable percept of two 3D parallelopipeds that alternate through time. Focusing spatial attention on a given line makes it look closer, and thus forces a consistent 3D interpretation. (c) Focusing spatial attention on one of the three disks makes it look both nearer and darker. See the text for details. Author created.

side of its receptive field, so it can either not respond at all or respond at best weakly. As a result, just as in the example of neon color spreading, an *end gap* is created in the stem boundary near where it intersects the top boundary, as illustrated in the top right image of Figure 2.60. Color can then spread through the end gap to both sides of this boundary, as it does during neon color spreading. This process enables figure-ground perception to begin when it interacts with processes leading to complementary consistency (Section 3.1), in a manner that I will now explain.

4.9 Complementary consistency: Surface contours and eye movements

Multiple boundary and surface representations are needed to represent a 3D scene during normal vision. Each of these boundary and surface representations can selectively respond to a different range of depths from an observer (Figure 2.19). The ensemble of all these form-sensitive and depth-sensitive filled-in surface representations, or FIDOs, gives rise to the 3D percept.

In particular, FACADE theory predicts how binocular boundary signals are topographically projected, from where they form in layer 2/3 of the interstripes of cortical area V2, to the monocular surface FIDOs within the thin stripes of cortical area V2 (Figure 2.61). These boundaries act as *filling-in generators* that initiate filling-in of surface brightness and color at positions where the boundary contour and feature contour signals are positionally aligned (Figure 2.18). After filling-in is initiated, the boundaries also act as *filling-in barriers* that prevent the filling-in of brightness and color from crossing object boundaries (Grossberg, 1994), as they do to generate all the percepts that have already been described.

If a boundary at a given depth is closed (Figure 2.18, top row), then it can contain the filling-in of an object's feature contour signals within it (Figure 2.18, bottom row, left column). If the boundary at a different depth has a sufficiently big gap in it, then surface brightness and color can spread through the gap and surround the boundary on both sides, thereby equalizing the contrasts on both sides of the boundary (Figure 2.18, bottom row, right column). Only a closed boundary can contribute to the final visible 3D percept. This last fact helps to explain how end gaps, such as those that occur during neon color spreading in Figure 2.36 and in response to the abutting rectangles in Figure 2.46, can contribute to figure-ground separation.

To complete our explanation of how this works, we need answers to the following two questions: How do closed boundaries help to form a visible 3D percept? How does this process also help to ensure complementary

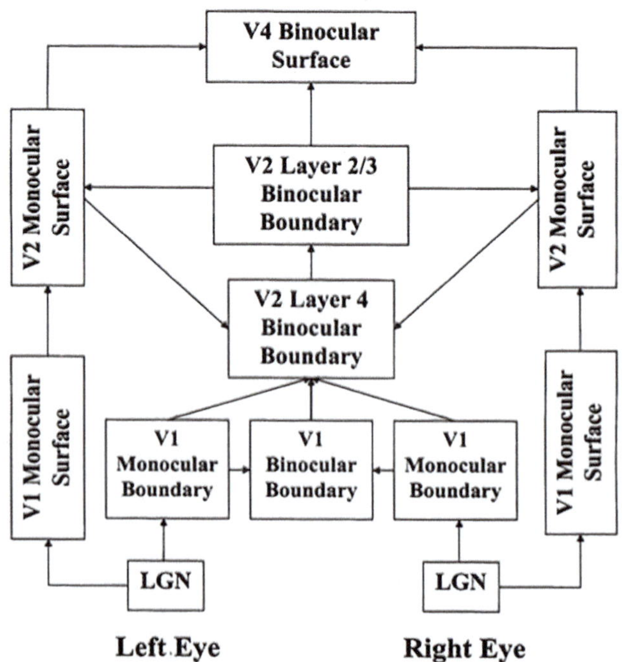

FIGURE 2.61 Macrocircuit of the main boundary grouping and surface filling-in processes stages from the lateral geniculate nucleus, or LGN, through cortical areas V1, V2, and V4. See the text for details. Author created.

consistency, and thereby, as a surprising consequence, contribute to figure-ground separation?

The answer follows from the fact that, in addition to the *boundary-to-surface* interactions that act as filling-in generators and barriers (Figure 2.19), there are also *surface-to-boundary* feedback interactions from filled-in surfaces in the V2 thin stripes to the corresponding depth-selective boundaries in the V2 interstripes (Figure 2.61). This feedback is carried out by *surface contour* signals in cortical area V2 (Figure 2.25). These signals are called surface contours because they are generated at the bounding contours of surface regions that fill-in brightness or color within a closed boundary (e.g., Figure 2.18, bottom left). As a result, the positions of surface contours and the closed boundaries that surround the region are the same (Figure 2.62).

Surface contours form at these positions because the outputs from the filled-in surface regions are generated by a *contrast-sensitive* on-center off-surround network. The inhibitory connections of this network's off-surround act *across positions* and *within depth*, and thus within each FIDO, to generate output signals only at positions where the filled-in contrasts change rapidly across space. These are the positions where boundaries block the further spread of the filling-in process.

Surface contour signals are not, however, generated at boundary positions near a big enough boundary gap, such as an end gap, because brightnesses and colors can then spread across the gap and equalize on both sides of the boundary, thereby causing zero contrast, which generates no output from the contrast-sensitive network (Figure 2.18, bottom right).

Surface contour output signals generate feedback signals to the boundary representations that induced them (Figure 2.62). These feedback signals are delivered to the boundary representations by another on-center off-surround network, whose inhibitory surface-to-boundary off-surround connections act *within position* and *across depth* (Figure 2.62). As a result, the on-center signals strengthen the boundaries that generated the successfully filled-in surfaces. They can do this because surface contours occur at the same positions as the boundaries that contain surface filling-in. The off-surround signals inhibit redundant boundaries at the same positions but farther depths (Figure 2.63). This inhibitory process is called *boundary pruning*.

Surface contour signals achieve complementary consistency by strengthening consistent boundaries and pruning redundant boundaries. The inhibited inconsistent boundaries can then contribute to neither seeing nor recognition in the final percept. Only boundaries and surfaces that can contribute to seeing and/or recognition remain. In addition, inhibiting the redundant boundaries prevents them from causing recognition of irrelevant contour fragments. Fang and Grossberg (2009) simulated these interactions to show how boundary pruning helps us to consciously perceive the emergent percepts hidden in random dot stereograms.

In addition to achieving figure-ground separation, the feedback process depicted in Figure 2.62 also enables an observer to move its eyes to selectively look at salient features in a scene or picture. This happens because surface contour signals are generated by a contrast-sensitive on-center off-surround network. The inhibitory signals of this network are weaker at high curvature points, such as

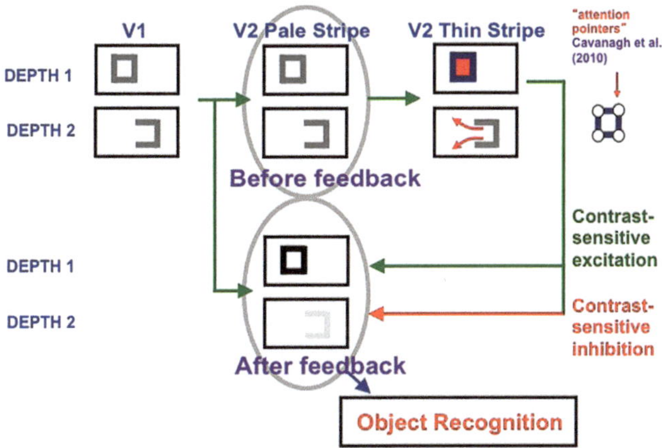

FIGURE 2.62 The same on-center off-surround feedback circuit within position and across depth that ensures complementary consistency between boundaries and surfaces also initiates figure-ground separation. See the text for details. Author created.

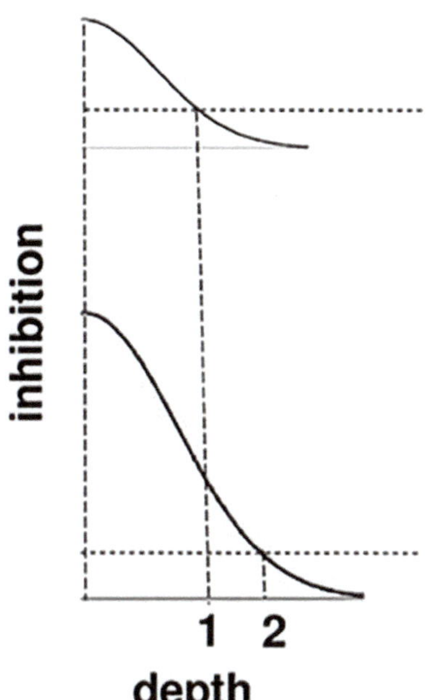

FIGURE 2.63 The on-center off-surround network within position and across depth that initiates figure-ground separation also helps to explain why brighter Kanizsa squares look closer. See the text for details. Author created.

the corners of the rectangle depicted in Figures 2.27 and 2.62. Such high curvature points are often the positions of salient features of an object. Using the circuit depicted in Figure 2.27, these salient feature positions are chosen one at a time, in order of signal strength, to compute target positions that control where the eyes will look next on an attended surface.

The ARTSCAN model hereby clarifies how several important processes during visual perception and recognition are coordinated: focusing spatial attention on an object of interest, enabling invariant object category learning and recognition of the object, and controlling scanning eye movements that sequentially foveate salient features, and thus multiple views, of an object while it is attended, so that invariant object categories can be learned.

4.10 From boundary pruning to figure-ground separation

Eliminating redundant boundaries at farther depths via boundary pruning allows figure-ground perception to begin. For example, in response to the image of three rectangles in Figure 2.46 (left panel), there are four T-junctions where end gaps will form. Only the horizontal rectangle is then surrounded by a closed boundary, so only the bounding contours of the horizontal rectangle will generate surface contour feedback signals (Figure 2.27). These surface contours use near-to-far inhibition to eliminate redundant copies of the boundaries of the horizontal rectangle at farther depths, using the network in Figures 2.62 and 2.63. The two pairs of collinear vertical boundaries due to the T stems then have no obstructing horizontal boundaries between them at farther depths. They can therefore use bipole grouping cells to complete a vertical boundary between them (Figure 2.46, right panel, dashed lines), thereby completing two vertical boundaries that are part of a vertically oriented rectangle. Because this boundary completion occurs in our brains at boundary representations that code farther depths (Figure 1.19), the completed vertical rectangle is perceived to lie "behind" the horizontal rectangle, even though the image that induces this 3D percept lies entirely in a 2D plane.

Why are the occluded parts of the vertical rectangle's boundary invisible? This is again easy to explain because *all* boundaries are invisible (Figure 2.16)! The harder question to explain asks why only the *unoccluded* surface regions of opaque objects are visible. These *visible*, or modal, surface percepts are predicted to occur in visual cortical area V4 (Figure 2.61). Cortical area V2 is, in contrast, proposed to generate amodal *recognition*, without visibility, of the occluded parts of the scene. Conscious, but amodal, recognition is supported by a feature-category resonance between prestriate visual cortex (V2/4) and inferotemporal cortex (IT) in the ventral, or What, cortical processing stream. This happens at the same time that conscious seeing, and visibility, is carried out by a surface-shroud resonance between prestriate visual cortex (V2/4) and PPC in the dorsal, or Where, cortical processing stream (Figure 2.29).

Cortical area PPC supplies top-down spatial *attention* to V2/4 as part of a surface-shroud resonance at the same time that it provides bottom-up sensory-motor signals that control the *intention* to look at and reach toward attended objects. Figure 2.64 hereby summarizes an answer to the question that was raised by images such as the ones in Figures 2.2, 2.3, and 2.4 (upper right panel), where we can consciously *recognize* properties of images without consciously *seeing* them. If seeing is not needed for recognizing, then why do we bother to see? Figure 2.64 indicates how *conscious seeing* due to a surface-shroud resonance controls *action*; notably, we can look at and reach toward surfaces that we can consciously see due to a surface-shroud resonance.

When the feature-category resonance within the What stream is eliminated by one or another kind of lesion (black X in Figure 2.64), *visual agnosia* occurs, where an individual can reach for an object (via PPC) without knowing (via IT) what the object is (Goodale et al., 1991).

The above discussion clarifies how basic mechanisms of perceptual grouping and complementary consistency can give rise to properties of figure-ground perception, notably

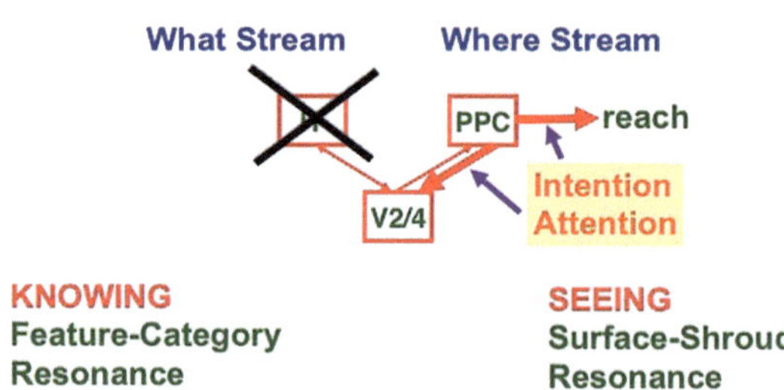

FIGURE 2.64 If a feature-category resonance, as depicted in Figure 2.29, cannot form, say due to a lesion in IT (see the black X across IT), then a surface-shroud resonance can still support conscious seeing of an attended object, and reaching for it. That can happen because the surface-shroud resonance supports both spatial attention (see thick diagonal downward red arrow) and motor commands that embody the intention to move toward the attended object (see thick horizontal red arrow). This can occur even if the individual doing so knows nothing about the object, as occurs during visual agnosia. Author created.

3D percepts of partially occluded objects, in response to 2D pictures. These mechanisms also help to explain temporally *bistable* 3D percepts, such as the 3D Necker cube percepts that oscillate through time in response to sustained observation of the 2D picture in Figure 2.60b (Necker, 1832). Computer simulations of bistable Necker cube percepts can be found in Grossberg and Swaminathan (2004). An explanation has also been given, using the same neural mechanisms, of the multi-stable percept wherein attending to one disk in a display of three overlapping disks (Figure 2.60c) makes that disk look both nearer and brighter, as was reported in psychophysical experiments of Peter Tse (Tse, 2005). Serendipitously, my article with Arash Yazdanbakhsh explaining how this happens was *in press* at the same time as Peter's article was, and in the same journal, *Vision Research* (Grossberg and Yazdanbakhsh, 2005)! Explanations and computer simulations of this and related percepts of figure-ground perception, and the control of eye movements, are provided in several articles: Grossberg and Yazdanbakhsh (2005), Grossberg, Srinivasan, and Yazdanbakhsh (2014), and Kelly and Grossberg (2000).

With this background in mind, I can now explain more examples of how particular artists have used figure-ground properties to create 3D percepts in response to viewing 2D paintings.

4.11 From boundary pruning to proximity-luminance covariance

The boundary pruning mechanism that is summarized in Figure 2.62 also helps to explain properties of proximity-luminance covariance as part of the figure-ground separation process. This is because of the way that the surface contour inhibitory signals in Figure 2.62 are delivered to the boundary system (Grossberg, 2014a, 2016). These inhibitory signals are part of an off-surround network whose strength decreases with the distance from the source cell, where "distance" is the difference in depth selectivity between multiple depth-selective boundary representations (Figure 2.63). The strength of this inhibitory surface contour signal *decreases* as the depth difference *increases* between the surface that generates the signal and its recipient boundaries.

In addition, a brighter surface generates a larger surface contour signal because the strength of the surface contour signal increases with contrast. This larger surface contour signal causes more inhibition to occur at every depth that the off-surround can inhibit. Thus, as surface brightness increases, so too does each inhibitory surface contour signal. As a result, boundaries at more depths can be inhibited, so the perceived depth difference increases between the figural surface and the depth of the nearest background surface that is not inhibited by these signals. Hence, the brighter a surface becomes, the closer it appears to be relative to its background. This explains "why brighter Kanizsa squares look closer" as a special case.

4.12 Frank Stella, occlusions, kineticism, and surface-shroud resonances

The spatial distributions of luminances and colors within a painting affect its grouping and stratification properties, which in turn affect the formation of attentional shrouds that determine where viewers focus their spatial attention. Section 4.7 discussed how larger stripe groupings within the paintings of Gene Davis (Figure 2.59) result, via proximity-luminance covariation, in the formation of surface-shroud resonances on nearer depth planes. This section discusses how surface-shroud resonances may form in response to viewing the Protractor paintings of Frank Stella (Figure 2.65). These paintings contain a number of brightly colored, interwoven, and/or overlapping figures. The series is called the Protractor series because the paintings in this series are based on the semicircular form of a protractor.

Figure 2.65 shows the painting titled *Firuzabad* above the painting titled *Khurasan Gate (Variation) I.* Stella wrote the following description of *Firuzabad*: "I was looking for a really symmetrical base and the protractor image provided that for me. Firuzabad is a good example of looking for stability and trying to create as much instability as possible. 'Cause those things are like bicycle wheels spinning around'" ("Frank Stella and the art of the protractor", 2000).

Stella also describes being inspired by decorative art in Iran in which patterns "doubled back on themselves" ("Collection: MOCA's first thirty years", 2010). Other

FIGURE 2.65 Two paintings of Frank Stella. See the text for details. © Frank Stella. ARS, NY and DACS, London 2025.

viewers have described these paintings as having an "engaging kineticism" ("Collection: MOCA's first thirty years", 2010) because the interwoven patterns in the paintings create a sense of movement when they are viewed.

FACADE theory, taken together with ARTSCAN model principles, can explain the sense of visual movement and rhythm within Stella's Protractor series paintings: In *Firuzabad*, circular forms occur in the outer protractor shapes as well as in the shapes of the interlocking rings. T-junctions occur where the surfaces of differently colored segments meet. The T-junctions trigger the figure-ground process whereby some shapes are seen to weave in front of other shapes. FACADE theory predicts that this percept arises due to the end gaps that form at T-junctions as a result of the long-range cooperation and short-range competition properties of boundary grouping using bipole cells (Section 4.8).

These end gaps create holes in stems of the T-junctions and lead to a percept of boundaries that are at a further depth than the nearer figure. This boundary discontinuity prevents filling-in of the "further" figure in the nearest FIDO, just as it does in response to the image in Figure 2.46. As a result of boundary pruning, surface-to-boundary feedback (Figure 2.62), and a process called surface pruning that is explained in Grossberg (2021), the boundary of the "further" figure is amodally completed, and its surface is filled-in within a FIDO that is behind the nearer figure. Only the non-occluded part of each figure is consciously seen, but whole protractor figures may nonetheless be amodally recognized.

The center region of *Firuzabad* contains a number of interwoven segments, some of which are recognized as whole protractors due to amodal completion, and others as smaller segments that are nested in the overall circular shape of the canvas and/or within other protractors. No single segment in this center portion of *Firuzabad* is "on top" of all the others. Instead, each segment weaves under and over other segments.

Similarly, in *Khurasan Gate (Variation) I*, no one protractor segment in the painting lies completely on top of all the others, even though each protractor segment has sufficiently large visible regions to allow the viewer to

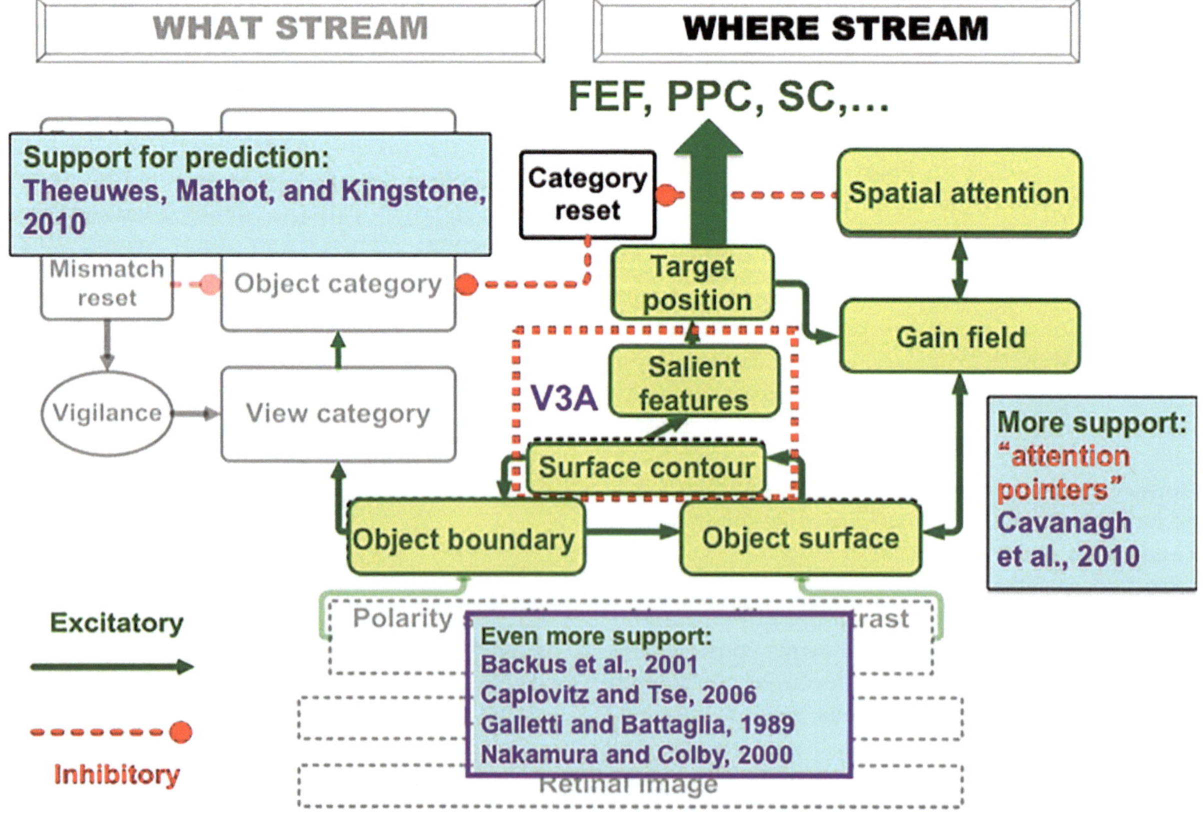

FIGURE 2.66 Once an attentional shroud forms on a protractor in a Frank Stella painting, a surface-shroud resonance directs eye movements on it to focus on salient features, while also guiding the eyes to move along surfaces in and out of different depth planes because no one protractor lies on top of all the others. See the text for details. Author created.

understand the protractor shapes. As in *Firuzabad*, these structures are perceived as overlapping segments in different depth planes. And once again, amodal boundary completion and surface filling-in of distinct segments divide the painting into separate, interlocking protractor forms that are tightly linked in a larger structure.

In all, amodal boundary completion allows perceptual completion of protractor boundaries that lay behind other forms, while consistent segment colors within amodally completed protractor surfaces allow disconnected segments to be understood as unified protractor forms. These boundary and surface completion processes enable attentional shrouds to spread across a whole protractor form, whether or not the entire form is consciously visible. Once an attentional shroud forms on a protractor, a surface-shroud resonance directs eye movements on it (see Figure 2.66) and guides these movements in and out of different depth planes because no one protractor lies on top of all the others. As noted in Figure 2.28, the ability of shrouds to form on the entire surface of figures, even when portions of the figure are occluded, combined with the interwoven nature of surface depths within the paintings shown in Figure 2.65, clarifies the "engaging kineticism" and simultaneous stability and instability within Stella's Protractor series paintings.

An attentional shroud that adheres to the surface of one figure inevitably collapses as a result of *inhibition of return*, which can be due either to inhibition of all fixation positions after they are foveated, or to activity-dependent habituation of the surface-shroud resonance itself (Chang, Grossberg, and Cao, 2014; Fazl, Grossberg, and Mingolla, 2009). When a shroud collapses, spatial attention is *disengaged* from that object surface, thereby enabling it to become *engaged* by a new surface of interest (cf., Posner, 1980). The formation of these successive shrouds enables spatial attention to shift through different depth planes and across different groupings, thus clarifying how Frank Stella achieves a sense of visual movement in paintings that are based on static interleaved patterns.

4.13 "Drawing directly in color": Monet, Matisse, Hawthorne, and Hensche

Monet directly painted the visible colors in a scene as he viewed it, rather than constructing a scene by drawing its boundaries and then filling them in. Monet's advice to painters was: "When you go out to paint, try to forget what objects you have before you, a tree, a house, a field, or whatever. Merely think, here is a little square of blue, here an oblong of pink...paint it just as it looks to you, the exact color and shape, until it gives your own naïve impression of the scene before you" (Perry, 1927, p. 120). This approach to painting inspired the entire Impressionist movement (Figure 2.47) and has had an enormous impact on the very concept of what a painting can be.

Matisse also struggled to define his own unique style. He wrote about "the external conflict between drawing and color...Instead of drawing an outline and filling in the color...I am drawing directly in color" (Matisse, 1947/1992). These words captured Matisse's realization that, if he painted directly with appropriately shaped and positioned color patches, these patches would induce the formation of amodal boundaries within the brain of the viewer, much as occurs when we perceive a Dalmatian in Snow (Figure 2.3). These amodal boundaries, in turn, capture the inducing colors to form surface representations whose form and color can be used by a viewer to understand a painting (Figure 2.67).

How does "drawing directly in color" change how a painting looks? If instead of "drawing directly in color," Matisse did "draw an outline" around his surfaces, and did so in a dark color, then these outlines could darken the surface colors of the entire scene via *color assimilation*. Color assimilation occurs due to mechanisms whereby boundaries control the filling-in of surface color. Grossberg and Todorovic (1988) and Kelly and Grossberg (2000) provide a unified explanation and computer simulations of situations in which assimilation occurs, as well as of situations that lead to brightness constancy, brightness contrast, contrast constancy, and many other visual percepts.

Matisse's method of "drawing directly in color" enabled him to create bright, glowing surface colors, without encountering the darkening and other distorting effects of visibly drawn outlines.

Matisse illustrated what he meant by "drawing directly in color" in his paintings from the Fauve period, such as his painting from 1905 called *The Roofs of Collioure* (Figure 2.68). In this painting, patches of pure color are arranged on the canvas in such a way that they are grouped by our brains into emergent boundaries, without the need for painted outlines. These emergent boundaries trigger and contain filling-in of the colors into surface representations of scenic objects. Because these emergent boundaries are invisible, or amodal, they do not darken the surface colors.

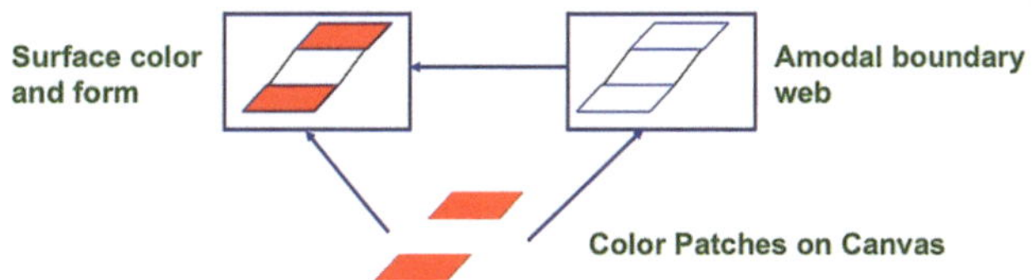

FIGURE 2.67 How "drawing directly in color" enabled Matisse to create conscious percepts of colored surfaces. Amodal boundary webs control the filling-in of color within these surface representations. See the text for details. Author created.

FIGURE 2.68 *The Roofs of Collioure* by Matisse is in the Hermitage in St. Petersburg. See the text for details.

FIGURE 2.69 The Matisse cut out called Les Codoma illustrates how he draws directly in color, without drawing an outline and filling in the color. See https://www.henrimatisse.org/cutouts.jsp#google_vignette.

Matisse had a life-long interest in these issues. It was in his book *Jazz*, which he published in 1947 toward the end of his life, that he wrote the above quote about the paper cut outs that he created during this period, when he could no longer manage the rigors of painting: "Instead of drawing an outline and filling in the color…I am drawing directly in color." Figure 2.69 shows the cut out called Les Codoma, which is one of my own personal favorites. It was done with gouache on paper, that was then cut and pasted on canvas.

Both Matisse and Monet hereby achieved their aesthetic goals by exploiting an intuitive understanding of the fact that "all boundaries are invisible."

Many other artists have also struggled with how to represent object surfaces without drawing explicit outlines, lines, or curves around them, including plein air painters of the Cape Cod School of Art. As I noted in Chapter 1, Charles Hawthorne (1938/1960) founded this influential school of painting in 1899 (see Figure 1.4). Hawthorne wrote: "Beauty in art is the delicious notes of color one against the other…all we have to do is to get the color notes in their proper relation" (p. 18). "… put down spots of color…the outline and size of each spot of color against every other spot of color it touches, is the only kind of drawing you need bother about…Let color make form—do not make form and color it. Forget about drawing…" (pp. 25–26).

Henry Hensche was Hawthorne's most famous student and a respected painter and teacher in his own right (Hensche, 1988). Hensche further developed and taught these concepts in his own Provincetown school starting in 1933. Hensche also acknowledged Monet's role in pioneering them: "When Monet came along…he revolutionized the 'art of seeing.'…it was the method of expressing light in color, and not value, to allow the key of nature to show clearly…The landscape helped Monet determine how color expressing the light key was the first ingredient in a painting, not drawing" (Robichaux, p. 27). "The untrained eye is fooled to think he sees forms by the model edges, not with color…Fool the eye into seeing form without edges" (Robichaux, p. 31). "Every form change must be a color change" (Robichaux, p. 33).

I have been lucky to know a student of Hensche, Hilda Neily, for a long time, and to have bought several of her beautiful plein air paintings. Hilda paints in Provincetown and has her own art gallery there at 364 Commercial Street. Hilda and I have gotten to know each other sufficiently well that she calls me "The Brain Guy"! The fact that I know Hilda, who is just one step removed from Hawthorne, illustrates how recently plein air painting began in America.

Many other artists have developed theories of painting to accommodate the aesthetic implications of the fact that "all boundaries are invisible."

4.14 Back to Monet: Gist and global-to-local spatial attention

Monet's painting process reflects how humans may initially perceive global information about a scene, such as its gist, before focusing attention on its finer details. As discussed in Section 2.7, ARTSCENE models how gist may be computed first as a large-scale texture category. Indeed, rapid and accurate classification of natural scenes can often be achieved using gist alone. Shifting attention via attentional shrouds to classify finer scenic textures then refines global scene classification hypotheses.

The principle that gist is the first available information in a scene is mirrored by Monet's aesthetic goal to preserve the first glance, or first impression, of a scene. Descriptions of Monet's painting process show that he started with distributed patches of local contrast and slowly added to them until long-range cooperation between his brushstrokes could more easily and unambiguously occur. Lilla Cabot Perry describes a canvas that Monet had painted only once: "it was covered with strokes about an inch apart and a quarter of an inch thick, out to the very edge of the canvas" (Perry, 1927, p. 120).

Perry (1927) went on to write: "[Monet] held that the first real look at the *motif* was likely to be the truest and most unprejudiced one, and said that the first painting should cover as much of the canvas as possible, no matter how roughly, so as to determine at the outset the tonality of the whole" (p. 120). Perry also describes one that was painted twice: "the strokes were nearer together and the subject began to emerge more clearly" (Perry, 1927, p. 120). Monet thereby carried out the kind of global-to-local process that many humans use to understand scenes, gradually building detail with more and more "brushstrokes," but never explicitly drawing fine scenic structures.

For example, in the paintings of the Rouen cathedral shown in Figures 2.48 and 2.49, the viewer can infer that there are sculptural elements on the facade of the cathedral, but the fine structure of these elements is not explicitly present in the paintings. Similarly, in the small patches within the white squares that are superimposed upon the Monet paintings in Figure 2.70, one cannot tell within the patches alone that they all contain trees. However, in the context of each complete scene, it is possible to recognize that these patches contain leaves on trees within the landscapes. By leaving out fine, scenic details in his paintings, Monet encourages viewers to experience the more global properties of a scene using larger-scale attentional shrouds. These shrouds capture the information acquired during the "first real look" at a scene and support Monet's desire to preserve the freshness of this first impression.

FIGURE 2.70 Three paintings by Monet in which the existence of trees and their leaves (surrounded by white squares) is contextually disambiguated, but not defined unambiguously by their local features. (Upper left) Grainstacks in the Morning, Snow Effect (1891) and detail. In the J. Paul Getty Museum, Los Angeles. (Bottom left) Mount Kolsaas, Rose Reflection (1895) and detail. In the Musée d'Orsay, Paris. (Right) Morning on the Seine, near Giverny (1896) and detail. Private collection. All images are freely available to download from the web.

4.15 Graffiti artists and Mooney faces

When a painting is not rendered on a smooth surface such as a canvas, additional constraints may influence an artist's technique. For example, the paintings of graffiti artists, such as Banksy (2005), were often made on walls. As Rubin (2015, p. 1) notes: "analysis of a large corpus of work by the graffiti artist Banksy suggests that the type and condition of the background wall significantly affected his artistic choices. To minimize on-site production time, Banksy renders his famous subjects (e.g., the rat) by applying single-color paint over pre-fabricated stencils. When the wall is smooth, Banksy leaves the regions previously covered by the stencil unpainted, relying on observers' perception to segregate figural regions from the (identically colored) background. But when the wall is patterned with large-scale luminance edges—e.g., due to bricks—Banksy takes the extra time to fill in unpainted figural regions with another color."

A rat wall painting by Banksy is shown in Figure 2.71 (top left). Rubin (2015, p. 2) discusses Banksy's paintings in terms of a "surface completion" process "whereby a whole surface is perceived when only fragments of its bounding contour are present in the image. The best known examples of surface completion are those of flat (2D) shapes bounded by illusory contours, such as Kanizsa's figures. . ."; see Figure 2.4. Rubin (2015) illustrates surface completion processes in response to "two-tone images obtained by binarizing the luminance levels of pictures of real-world objects or scenes" using "Mooney faces" (Mooney, 1957).

Mooney faces depict heads and faces in black and white where only salient shadows or highlights are shown, as they would appear in strongly lit photographs (Figure 2.71, top middle). These stimuli have been used to study the development of "closure" in schoolchildren; namely "the perception of an object or event which is not completely or immediately represented" (Mooney, 1957, p. 219). Rubin (2015) identifies closure with the process of surface completion. Rubin also discusses how surface completion may work by modifying Mooney faces with a lattice much like the bricks on a wall (Figure 2.71, top right) and by discussing how such backgrounds may influence the formation of Kanizsa square percepts (Figure 2.71, bottom row).

Using the term "surface completion" does not, by itself, explain how our brains process these images. A more complete explanation invokes how the computationally

FIGURE 2.71 (top row) Graffiti art by Banksy exploits properties of amodal boundary completion and spatial impenetrability. (bottom row) Pac-man figures that are not aligned with the checkerboard background (left image) versus those that are (right image) created different percepts of figural objects and their relative depths, with the checkerboard appearing nearer in the left image and in a planar percept in the right image. See the text for details. Top row: (Left image) Photo of a graffiti painting by Banksy. (Middle image) Reprinted from Face #13 in: Mooney, C. M. (1957). Age in the development of closure ability in children. *Canadian Journal of Psychology*, 11, 219–226. (Right image) Mooney Face #13 with brick pattern added. Entire figure reprinted from p. 2 in Rubin, N. (2015). Banksy's graffiti art reveals insights about perceptual surface completion, *Art & Perception*, 3, 1–17. Bottom row: Reprinted from Figure 2 in Ramachandran, V. S., Ruskin, D., Cobb, S., Rogers-Ramachandran, D., and Tyler, C.W. (1994). On the perception of illusory contours, *Vision Research*, 34, 3145–3152.

complementary processes of boundary completion and surface filling-in (Figure 2.16) individually work and interact. From this perspective, consider the Mooney face image in Figure 2.71 (top middle). Recognition of this face is facilitated when an illusory contour forms between the chin of the face at the picture's bottom and the cheek of the face at the picture's middle right. This illusory contour occurs obliquely upwards and to the right from the chin to the cheek. The illusory contour organizes the surface filling-in process by separating the white of the face from the white of the background, just as when we recognize the Dalmatian in Snow (Figure 2.3).

How does the brickwork pattern in Figure 2.71 (top right) interfere with this percept? This interference is due to the way in which at least two neural mechanisms react to the bricks:

The first mechanism activates the property of spatial impenetrability that was described in Section 3.4; namely, the horizontally oriented hypercomplex cells that are activated by the brick horizontal edges inhibit the (almost)-vertically oriented bipole cells that would otherwise create the illusory contour between the chin and the cheek of the face (Figure 2.11). Because this illusory contour cannot form, it cannot separate the face from its background during the surface filling-in of white in the right half of the percept. This problem is triggered by the *boundary* system, not the *surface* system, as Rubin (2015) suggested. Surface filling-in ("surface completion") is the result, not the cause, of this property.

The second mechanism causes the amodal completion of the horizontal boundaries of the bricks "behind" the black shapes in the image, in the same way that the vertical boundaries in response to the image in Figure 2.46 are completed behind the horizontal rectangle there. All the white parts of the face are therefore relegated to the background of the percept, and thereby prevented from being part of the facial figure. This process is again triggered by the boundary system, which then influences surface filling-in.

A similar analysis can be immediately applied to the two Kanizsa square images in Figure 2.71 (bottom row). In response to the left image, spatial impenetrability causes the horizontal boundaries of the background squares to interfere with vertical boundary completion by vertically oriented bipole cells between the pairs of vertically oriented collinear pac-man edges. Likewise, the vertical boundaries of the background squares interfere with horizontal boundary completion by vertically oriented bipole cells between the collinear horizontally oriented pac-man edges. As a result, a percept of four separate pac-men in front of a partially occluded checkerboard background is seen, much as the percept in response to Figure 2.14.

In contrast, when we view the right image, the vertical boundaries of the background squares are collinear with the vertical pac-man inducers, and the horizontal boundaries of the background squares are collinear with the horizontal pac-man inducers. Both collinear pairs of inducers contribute to formation of the Kanizsa square boundaries by bipole grouping. A percept of a square checkerboard in front of four partially occluded black disks results.

Why the Kanizsa squares in Figure 2.71 (bottom right), as in Figure 2.4 (top left), appear to be in front of four partially occluded circular disks, can be understood using FACADE theory figure-ground feedback interactions between the complementary boundary and surface cortical streams (Figure 2.16).

5 Brain processes that enable us to consciously see 3D shapes of sculptures

5.1 Sculptures of Bernini, Michelangelo, and Rodin

Some of the greatest artistic creations are sculptures. I will only discuss a few of the greatest examples by European sculptors of this inspiring type of art, notably famous sculptures by *Michelangelo* di Lodovico Buonarroti Simoni (Figure 2.72), Gian Lorenzo *Bernini* (Figure 2.73), and Auguste René *Rodin* (Figure 2.74). My discussion of these masterpieces will not comment upon their aesthetic or historical significance, about which many volumes have already been written. I just want to use them to illustrate how we see complex 3D shapes. Similar comments apply to sculptures as different as those created by European artists like Constantin Brâncuşi, Alberto Giacometti, Henri Matisse, Henry Moore, and Pablo Picasso, or the no less great creators of African, New Guinea, and Asian masks, statues, and shields (e.g., Figures 1.4F–1.7), many of whose identities are unknown.

Michelangelo lived between March 6, 1475, and February 18, 1564. As noted in Wikipedia (https://en.wikipedia.org/wiki/Michelangelo), Michelangelo has been praised by many biographers for being the most accomplished and versatile artist of his era, a true Renaissance Man who was equally at home doing architecture, painting, and sculpture. Even during his lifetime, he was often called *Il Divino* ("the divine one"). His David (Figure 2.72) and Pietà were sculpted before the age of thirty and brought him early and enduring fame. Michelangelo's David became a symbol of the Renaissance and a treasure of the city of Florence where it is housed in the Accademia Gallery of Florence (Galleria dell'Accademia di Firenze).

FIGURE 2.72 The statue of David by Michelangelo. See the text for details. This image is available on the web from the *Wikipedia* article entitled David (Michelangelo).

FIGURE 2.73 The statue of David by Bernini. See the text for details. This image is available on the web from SMARTHISTORY: The Center for Public Art History, https://smarthistory.org/bernini-david-2/.

Bernini is broadly acknowledged to be the greatest European sculptor of the seventeenth century. He was born in 1598 in Naples, Italy, and died in 1680 in Rome, where he created the Baroque style of sculpture. His sculptures are often large, include elaborate details, and often depict the midst of a dramatic movement. For example, in Bernini's David (Figure 2.73), David's arms, legs, and drapery describe an action during which David has twisted his body as he prepares to throw a rock at Goliath. I vividly remember standing at the foot of this David while marveling at how beautifully Bernini sculpted the finest details of David's body in action.

FIGURE 2.74 Statue of the Thinker by Rodin. See the text for details. This image is available on the web from Dreamstime Stock Photos, photo by S10001 on dreamtime. Dennis MacDonald/Shutterstock.

Rodin lived in France between November 12, 1840, and November 17, 1917. His work departed from the tradition of figurative sculptures that were merely decorative, by modeling the human body naturalistically to celebrate individual character and physicality. Although criticized at first, by 1900, he was acknowledged to be the pre-eminent French sculptor of his time, and world-renowned as the father of modern sculpture.

My main comments about sculptures, including the ones in Figures 2.72–2.74, are that the same basic processes of 3D vision that enable us to understand 2D paintings as representations of 3D objects and scenes, also apply, even more directly, to sculptures that are already rendered in 3D.

5.2 Seeing the 3D shapes of sculpture: Multiple-scale boundary webs again

The main brain process that I will remind the reader of here is how multiple-scale boundary webs are formed, and lead to percepts of 3D shape, as well as of perspective (see Sections 3.6 and 3.7). The 2D picture of the shaded ellipse in Figure 2.40 (left panel) elicits a compelling conscious percept of a 3D ellipsoid by generating in our brains the kind of multiple-scale boundary web that is summarized in Figure 2.42. It is clear that a 3D ellipsoid in the real world will also generate such a multiple-scale boundary web and, thus, a corresponding conscious percept of its 3D shape.

Let me now explain in greater detail the brain processes that transform the 2D image in Figure 2.40 (right panel), that is built up from spatially separated black shapes, into a conscious percept of a continuously curved hemispherical 3D surface. This image can be thought of as a 3D analog of the 2D Dalmatian in Snow (Figure 2.3), in the sense that the percepts they both induce arise from the spatial organization of completed invisible boundaries.

When we can explain how both shaded and textured objects are consciously seen as 3D shapes, either in 2D pictures or in the 3D world, we will have a good foundation for explaining how we see the 3D shapes of sculptures as well.

Percepts that are induced by images like the one in Figure 2.40 (right panel) were studied in 1987 by James Todd and Robin Akerstrom (Todd and Akerstrom, 1987). Unlike the Dalmatian-in-Snow image, the different sizes, relative orientations, and spatial arrangements of the discrete black shapes in such a texture induce a 3D surface percept. Importantly, the spatial scale of the shapes is smaller near the bounding contour of the figure. This effect of perspective activates boundary cells of different sizes, or scales, in response to different parts of the image, with the smallest boundary cells responding nearest to the bounding contour, just as in the case of the shaded ellipse (Figure 2.42). As in all cases of the size-disparity correlation (Figures 2.41 and 2.43), where smaller (larger) scales tend to code for farther (nearer) depths, the collection of all these multiple-scale boundary webs can generate a 3D percept of a curved surface.

Because at least one edge of each black shape tends to be (approximately) parallel to the bounding circular edge of the picture in Figure 2.40 (right panel), each spatial scale of boundary cells generates a boundary web that also tends to be aligned with the bounding circular edge. These depth-selective boundary webs project to depth-selective surface representations, where filling-in of surface brightness and color occurs (Figure 2.19), again as in the case of the shaded ellipse.

Todd and Akerstrom (1987) studied the perceptual judgments induced by images generated using five different statistical rules (Figure 2.75): High Perspective (HP), Low Perspective (LP), Constant Compression Elongated (CCE), Constant Compression Square (CCS), and Randomly Oriented (RO) elements. Figure 2.75 shows that the first three ways of generating images elicited depth percepts, as reflected by the increasing data curves

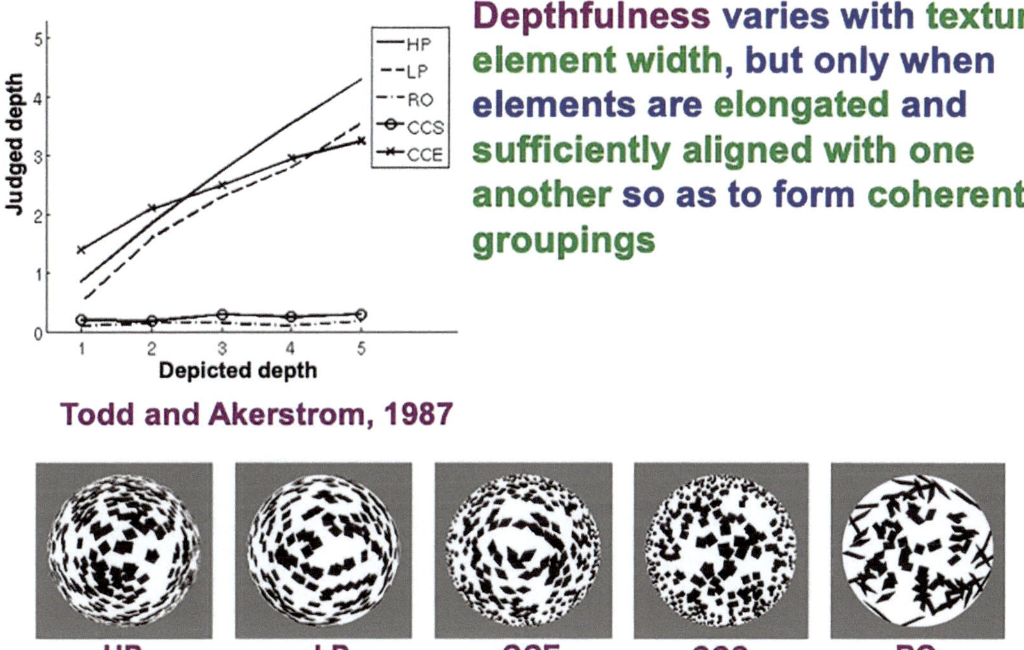

FIGURE 2.75 Todd and Akerstrom (1987) studied human perceptual judgments induced by images generated using five different statistical rules: High Perspective (HP), Low Perspective (LP), Constant Compression Elongated (CCE), Constant Compression Square (CCS), and Randomly Oriented elements (RO). Their data showed how well they could generate a conscious percept of surface depth. See the text for details. Author created.

as a function of depicted depth, with HP generating the most veridical judgments. The last two ways, CCS and RO, did not generate any significant percept of depth, as reflected by the horizontal data curves.

I worked with my PhD student, Levin Kuhlmann, and colleague Ennio Mingolla to simulate these percepts using three different variants of the same model heuristics (Grossberg, Kuhlmann, and Mingolla, 2007). Figure 2.76 shows the data in the upper left and the three model variants that we simulated on the computer in the other three images. We often simulated variants of a neural model wherein each variant embodies similar design principles. This approach demonstrates the robustness of the model principles.

The model's acronym LIGHTSHAFT stands for LIGHTness-and-SHApe-From-Texture. As shown in the model microcircuit depicted in Figure 2.77 and summarized in the article's Abstract, "changes in the statistical properties of texture elements across space induce the perceived 3D shape...through multiple-scale filtering of a 2D image, followed by a cooperative-competitive grouping network that coherently binds texture elements into boundary webs at the appropriate depths using a scale-to-depth map and a subsequent depth competition stage. These boundary webs then gate filling-in of surface lightness signals in order to form a smooth 3D surface percept. The model quantitatively simulates challenging psychophysical data about perception of prolate ellipsoids."

The LIGHTSHAFT model microcircuit in Figure 2.77 summarizes familiar model processing stages (Figure 2.11) that are adapted in LIGHTSHAFT to a multiple scale, and thus multiple depth, setting. These stages include filtering by oriented simple cells and complex cells, spatial competition across position and within orientation, followed by orientational competition within position, and bipole grouping. Because of the multiple scales, interactions are needed across scales (Figure 2.43) to achieve a *scale-to-depth* transformation because multiple scales contribute to the computation of each depth, and a *depth-to-scale* transformation because each depth generates feedback to the multiple scales that contributed to that depth.

Todd and his collaborators have described many ingenious variations and extensions of the types of displays that were discussed in the original Todd and Akerstrom (1987) article. Some of these are found in the articles Todd et al. (2004) and Todd and Thaler (2010). The authors also propose algebraic form factors to fit data

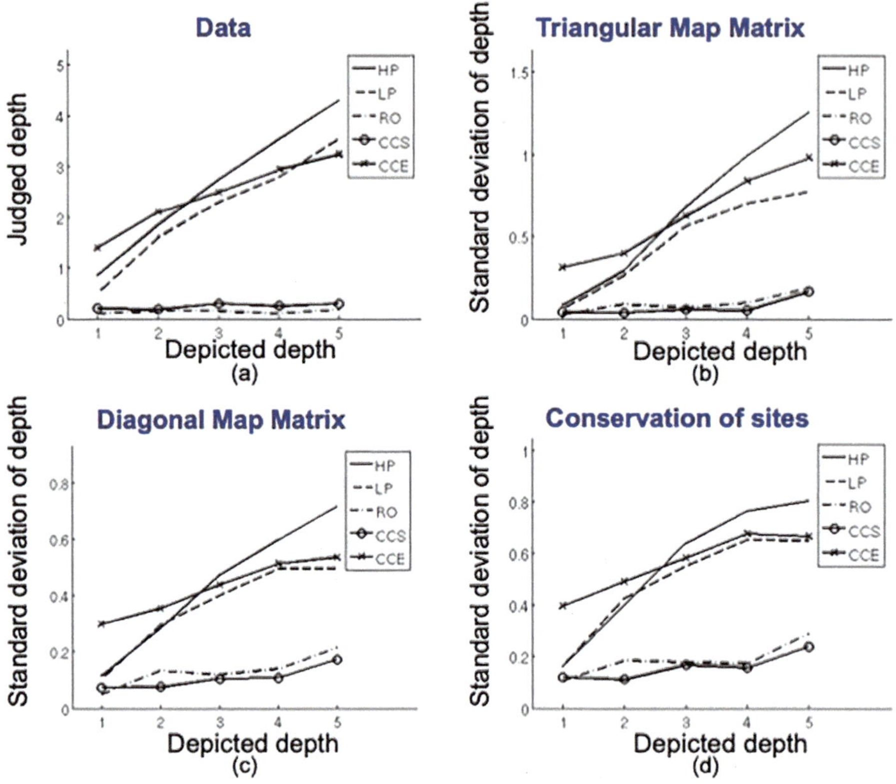

FIGURE 2.76 Comparison of the Todd and Akerstrom (1987) depth percepts with the simulated depth percepts of variants of the LIGHTSHAFT neural network model. See the text for details. Author created.

that are generated by their displays, and to thereby at least partially explain properties of the psychophysical judgments that they induce. Such form factors, however, do not explain the underlying brain processes or how they generate the 3D conscious percepts that we see when we look at these displays.

When I view these various Todd et al. displays, the conscious percepts of 3D-shape-from-texture that I experience can be explained by using the concepts and mechanisms that I have already reviewed, although not all of these concepts and mechanisms are needed to explain percepts of the different displays or even parts of a single display.

One crucial concept is that the multiple-scale *global* boundary grouping which is chosen from among many possible *local* groupings is sensitive to the context of all groupable inducers. Only the grouping with the most evidence is chosen (Figures 2.38, 2.39, 2.41, and 2.43). Although the *size-disparity correlation* (see Section 3.6) is one factor that may influence a 3D percept, it is not the only one and may be put into conflict with other factors, such as the *proximity-luminance covariance* (see Section 4.2).

Another crucial concept is that, after multiple-scale surface filling-in occurs, it feeds back to the boundary system, possibly re-organizing the multiple-scale boundary grouping to achieve complementary consistency (Section 2.5). This happens, for example, to make brighter Kanizsa squares look closer (e.g., Figure 2.4, left column).

Although our modeling concepts and mechanisms have been used to successfully simulate some of the data about perception of shape-from-texture from the Todd laboratory, additional work will need to be done to simulate all of the Todd et al. percepts, as well as the percepts generated by the great variety of shape-from-texture displays.

5.3 Distinguishing shadows from material properties of objects

In both the real world and paintings that represent it, observers can usually readily distinguish objects from the shadows that they cast. Several articles and books describe properties of shadows and how we see them, but

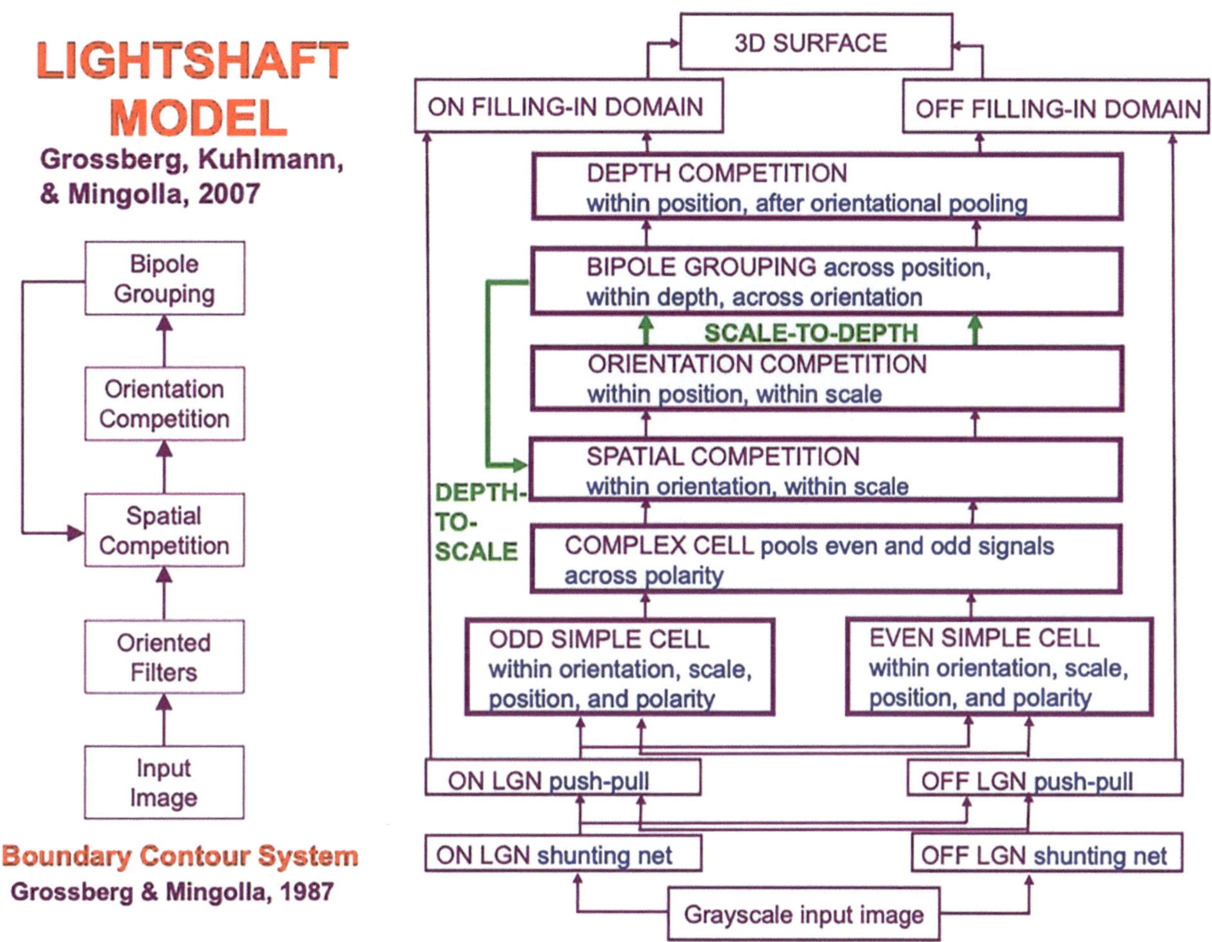

FIGURE 2.77 Macrocircuit of the LIGHTSHAFT model. Adapted from Grossberg, S., Kuhlmann, L., and Mingolla, E. (2007). A neural model of 3D shape-from-texture: Multiple-scale filtering, boundary grouping, and surface filling-in. *Vision Research*, 47, 634–672. Reprinted with the author's permission.

not how they generate our conscious percepts of them; e.g., Casati (2004), Dee and Santos (2011), and Casati and Cavanagh (2019). Figure 2.78 depicts the shadow cast by an umbrella pole on the table where I was having lunch one afternoon on the patio of one of my favorite restaurants in Provincetown. Note that the shadow uniformly changes the contrast of the table's textured colors. In particular, the shadow multiplies these contrasts by the same number. How does our brain determine that the shadow is not just a change in the contrasts of the table's colors?

One contributing factor is that the two bounding edges of the shadow have a much larger spatial scale than the individual textural elements of the table surface. The shadow edges also tend to have the same orientation over a larger spatial scale than the individual texture elements or change their orientation over a larger scale than the individual texture elements. The size-disparity correlation hereby enables us to perceive the shadow boundaries as lying on a slightly nearer depth plane than the table surface itself.

On this separate depth plane, the bounding edges generate boundary contours, as well as abutting feature contours whose strength encodes the relative contrasts of the shadow with the table surface textures, much as in Figure 2.18 (top row). When the feature contours fill-in within the texture boundaries, they do so with the uniform average relative contrast of the shadow and the table texture, as in the simpler case in Figure 2.18 (bottom left). The net percept is of a shadow lying on the textured table surface.

6 Concluding remarks

This chapter illustrates how artists have intuitively understood and exploited different combinations of brain processes to achieve their aesthetic goals. There are, of course, many more artists and artistic styles than a single chapter can review, as well as additional processes whereby

FIGURE 2.78 A shadow cast by a vertical umbrella pole on a table where I was having lunch in Provincetown, MA. See the text for how it can be explained.

humans see the world around them that have been incorporated into artistic creations.

I have described in the chapter some key brain processes that I find useful to explain how we consciously see various of the paintings and sculptures that I like a lot, as well as other artistic works whose properties lend themselves to clear explanations of how these brain processes work. To this end, the chapter reviews how neural models have shed light on the aesthetic effects that are achieved by paintings and painterly methods of Jo Baer, Banksy, Ross Bleckner, Gene Davis, Charles Hawthorne, Henry Hensche, Henri Matisse, Claude Monet, Jules Olitski, Rembrandt van Rijn, Graham Rust, Frank Stella, and Leonardo da Vinci, as well as by sculptures of Michelangelo di Lodovico Buonarroti Simoni, Gian Lorenzo Bernini, and Auguste René Rodin.

These processes range from discounting the illuminant and lightness anchoring, to multiple-scale boundary and texture grouping and classification, through filling-in of multiple-scale surface brightness and color, to the allocation of spatial attention and eye movement control. The chapter also clarifies the role of feature-category and surface-shroud resonances in supporting conscious visual recognition and perception of paintings and sculptures, how such resonances control where people may shift their attention to view paintings and sculptures, and how the attention shifts and eye movements that are regulated during such resonances can be used to better understand and appreciate paintings and sculptures. In this way, the chapter clarifies how humans consciously see paintings and sculptures, while also shedding light on how these works of art illuminate how humans see.

Music

How humans learn and consciously perform musical lyrics and melodies

Theme from Schindler's List
https://www.youtube.com/watch?v=cLgJQ8Zj3AA
Music composed and conducted by John Williams
Violin solo by Itzhak Perlman

Nessun Dorma
https://www.youtube.com/watch?v=ViOIImq_84g
Aria from the opera Turandot by Giacomo Puccini
Sung by Luciano Pavarotti
Zubin Mehta conducted the New York Philharmonic

Misty
https://www.youtube.com/watch?v=lJXLqAutql4
Music composed by Erroll Garner, lyrics by Johnny Burke
Sung by Sarah Vaughan

Rachmaninov Piano Concerto No. 3
https://www.youtube.com/watch?v=D5mxU_7BTRA
Music composed by Sergei Rachmaninov
Pianist: Vladimir Horowitz
Zubin Mehta conducted the New York Philharmonic Orchestra

Let It Be
https://www.youtube.com/watch?v=bVeVMFlwUb8
Composed and performed by The Beatles

Beethoven's Ninth Symphony
https://www.youtube.com/watch?v=dr2mcEtCxnY
Music composed by Ludwig van Beethoven
Arturo Toscanini conducted the NBC Symphony Orchestra

Your Creative Brain and AI. Stephen Grossberg, Oxford University Press. © Oxford University Press (2026).
DOI: 10.1093/9780198965367.003.0003

1 Overview

Just like visual art, music is a vast enterprise that has developed in multiple societies over thousands of years. The History Channel (https://www.history.com/news/what-is-the-oldest-known-piece-of-music) notes that, in fact:

> "The history of music is as old as humanity itself. Archaeologists have found primitive flutes made of bone and ivory dating back as far as 43,000 years, and it's likely that many ancient musical styles have been preserved in oral traditions.
>
> "When it comes to specific songs, however, the oldest known examples are relatively more recent.
>
> "The earliest fragment of musical notation is found on a 4,000-year-old Sumerian clay tablet, which includes instructions and tunings for a hymn honoring the ruler Lipit-Ishtar."

As this last sentence illustrates, the development of music has been influenced by cultural, social, historical, religious, and technological factors, among others. This chapter will not try to summarize such factors, which have been expertly described in many articles and books, for example, Damm et al., 2020; Deutsch, 1992, 2013; Gjerdingen, 1989; Hesmondhalgh (2013); Howell, West, and Cross, 1991; Krumhansl, 2000; Large, 2010; Large, Herrera, and Velasco, 2015; Levitin, 2006; Nguyen, Gibbings, and Grahn, 2018; Patel and Iversen, 2014; Rajendran, Teki, and Schnupp, 2018; Repp, 2005; Thompson, 2009; and Zatorre, Chen, and Penhune, 2007.

Instead, this chapter complements those contributions by summarizing a neural network model that proposes explanations of how human brains consciously attend, perceive, learn, and perform musical lyrics and melodies with variable rhythms and beats. The model accomplishes this by applying and specializing brain design principles and mechanisms that evolved earlier than human musical capabilities, and that have been used to explain and predict many other kinds of psychological and neurobiological data. This fact raises questions about whether music is "special" or is the outcome of an opportunistic evolutionary combination of capabilities that evolved to achieve more basic survival skills.

The remainder of this section summarizes highlights of these principles and mechanisms that will be explained in greater detail in subsequent sections. The exposition is non-technical and illustrates its proposals with analyses of specific melodies and songs.

One principle is called *factorization of order and rhythm*. We already know from Chapter 2 that a *working memory* temporarily stores information about sequences of visually perceived events (see Figure 2.30). A different working memory stores sequences of auditorily perceived events in a *rate-invariant* and *speaker-invariant* way in order to enable learning of speech, language, and music perception and recognition. These stored invariant representations can then be flexibly *performed* in a *rate-dependent* and *speaker-dependent* way under volitional control. This dissociation between invariant storage and variant performance is accomplished by the factorization of order and rhythm, where the "order" is the invariant information, and the "rhythm" is its variable way it can be performed.

I will explain in this chapter how a *canonical* working memory design stores linguistic, spatial, motoric, and musical sequences, including sequences with repeated words in lyrics, or repeated pitches in songs. This canonical working memory is reminiscent of the *canonical* laminar cortical circuit design that I discussed in Chapter 2 (see Section 2.2), and whose variations support vision, audition, speech, language, and cognition.

The canonical working memory design shows how this laminar cortical design can temporarily store *sequences* of events. The fact that the canonical cortical circuit can be realized as a working memory for temporarily storing sequences of multiple kinds of events, and for enabling these sequences to be learned and stably remembered, illustrates how our brains achieve a great diversity of function using variations of a small number of canonical cortical circuits.

Sufficiently many presentations of individual words trigger learning of recognition categories that respond selectively to them. These word categories are also called word *chunks* for short, and the categorization process is also called *chunking*. The same kind of category learning occurs in response to presentations of individual pitches. Pitch chunks are hereby learned that categorize the harmonics of each pitch.

Sequences of word chunks can be stored in working memory in response to a heard phrase or sentence. Such a sequence triggers learning of a recognition category, called a *list chunk*, that responds selectively to the whole sequence. When the sequence consists of the words in a song, the list chunk is called a *lyrics chunk*. When the sequence consists of the pitches in a song, the list chunk is called a *pitches chunk*. It is because pitches chunks respond selectively to stored sequences of pitch chunks that categorize harmonics that *tonal music* can be stored and learned. *Songs* are learned by associatively linking sequences of lyrics and pitches chunks.

Bottom-up and top-down learned interactions between working memory and chunking networks dynamically stabilize learned memories of music. These interactions are modeled by Adaptive Resonance Theory, or ART, which was summarized in Section 2 of Chapter 2. ART applies equally to visual and auditory information processing because, as shown by the thought experiment from which ART was derived, it embodies *universal* processes whereby humans and other primates learn to attend, recognize, and predict objects and events in a changing world.

Performance of a song begins when list chunks read word and pitch chunk sequences into working memory. Such a read-out occurs via ART top-down expectations from learned list chunks to their corresponding sequences of item chunks.

Learning and performance of *regular rhythms* exploits cortical interactions with *beats* that are generated in the subcortical brain nuclei called the *basal ganglia*. *Arbitrary rhythms* are learned and performed by *adaptive timing* circuits in the *cerebellum*, which regulates timed performance via interactions with the *prefrontal cortex* and *basal ganglia*.

I will explain below how the same network design that controls *walking*, *running*, and *finger tapping* also generates beats and the urge to move with a beat.

1.1 Lyrics, melodies, rhythm, and beat

In summary, this chapter explains how our ability to store, learn, and perform music builds upon brain mechanisms that are used in multiple perceptual, cognitive, and motor processes. I will also explain how variations of the same types of neural circuits that can store lyrics or melodies can be used to oscillate with a beat. These unifying mechanistic insights contribute to understanding how music may have emerged through evolution from brain processes that earlier evolved to carry out more basic psychological functions. Indeed, these variations have elsewhere been used to qualitatively explain and quantitatively simulate on the computer many kinds of psychological and neurobiological data. My Magnum Opus Grossberg (2021) provides a self-contained and non-technical exposition and synthesis of many of these explanations.

1.2 Bach's Partita No. 1 for piano illustrates that musical phrases are short

I will first introduce musical notation to illustrate that many musical phrases are short. I will then explain how short musical phrases make possible how we learn and perform music.

Figure 3.1 copies the first page of Johann Sebastian Bach's Partita No. 1. The incomparable Glenn Gould plays it here: youtube.com/watch?v=7pj5r8anMdc. Figure 3.1 illustrates the information that musical notation embodies. A brief review of musical notation is included to make the chapter accessible to those who do not read music.

Piano sheet music is organized into two separate rows of notes. The *treble clef*, also called the G clef, describes higher sounding notes, which are usually played with the right hand. The *bass clef*, also called the F clef, describes lower sounding notes, which are usually played with the left hand.

Consider the notes to the right of the treble clef in the top row of the music. They are grouped into several phrases that are defined by horizontal bars either above or below the notes that they link. These phrases influence how the piece is practiced and encoded in long term memory.

The first horizontal bars lie above the first four notes printed in the treble clef. The next horizontal bars lie above the fifth through eighth notes. Each bar also indicates the relative speed with which these grouped notes are played. Two parallel bars command the right hand to play these notes twice as fast as notes that are linked by one bar, such as the notes below them that are played by the left hand. Finer structure is depicted among the second and third notes of the treble clef that are linked by three parallel bars, and are therefore played even faster. The jagged line, or chevron, above the second note commands that note to be played even more quickly as part of a trill.

The two ♭-shaped symbols that are printed to the right of the clefs define the *key*: They require that notes b and e be played on the "flat" black keys that lie on the piano keyboard just before the white keys for b and e.

The music is further divided by vertical lines, which separate the music into bars. The C shapes to the right of the clefs and the ♭ symbols denote 4/4 time. The upper 4 means that there are four beats in a bar, and the bottom number 4 says that there are four quarter notes in a bar. The first black note to be played by the left hand is a quarter note. Thus, the notes within each bar are played in four beats that are equally spaced through time.

This segment of music illustrates that musical groupings are often short, here four or five notes long, with more notes playable in a given amount of time at a faster speed. These facts raise basic questions, including the following:

- Why are musical phrases so short?
- How do musical key and harmonic relationships constrain the notes that are played in musical phrases?
- How do variable numbers of notes get fit to an underlying beat?

1.3 Musical grouping: Harmonics, pitch, streams, arpeggios, and tonality

The kind of grouping that is marked in a piece of music like Bach's Partita is influenced by several different kinds

FIGURE 3.1 Page 1 of the score of J. S. Bach's Partita #1 for piano, BWV 825.

of grouping constraints that are due to the physics of sound and the properties of hearing.

The physics of sound determines one crucial source of grouping, namely, the *pitch* of a sound, or note, in music. The pitch of a sound determines how high or low the sound is consciously heard in any piece of music. The perceived pitch typically depends on the fundamental frequency of the sound, with higher fundamental frequencies sounding like higher pitches. Harmonics, or frequencies that are integer multiples of the fundamental frequency, are overtones that all contribute to the percept of pitch. A pitch percept is the result of another form of grouping. In particular, the harmonics that a pitch percept represents are bound together to learn a *pitch category*.

Figure 3.2 (left panel) summarizes the Spatial Pitch NETwork, or SPINET, model that I developed with Michael Cohen and Lonce Wyse (Cohen, Grossberg, and Wyse, 1995) to explain the brain processing stages that end with pitch categories. The SPINET model was used to quantitatively simulate many human psychophysical data about pitch perception, including data about the phase of mistuned components, shifted harmonics, dominance region, octave shift slopes, pitch shift slopes, pitch of narrow bands of noise, rippled noise spectra, tritone paradox, edge pitch, and distant modes. The amount of psychophysical data that the model explains and simulates assured that SPINET could confidently be used to provide inputs to auditory streaming processes.

Auditory streams are another source of grouping that enables the Bach Partita, and indeed all other music, to sound like a continuous flow of sound, even though we consciously hear only discrete notes through time. Gjerdingen (1994, p. 335) has discussed the conscious perception of auditory streams during music, noting that

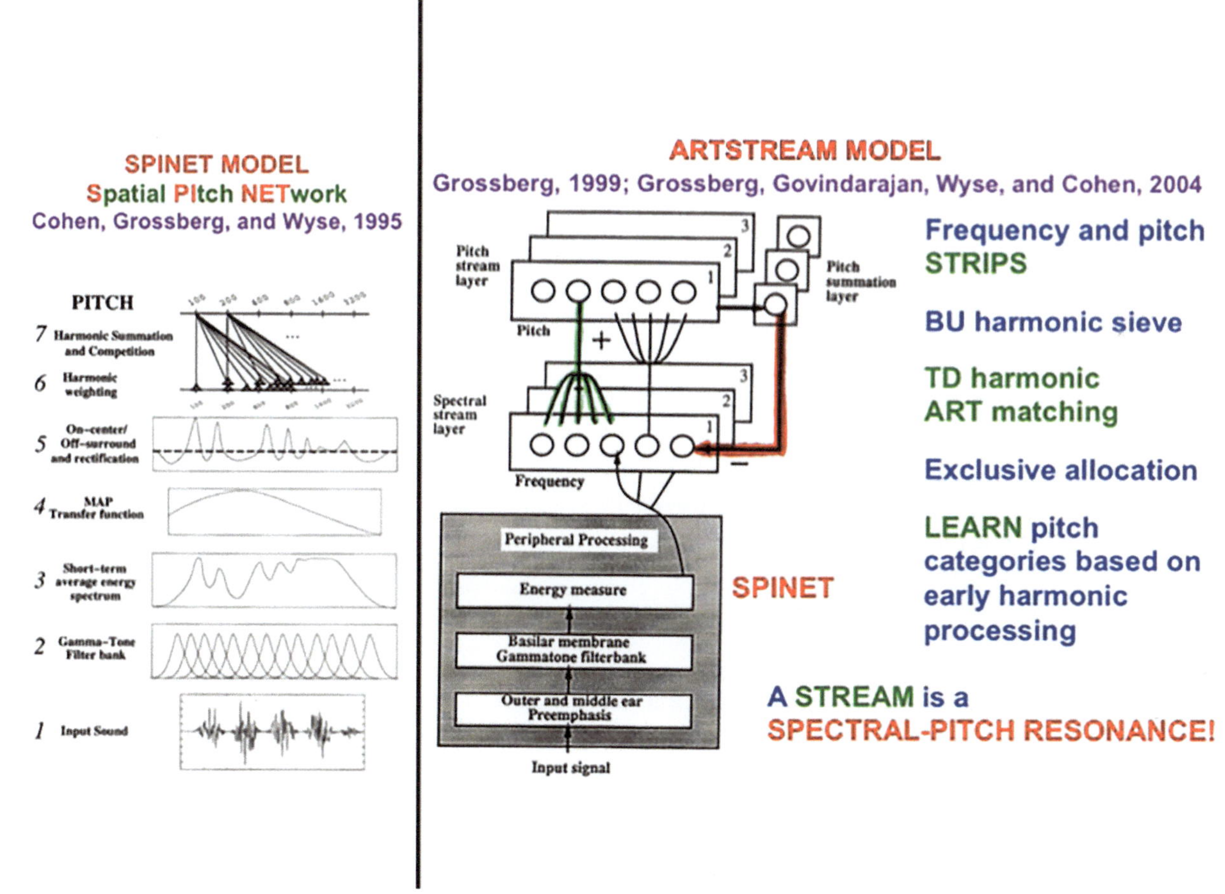

FIGURE 3.2 (left panel) The Spatial Pitch Network, or SPINET, model (Cohen, Grossberg, and Wyse, 1995) shows how a log polar spatial representation of the sound frequency spectrum can be derived from auditory signals occurring in time. This spatial representation allows the ARTSTREAM model to compute spatially distinct auditory streams. (right panel) The ARTSTREAM model explains and simulates the auditory continuity illusion as an example of a spectral-pitch resonance. Interactions of the ART Matching Rule and asymmetric competition mechanisms in cortical strip maps explain how the tone selects the consistent frequency from the noise in its own stream while separating the rest of the noise into another stream. Reprinted with the author's permission from Grossberg et al. (2004).

"a great deal of the motion perceived in music is apparent rather than real. On the piano, for example, no continuous movement in frequency occurs between two sequentially sounded tones," say in an arpeggio, "though a listener may perceive a movement from the first tone to the second." Properties of arpeggio playing were modeled by Gjerdingen (1994) using a neural model of apparent motion in vision that was introduced and developed by Grossberg and Rudd (1989, 1992).

Apparent motion in vision occurs when a discrete series of lights that are placed in a linear row turn on sequentially through time. When the spacing of the lights, and the timing with which they are sequentially lit, are within an appropriate range, one perceives a continuous motion between them in the order in which they are lit.

Apparent motion also occurs in music, when a discrete series of tones that are organized tonotopically in a linear row turn on sequentially through time within an appropriate range of rates. The brain mechanism that causes both percepts is the same, namely, when each light or tone is turned on, it activates a *Gaussian receptive field* that is centered at that light or tone. Gaussian receptive fields are ubiquitous in our brains. Successive activations, within an appropriate range of rates, of lights or tones whose Gaussian receptive fields overlap across space can cause a traveling wave of continuous activation to flow across each network from the first to the second light or tone. I call this traveling wave a *G-wave*, which is short for Gauss-wave (Figure 3.3).

Remarkably, a simple process like a G-wave has psychophysical properties that are observed during long-range apparent motion in vision, and during arpeggio playing in music. For example, if the ISI, or interstimulus interval, between the first and second tone decreases, then the traveling wave speeds up to smoothly interpolate the two tones. If the frequency difference between the two tones increases, but the ISI stays fixed, the traveling wave again speeds up to smoothly interpolate the two tones. Since the second tone turns on only after the first tone turns off, these traveling wave properties raise interesting conceptual and philosophical questions that are settled by how a G-wave works.

During apparent motion in music, the apparent motion from one tone to another allows us to enjoy music, as it does when we experience the playing of arpeggios or the singing of a song. In both vision and audition, apparent motion has an important survival function: In vision, it enables our brains to continuously track a moving object, such as a prey or predator, as it runs with variable speed in a forest, while intermittently disappearing behind occluding cover. In addition, it enables our brains to continuously track a temporally discrete sequence of acoustically similar sounds, either emitted by a prey or predator in the forest or, much more recently, during the performance of a piano sonata or string quartet. My Magnum Opus Grossberg (2021) provides a thorough review, explanation, and computer simulations of the main experimental facts about apparent motion in its Chapter 8.

TRAVELING WAVE (G-WAVE): LONG-RANGE MOTION

If the Gaussian activity profiles of two flashes overlap sufficiently in space and time, then the sum of Gaussians produced by the waning of the first flash added to the Gaussian produced by the waxing of the second flash can produce a single-peaked traveling wave from the position of the first flash to that of the second flash

The wave is then processed through a WTA choice network.

The resulting continuous motion percept is both long-range and sharp

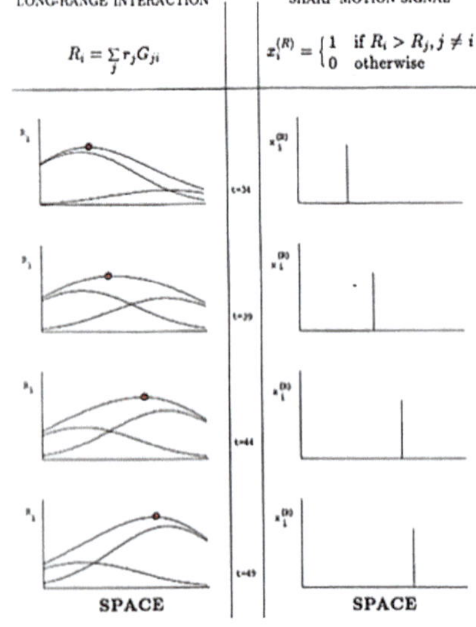

FIGURE 3.3 The sum of the waning Gaussian activity profile due to the first flash and the waxing Gaussian activity profile due to the second flash has a maximum that moves like a traveling wave from the first to the second flash. Reprinted with the author's permission from Grossberg (2021).

Our conscious recognition of a pitch percept is only partly modeled by the SPINET model. It is more fully modeled by the ARTSTREAM model (Figure 3.2, right panel) of how our brains can track multiple streams of sound through time, such as voices or instruments during music (Grossberg, 1999, 2021; Grossberg et al., 2004). The ARTSTREAM model incorporates the SPINET model as the front end of a larger neural architecture with enhanced capabilities, including conscious recognition of changing pitch sounds in an *auditory stream*, a concept that I will explain more fully now.

In ARTSTREAM, the event that supports conscious recognition of a pitch percept is modeled by a *spectral-pitch resonance* that creates an emergent *bound state* between a pitch category and the harmonic spectrum of sounds that it categorizes. Such a resonance emerges when the bottom-up adaptive filter that activates a learned pitch category within the Pitch Stream Layer triggers read-out of a top-down learned expectation back to the pitch's frequency spectrum across the Spectral Stream Layer (see Figure 3.2). When these bottom-up and top-down signals continue to cycle, they give rise to a resonant state between the pitch category and its harmonics. As in any adaptive resonance, this particular resonant state can drive learning in the bottom-up adaptive filter, leading to selective activation of its pitch category, and in the top-down expectation, leading to selective activation by the pitch category of the harmonics that support its activation.

The ARTSTREAM model also explains how a top-down expectation *focuses attention* upon the pitch's harmonics while synchronizing and gain-amplifying their activation. It does this because each top-down expectation obeys the ART Matching Rule. The ART Matching Rule is embodied within a circuit that has been mathematically proved necessary to stabilize the learning and memory of any recognition category, including a pitch category (e.g., Carpenter and Grossberg, 1987a, 1991). As I noted in Chapter 2, this circuit has a *top-down, modulatory on center, off-surround anatomy*. Top-down signals to the modulatory on-center can sensitize, or *prime*, the cells that its learned signals excite. The top-down signals, by themselves, cannot drive the activities of these cells to suprathreshold levels that can trigger output signals. Primed cells are, however, ready to respond to sufficiently well *matched* bottom-up signals more quickly and efficiently than they would without priming. The off-surround inhibits its target cells.

A striking example of the ART Matching Rule acting through time during audition occurs during the *auditory continuity illusion* (Figure 3.4; Bregman, 1990). This illusion is a simple version of our ability to complete percepts of speech and music during intervals of occluding noise, whether it is a jumble of other speakers' voices at a cocktail party, or coughs by other members of an audience during a concert.

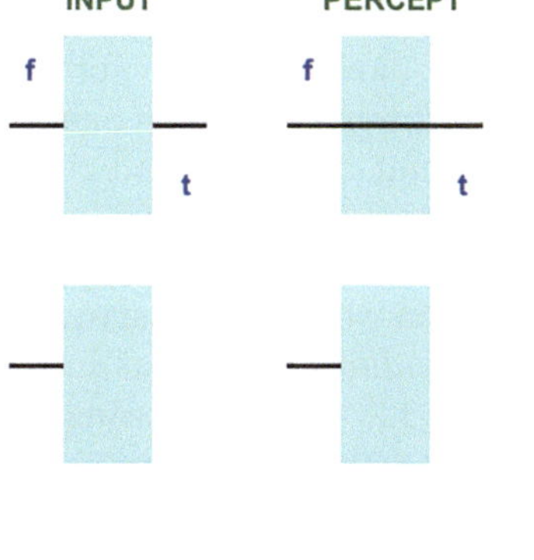

FIGURE 3.4 The auditory continuity illusion is an example of auditory streaming that illustrates the ART Matching Rule operating in the auditory system. The effect of future context on past conscious perception illustrates how long it takes for a resonance to enter consciousness. Author created.

The auditory continuity illusion can be experienced when a steady tone occurs right before and after a burst of broadband noise. Under appropriate conditions, a percept is consciously heard of the tone continuing through the noise. The residual noise then is heard in a different auditory stream.

If second tone does not follow the first tone-followed-by-noise, then the first tone is not heard continuing through the noise. The brain hereby distinguishes between a tone that is partially occluded by noise—which it completes through the noise—and one that ends before the noise begins—which it does not complete through the noise.

This is a remarkable fact because the first tone-followed-by-noise combination is the same in both cases. Somehow the second tone operates "backwards-in-time" to create a conscious percept of the first tone continuing through the noise. Without the second tone, this does not occur.

There is another way to prevent the backwards-in-time percept from occurring: If there is no noise, this backwards-in-time percept does not occur. In particular, if two tones are played with the noise replaced by an interval of silence, then silence is heard between them.

Although these percepts may seem strange in isolation, they can be explained as properties of the ART Matching Rule (Figures 1.34–1.36, 2.21, and 2.22) acting through time in the auditory system, and thus as manifestations of how ART explains our brains' ability to learn quickly without experiencing catastrophic forgetting.

In particular, if a second tone follows the noise, then ART can complete the first tone through the noise because the conscious percept lags the stimulus until a top-down expectation can match the incoming auditory signals.

An ART pitch category of the tone can then select the frequency of the tone from the noise to generate a spectral-pitch resonance of the tone continuing through the noise. If there is no noise, however, then nothing happens because the ART Matching Rule, being defined by a *modulatory* on-center, cannot create a sound out of nothing at all.

In more general situations, a spectral-pitch resonance can flow through time between successively activated pitch categories as part of an *auditory stream*, as in the groupings described in Section 1.2 while discussing the Bach Partita.

An auditory stream is caused when a G-wave flows across Gaussian receptive fields interacting across a topographically organized map of pitch categories in response to inputs to displaced positions across the map through time. One difference between apparent motion in vision and apparent motion in music is illustrated by the interactions between harmonics that occur during arpeggio playing. Due to these harmonic interactions, arpeggio playing of, say, the notes C E G C of increasing frequency, illustrates tonal music in which C E and G together form a "tonic triad" and end in the "tonic," or key note, C.

1.4 Grouping by working memories and learned plans

In addition to grouping by pitch categories and auditory streams, our brains can temporarily store the sequences of words and notes that make up the lyrics and melodies of music, in order to enable us to learn and perform them from memory.

As noted above, a *working memory* is the type of brain circuit that can temporarily store a sequence of events. Because a working memory can perform a stored sequence in the order that it occurs, it embodies a third kind of grouping. To date, it has been shown that a *single* canonical circuit design, suitably specialized, can store auditory, linguistic, spatial, and motor sequences in working memory, including sequences with repeated items, as in the sequence ABACBD. This type of Item-Order-Rank, or IOR, working memory will be described in more detail in Section 3. Lyrics and melodies, possibly with repeated words and notes, can both be stored in suitably specialized IOR working memories.

Sequences that are temporarily stored in working memory can be learned using list categories that I call *list chunks*. List chunks are a fourth kind of musical grouping. Just as harmonics can resonate with pitch categories in a spectral-pitch resonance, items that are stored in working memory can resonate with their list chunks in an *item-list resonance*. An item-list resonance supports learning and conscious recognition of the list chunk that selectively categorizes the resonating list.

Although a spectral-pitch resonance supports *conscious recognition* of a pitch, it does not support *conscious hearing* of it. Conscious perception and recognition are each supported by different resonances. As I explained in Section 2.19 of Chapter 2, a *surface-shroud resonance* supports conscious seeing of a visual object or scene. In a similar way, a *stream-shroud resonance* supports conscious hearing of an auditory object or stream. During vision, when a feature-category resonance and its corresponding surface-shroud resonance are simultaneously active, we can both consciously recognize and see the corresponding object. During audition, when a spectral-pitch resonance and the corresponding stream-shroud resonance are simultaneously active, we can both consciously recognize and hear the corresponding pitch.

These various resonances are part of a classification of the resonances that support conscious seeing, hearing, feeling, and knowing (or recognition) that are explained in Grossberg (2017c, 2021).

1.5 Invariant learning versus variant performance: Factorization of order and rhythm

These and related issues and processes are illustrated by the following two songs: The Alphabet Song (https://www.youtube.com/watch?v=dpGYmYC3p0I), and the song Smoke Gets In Your Eyes, with music by Jerome Kern and lyrics by Otto Harbach (https://www.youtube.com/watch?v=57tK6aQS_H0).

English-speaking children typically learn the Alphabet Song: A B C D E F G (pause) H I J K (L M) (N O) P (pause), ... This notation connotes that each of the letters A, B, ... , F and H, I, J, K is typically performed on a single beat, G and P are followed by a pause—that is, a silent beat with no letter performed—and the pairs of letters L, M and N, O are performed within a single beat. Moreover, the speed of performance can be volitionally increased or decreased without disrupting the relative timing of the letters. The letters can also be performed under volitional control with a different melody and/or rhythm. The Alphabet Song hereby raises questions about how a sequence of items is stored, learned, and performed in a given order, and with a prescribed melody and beat.

The words of the Alphabet Song can be performed with different rhythms due to our brain's ability to *factorize order and rhythm information*. The phrase *factorization of order and rhythm* denotes the fact that a sequence which is stored invariantly in an IOR working memory can be flexibly performed at a variety of rhythms under volitional control (Grossberg, 1986, 2003). Because of the factorization of order and rhythm, rate-invariant and speaker-invariant working memory representations (the "order") can be flexibly performed in a rate-dependent and speaker-dependent way (the "rhythm") that is under volitional control. In the special case of music, one must then explain how the lyrics that are stored in invariant working memories may be performed with different learned rhythms.

Using an invariant working memory representation greatly reduces the amount of memory that is needed for storage, and makes it possible to learn the stored sequence's *meaning*, which is coded by list chunks and their many learned associations throughout our brains, including to brain regions that support perception and emotion. Chapter 4 explains how this may happen.

In contrast, were every language utterance stored in a rate-dependent and speaker-dependent way, then learning the meaning of one such representation would not generalize to any other representation. Indeed, learning from one teacher whose words are uttered with a given rate using a given frequency range (e.g., female) could not be understood when another teacher said the same words at a different rate or used a different frequency range (e.g., male). Language learning, among other skills, would then become impossible.

To avoid this catastrophe, order information is temporarily stored using a canonical IOR working memory circuit design whose specializations are capable of temporarily storing auditory, linguistic, spatial, or motoric sequences, including sequences with repeated sequences of letters such as ABACBD, repeated words in lyrics, or repeated pitches in songs. IOR working memories are used ubiquitously in our brains because they can quickly store, stably learn, and flexibly perform sufficiently short sequences of arbitrary kinds of information, including sequences that include repeated items (Grossberg, 1978a, 2017c, 2021; Grossberg and Pearson, 2008; Silver et al., 2011).

An example from speech illustrates the main idea that the process of factorization of order and rhythm embodies. You can ask: "How ARE you today?" or, just as easily, "HOW are YOU today?," where the capital letters indicate a different rhythmic emphasis and duration. In past modeling analyses, this factorization property has been used to explain how speech can be performed with different rhythms (e.g., Boardman et al., 1999; Grossberg, Boardman, and Cohen, 1997). These studies immediately apply to performing the speech that constitutes lyrics with different rhythms. The current article extends this analysis to propose how lyrics and pitches can synchronously be performed with different rhythms, including a regular beat that is generated in the basal ganglia; how learning and performance of more general, but still regular, rhythms is regulated by prefrontal cortical modulation of the beat; and how arbitrary performance rhythms are learned by adaptive timing circuits in the cerebellum as they interact with prefrontal cortex and basal ganglia. Moreover, the same type of circuit which controls beats also controls such basic motor skills as walking, running, and finger tapping.

1.6 Regular rhythms, counting, and storing repeated words and notes

The song *Smoke Gets in Your Eyes* begins with the phrase: "They asked me how I knew my true love was true." The melody of this phrase poses at least two challenges to understanding how each brain learns and controls the performance of music. The first challenge is due to the fact that different words in the lyrics are performed with different timing. For example, the words "they," "knew," and the second occurrence of "true" are all held for four beats, while the remaining words are performed within one beat. How do we do this? When learning to play this piece on an instrument like the piano, one strategy that piano teachers use is to ask their students to count the number of beats before the next note is played. That leads to the basic question: How do we count? A review of how humans count will be given in Sections 3.14 and 3.15.

A second challenging feature of the lyrics for *Smoke Gets in Your Eyes* is that the word "true" is repeated in two different places, each performed with different timing. These lyrics hereby illustrate the general cognitive problem of storing, learning, and performing sequences of items or events with repeated elements, such as ABACBD, and to do so with their own timing. Performing the same word with different timing in a single song is a good example of why I call our ability to do this *factorization* of order and rhythm.

1.7 Lyrics and pitches working memories

Multiple working memories exist in our brains to temporarily store sequences of different kinds of information. Figure 3.5 summarizes the ARTSPEECH model of two working memories that are needed for speech perception and recognition (Ames and Grossberg, 2008). ARTSPEECH illustrates the fact that working memories occur after a series of preprocessing stages.

SOLVING THE COCKTAIL PARTY PROBLEM

One of these processing stages separates acoustic sources into distinct *processing streams* to which our attention can be paid. This ability enables humans to solve a problem that is often called the Cocktail Party Problem, because it helps us to understand the speech of the person with whom we are conversing in a busy cocktail party at which multiple voices are simultaneously speaking using overlapping frequencies. Albert Bregman wrote eloquently wrote about this problem in his classical book called *Auditory Scene Analysis: The Perceptual Organization of Sound* (Bregman, 1990).

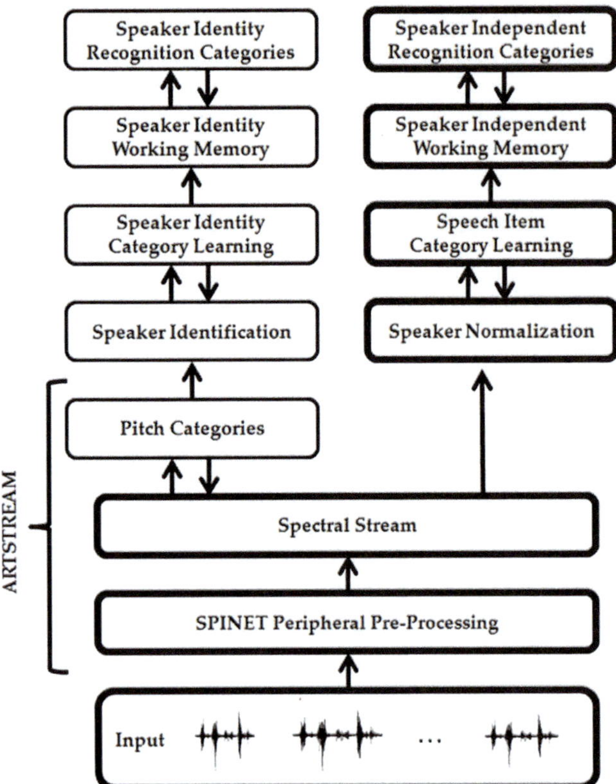

FIGURE 3.5 The ARTSPEECH architecture. ARTSPEECH consists of two parallel cortical processing streams, one devoted to speaker identification and the other to speaker meaning. Speaker identification can be learned from a speaker-dependent and rate-dependent representation of speech prosody. Speaker meaning can be learned from a speaker-normalized and rate-normalized representation of speech Item-Order-Rank information. Both streams process their distinct representations in working memories and link chunking networks of similar design. Reprinted with the author's permission from Ames and Grossberg (2008).

SOLVING THE SPEAKER NORMALIZATION PROBLEM

Separating acoustic sources, including voices, into separate auditory streams is necessary, but not sufficient, to understand the meaning of speech and language, including musical lyrics. That is because babies, women, and men typically speak using different ranges of acoustical frequencies. Their speech is thus speaker-*dependent*. In contrast, as I already noted, representations of language meaning are speaker-*independent*. Were this not the case, and we had to separately learn the meaning of every language utterance for each speaker, then we could never transfer the language meanings that we learned from one speaker to understand another one.

The process whereby a speaker-independent representation is formed is called *speaker normalization*. Speaker normalization occurs in our brains at a processing stage *after* the selection and attentive tracking of an auditory stream, but *before* the stage where the meaning of what the speaker is saying is extracted. As a result, we can learn to understand the meanings of many speakers by learning language from just a few, notably our parents and teachers.

The very fact that speaker normalization is effortlessly achieved by very young children raises basic questions, not the least of which is: How was brain evolution smart enough to discover what appears, at least at first, to be a process as complicated as speaker normalization?

I propose a detailed explanation of how this happens in Grossberg (2021). For present purposes, it sufficed to note that the neural network that carries out speaker *normalization* is proposed to use a variant of the same network mechanisms that accomplish auditory *streaming*. This proposal immediately clarifies how the speaker normalization processing stage occurs right after the stage where multiple auditory streams are formed and separated: Just replicate the streaming stage and specialize it for speaker normalization. Once speaker normalization is accomplished, speech and language classification and meaning extraction can begin.

Accordingly, the rightmost stream in Figure 3.5 includes a Speaker Normalization stage that converts the different frequency ranges of child and adult speech into a shared frequency range, or speaker-independent form, before speech item categories are learned and stored in working memory.

SOLVING THE RATE-INVARIANCE PROBLEM

Speaker-independence is not the only preprocessing that is needed to create an invariant working memory that can represent language meanings. *Rate-invariance* is also needed. I modeled how rate-invariant speech representations are formed from rate-variant speech tokens in the articles Boardman et al. (1999) and Grossberg, Boardman, and Cohen (1997) and reviewed them in Chapter 12 of Grossberg (2021) that reviews models with names like PHONET and ARTPHONE.

The following kinds of transformations were modeled in these articles to transform rate-variant speech inputs into a rate-independent representation of speech in our brains:

SUSTAINED AND TRANSIENT CORTICAL PROCESSING STREAMS

Just as in vision, where the Parvo and Magno cortical processing streams emphasize sustained and transient properties of visual inputs (Figure 2.17), parallel cortical streams exist in the auditory system for processing sustained (S) versus transient (T) acoustic features to represent all the sounds that occur during audition, speech, and music. Transient sounds include the sounds of *consonant* letters such as T and P. Sustained sounds include the sounds of *vowel* letters such as A and E.

INTRA-WORD RATE INVARIANCE

As the rates of uttering speech or lyrics speed up, non-invariant changes occur in how certain sounds *within* syllables are uttered, such as pairs of consonants (C) and vowels (V), also called CV pairs. Understanding of speech or lyrics is greatly facilitated by a transformation that creates a rate-invariant representation of such sounds within syllables. In particular, excitatory signals, also called *gain control* signals, occur from the T stream to the S stream to compensate for non-invariant changes within syllables. This T-to-S stream gain control process creates more invariant ratios of sustained to transient sounds (also called S/T ratios) across rates.

INTER-WORD RATE INVARIANCE

A different rate-dependent gain control process operates *across* syllables to modify the integration rates of working memory and list chunking networks as the *overall* speech rate changes. This process generates more rate-invariant speech and language representations in working memories and list chunking networks.

SPEAKER-DEPENDENT PROCESSING: SPECTRAL-PITCH-AND-TIMBRE CATEGORIES IN AUDITORY STREAMS

The leftmost stream in Figure 3.5 uses pitch categories to categorize the speaker-dependent and rate-dependent voices of different speakers, including when they sing different melodies. The preprocessing stages that are used to categorize pitch harmonics are illustrated in Figure 3.2. Starting with an acoustic spectrum, the six subsequent processing stages in Figure 3.2 (left panel) are sufficient to compute pitch categories using the SPINET model that I developed with Michael Cohen and Lonce Wyse (Cohen, Grossberg, and Wyse, 1995). This SPINET front end input to multiple *spectral streams*, that separate different speaker or instrument voices, is explained by the ARTSTREAM model in Figure 3.2 (right panel) that I developed with Michael Cohen, Krishna Govindarajan, and Lonce Wyse (Grossberg, Govindarajan, Wyse, and Cohen, 2004). Sequences of Pitch Categories are activated by the Spectral Streams in Figure 3.5.

A sound's pitch and loudness are, however, not sufficient to identify all naturally important sounds. *Timbre* describes more complex sound qualities, such as a sound's entire *frequency spectrum and envelope*. For this reason, I have shown how Spectral-Pitch-and-Timbre categories, not just Spectral-Pitch categories, are the units that are categorized.

A sound's *frequency spectrum* represents the amount of sound pressure, or amplitude, in each sound frequency that is active during the sound's occurrence, where each frequency measures the number of vibrations per second (called hertz, or Hz) or thousands of vibrations per second of the sound (called kilohertz, of kHz). The *envelope* of a sound is a smoothed total amplitude of a sound wave. Listeners use properties of a sound's spectrum and envelope to distinguish different voices and musical instruments, even sounds that may have the same pitches and loudnesses. The ARTSTREAM model explains how a listener can track these important cues to a sound source's identity, even when they are interrupted by occluding noise bursts. Such model details are reviewed in Grossberg (2021).

The above multiple processing stages are needed before sequences of acoustic events are ready to be temporarily stored in a working memory. I will summarize below how conscious hearing of a song is achieved when these working memories read out production commands at the same time as they read out top-down sensory expectations of the sounds that are predicted to be heard. When these expectations are matched by the auditory feedback of the productions, a song is consciously heard by coordinated *spectral-pitch-and-timbre resonances*, for a song's lyrics, and *stream-shroud resonances*, for its melodies.

LYRICS AND PITCHES WORKING MEMORIES AND THEIR LIST CHUNK NETWORKS

The IOR working memory (see Section 3.5) that stores the sequences of words in a song is called a *lyrics working memory*. *List chunks* are learned from sequences of words that are stored in the lyrics working memory. As I noted above, list chunks are recognition categories that are selectively activated by specific sequences of words that are stored in working memory. This selective activation is possible due to learned changes in the adaptive filter from the working memory level to the list chunk level (Figure 3.6). The corresponding list chunks during speech perception are denoted by Speaker Independent Recognition Categories in Figure 3.5. Each list chunk can, in turn, learn to activate the sequence of lyrics words that it codes via learned changed in the adaptive weights at the ends of the top-down pathways from the list chunk level to the working memory level. These pathways are called top-down learned expectations (Figure 3.6), and are modeled by Adaptive Resonance Theory, or ART, circuits and dynamics (see Section 2 of Chapter 2). As each lyrics list chunk is activated in its turn, it can use its top-down learned expectation to read-out its lyrics into working memory, from which they can be performed under volitional control.

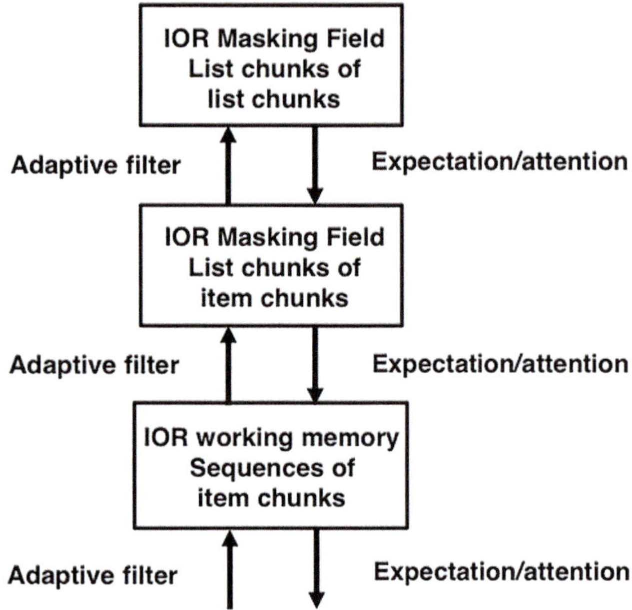

FIGURE 3.6 Hierarchy of speech processing levels. Interactions among three speech processing levels are capable of working memory storage, learning, stable memory, and performance of word sequences with repeated words. Each level consists of an Item-Order-Rank working memory. The second and third levels are, in addition, multiple-scale Masking Fields that can store sequences of variable length. All the levels are connected by Adaptive Resonance Theory bottom-up adaptive filters and top-down learned expectations and their attentional focusing and memory stabilization capabilities. The first level stores sequences of item chunks. Its inputs to the second level enable that level to store list chunks of item chunks. The inputs of the second level to the third level enable it to store list chunks of list chunks, in particular sequences of words that may include repeated words. Reprinted with the author's permission from Grossberg and Kazerounian (2016).

A different working memory temporarily stores the sequence of pitches that constitute the melody of the song. This is the *pitches working memory*. The lyrics working memory and the pitches working memory are activated in parallel when listening to someone singing a song with those lyrics and melody. Just as the right-hand stream of Figure 3.5 can represent lyrics, the left-hand stream of Figure 3.5 can represent pitches. As in the case of the lyrics working memory, a bottom-up adaptive filter from the pitches working memory can learn to activate pitches list chunks (cf. Speaker Identity Recognition Categories in Figure 3.5) and top-down pathways from a pitches list chunk to the pitches working memory can learn a top-down expectation whereby to read-out the sequence of pitches that it codes across the pitches working memory.

How these various processes work will be explained in Section 4. Additional learned associations enable the name of the piece of music, whether seen via vision or heard via audition, to activate temporally ordered series of lyrics and pitches list chunks under volitional control. Volition also controls finer aspects of performance. For example, the words that take up four beats in *Smoke Gets in Your Eyes* can be sung quickly—within a single beat that is followed by three beats filled with silence—or can be sustained throughout all four beats. This fact illustrates the distinction between the circuitry that controls the timed performance of the song as a whole, and the circuitry that modulates each word's performance during the allotted timing using volitionally regulated breath control and emphasis. How such volitional signals are generated by the basal ganglia of our brains is discussed in Sections 3–5.

2 Is Music Special?

2.1 An analysis based on shared brain designs

A great deal has been written about how music may have emerged during evolution, how it compares with language, and how it has contributed to the development of social cognition, among other topics (e.g., Deutsch, 2013; Dowling and Tighe, 2014; Howell, West, and Cross, 1991; Levitin, 2006; Schulkin and Raglan, 2014; Tan, Pfordresher, and Harré, 2018; Thompson, 2009). Essentially all of these observations have described psychological or neurobiological data about what happens during musical experiences.

The current chapter supplements this kind of descriptive knowledge with mechanistic neural explanations of how we learn and perform music. It hereby provides new insights about issues such as the following:

- Is music special to humans? If so, how?
- What similarities and differences exist between the neural mechanisms that control music perception and production versus those that control language perception and production?
- How does musical rhythm and beat compare with other rhythmic activities?
- How can we perceive and perform music in different musical keys?

The discussion in Section 1 already discussed some of these issues. The remainder of this chapter will propose answers by describing brain design principles, mechanisms, and architectures that are needed to learn and consciously perform lyrics and melodies with variable rhythms and beats. It can thereby demonstrate how variations of the same brain design principles and mechanisms that control musical experiences are also used to accomplish other perceptual, cognitive, and motor competences than music. Although music has a unique place in our personal and cultural experiences—and has special properties due to its underlying harmonic structure—it also builds upon variations and specializations of other mental capabilities. It is because of that fact that this chapter was able to apply and specialize biological neural models that I developed since the 1980s with my colleagues to explain perceptual, cognitive, and motor processes other than music. My Magnum Opus Grossberg (2021) explains these concepts and models in greater detail and with many more scholarly references to other models and relevant data than I can present here. My current exposition will supply enough information to explain how these ideas can be applied to understand music.

3 Storing and Learning Sequences with Repeats: Working Memory and List Chunks

This section will review neural models of how working memories are designed and work in the brain. For reasons that will be explained below, the same kind of model circuit has been used in the past to successfully explain data about linguistic, motor, and spatial working memories. The lyrics of a song are an example of a linguistic working memory. The present chapter adds a pitches working memory to this list since pitches, no less than lyrics, obey the same laws that govern all of these types of working memory. This shared design of the working memories that encode lyrics and melodies enables our brains to coordinate the performance of a song's words and pitches using a prescribed rhythm.

3.1 Linking working memory items and list chunks

The Item-and-Order neural model of working memory (WM) proposes that an incoming sequence of inputs that is received through time by our brains is stored an evolving *spatial pattern* of item activities (Figure 3.7; Grossberg, 1978a, 1978b). This name Item-and-Order model summarizes that its individual nodes, or cell populations, represent list *items*, while the temporal *order* of the items is stored by the spatial pattern of activity across the nodes. Item activities are sustained through time by recurrent excitatory neural signals from the item representations to themselves (vertical curved green arrows in Figure 3.7), balanced by recurrent inhibitory signals across multiple items (horizontal red arrows in Figure 3.7); see Section 3.12 for further discussion of this kind of circuit. These stored

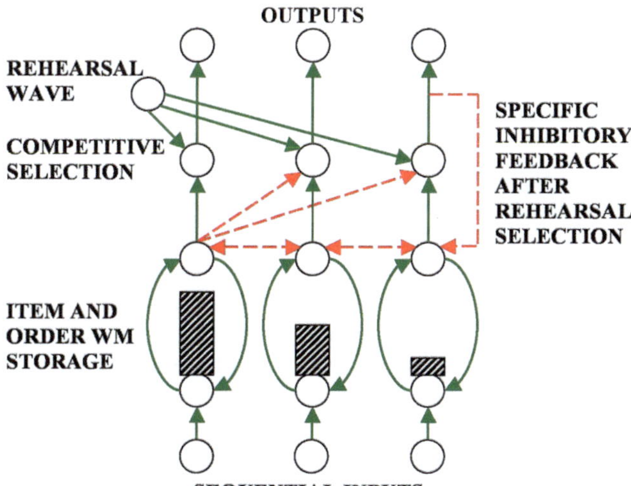

FIGURE 3.7 An Item-and-Order working memory is defined by a recurrent on-center off-surround network whose cells obey the membrane equations of neurophysiology, also called shunting laws (Carpenter, 1977a, 1977b, 1979; Grossberg, 1973, 1980a; Hodgkin, 1964; Hodgkin and Huxley, 1952). Excitatory connections are in green. Inhibitory connections are in red. A primacy gradient of activity is stored in working memory in this figure (dashed rectangles denote relative cell activities). Two simultaneously converging inputs are needed to fire a competitive selection cell. One input is a specific input from the corresponding working memory cell. The other input is a nonspecific input called a rehearsal wave. The working memory cell with the largest activity can fire the corresponding competitive selection cell when a rehearsal wave is on. Its output signal also activates a specific inhibitory feedback signal that shuts the competitive selection cell off, and thus allows the next most active working memory cell to be rehearsed next. Competitive selection cells are called polyvalent cells in the subsequent exposition. Author created.

spatial patterns of item representations are, in turn, unitized through learning into list chunk representations at the next processing level (Figure 3.6).

An item, or more precisely *item chunk*, selectively responds to prescribed patterns of activity across the distributed feature detectors within a prescribed time interval (e.g., a phoneme, musical note, or musical chord). A *list chunk* selectively responds to prescribed sequences of item chunks that are stored in working memory (e.g., a word or familiar phrase in the lyrics of a song). Thus, the item chunks of an Item-and-Order WM mediate between distributed feature patterns and list chunks. Properties of these functional units, interacting via bottom-up and top-down interactions, have been supported by their successful explanations and predictions of psychophysical data about speech perception, including immediate serial recall; immediate and delayed free recall; continuous distracter free recall; long-term recency, word frequency, and word superiority effects; list length and list strength effects; presentation variability; phonemic similarity; and non-word lexicality (Grossberg, 1978a, 1978b, 1984, 1986,

2003; Grossberg and Pearson, 2008). These functional units and their interactions will herein be used to explain how musical information is temporarily stored.

Remarkably, as will be explained below, just the three interacting processing levels shown in Figure 3.6 can store, learn, and perform lyrics that include repeats, such as "our true love was true." This kind of example illustrates that our brains do not need, nor do they possess, many processing levels to store, learn, and perform sequential behaviors. Figure 3.8 summarizes a macrocircuit of the predictive Adaptive Resonance Theory, or pART, model neural architecture of many of the brain's most important processing regions (Grossberg, 2018). Regions like the ventrolateral prefrontal cortex (VLPFC) and dorsolateral prefrontal cortex (DLPFC) contain the main working memories that our brains use for the storage of event sequences and their list chunks.

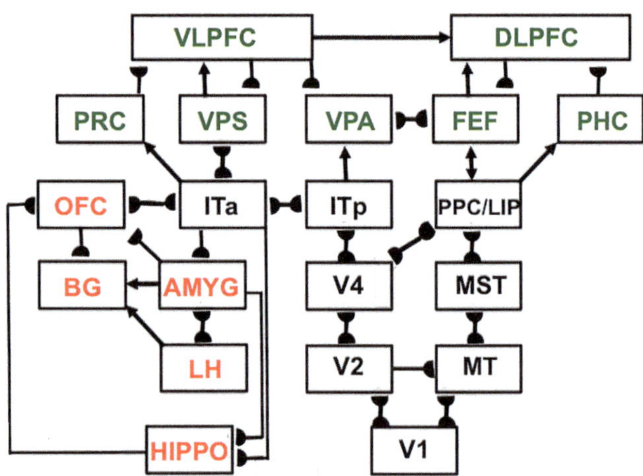

FIGURE 3.8 Macrocircuit of the main brain regions, and connections between them, that are modeled in the unified *predictive Adaptive Resonance Theory* (pART) of cognitive-emotional and working memory dynamics. Abbreviations in red denote brain regions used in cognitive- emotional dynamics. Those in green denote brain regions used in working memory dynamics. Black abbreviations denote brain regions that carry out visual perception, learning and recognition of visual object categories, and motion perception, spatial representation, and target tracking. Arrows denote non-adaptive excitatory synapses. Hemidiscs denote adaptive excitatory synapses. Many adaptive synapses are bidirectional, thereby supporting synchronous resonant dynamics among multiple cortical regions. The output signals from the basal ganglia that regulate reinforcement learning and gating of multiple cortical areas are not shown. Also not shown are output signals from cortical areas to motor responses. V1: striate, or primary, visual cortex; V2 and V4: areas of prestriate visual cortex; MT: middle temporal cortex; MST: medial superior temporal area; ITp: posterior inferotemporal cortex; ITa: anterior inferotemporal cortex; PPC: posterior parietal cortex; LIP: lateral intraparietal area; VPA: ventral prearcuate gyrus; FEF: frontal eye fields; PHC: parahippocampal cortex; DLPFC: dorsolateral hippocampal cortex; HIPPO: hippocampus; LH: lateral hypothalamus; BG: basal ganglia; AMGY: amygdala; OFC: orbitofrontal cortex; PRC: perirhinal cortex; VPS: ventral bank of the principal sulcus; VLPFC: ventrolateral prefrontal cortex. Reprinted with the author's permission from Grossberg (2018).

3.2 Correct temporal order is stored in working memory by a primacy gradient

How does a spatial pattern that is stored in an Item-and-Order WM get performed in its correct temporal order? Performing musical notes in their correct order is, of course, essential in all musical performance. Correctly ordered performance occurs if the items in working memory are stored by a *primacy gradient* (Figure 3.7). For example, when a sequence "A-B-C" of items is stored by a primacy gradient, cells that store item "A" have the highest activity, cells storing "B" have the second highest activity, and cells storing "C" have the least activity. Then the list ABC can be performed in the correct order because the item chunk with the highest activity is performed first, the item chunk with the second highest activity is performed second, and so on, until all items in the sequence are performed.

3.3 Rehearsal waves and inhibition of return

A primacy gradient that is stored in working memory does not have to be immediately performed. Performance occurs in response to a volitional signal that is called a *rehearsal wave*. The basal ganglia (BG) where the volitional signal originates has no knowledge about what is stored in working memory in prefrontal cortex, or PFC (Figures 3.8 and 3.9). A *rehearsal wave* is therefore delivered uniformly, or nonspecifically, with equal activity from the basal ganglia to the entire working memory in the PFC (Figure 3.7). The rehearsal wave enables read-out of stored activities by opening a rehearsal gate. The item chunk with the highest activity is read out fastest because it exceeds its output threshold fastest. Its output signal also self-inhibits its WM representation via a specific inhibitory feedback pathway (Figure 3.7), leading to *inhibition-of-return* that prevents perseverative performance of the most active item (Grossberg, 1978a, 1978b; Klein, 2000; Posner and Cohen, 1984; Posner et al., 1985). Each item representation self-inhibits as it is rehearsed until no active items are left in working memory.

3.4 Item-and-Order, competitive queuing, and primacy models

Since the Item-and-Order model was introduced, many modelers have applied it (e.g., Boardman and Bullock, 1991; Bohland, Bullock, and Guenther, 2010; Bradski, Carpenter, and Grossberg, 1994; Bullock and Rhodes, 2003; Grossberg and Pearson, 2008; Houghton, 1990; Page and Norris, 1998). For example, Page and Norris (1998) have used a

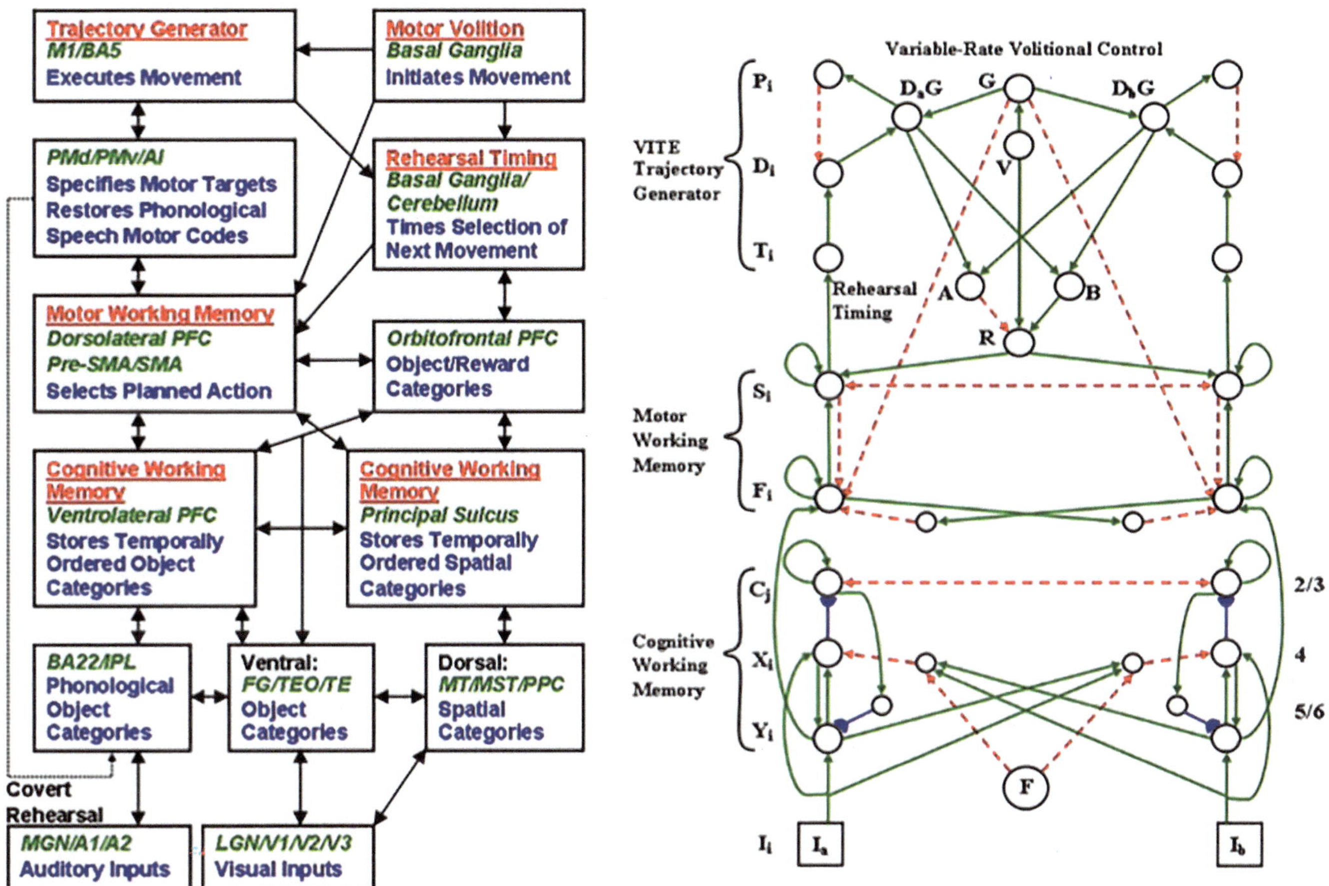

FIGURE 3.9 (left panel) The LIST PARSE laminar cortical model of working memory and list chunking includes circuits to model the brain regions that are marked in red. (right panel) The model's Cognitive Working Memory circuit is proposed to occur in ventrolateral prefrontal cortex. The Motor Working Memory, VITE Trajectory Generator, and Variable-Rate Volitional Control circuits model how other brain regions, such as dorsolateral prefrontal cortex, motor cortex, cerebellum, and basal ganglia, interact with the Cognitive Working Memory to control working memory storage and volitional control of variable-rate performance of item sequences. Reprinted with the author's permission from Grossberg and Pearson (2008).

variant of it, that they call the Primacy Model, to explain their cognitive data about word and list length, phonological similarity, and forward and backward recall effects. Houghton (1990) called the model Competitive Queuing when he also used it to explain other cognitive data.

3.5 Item-and-Order model explains psychological and neurophysiological data

Subsequent psychophysical and neurophysiological experiments confirm that, as predicted, item order is encoded by relative activity levels and is reset by self-inhibition. For example, Farrell and Lewandowsky (2004) studied the latency of human responses that follow serial performance errors. They found that (p. 115):

> Several competing theories of short-term memory can explain serial recall performance at a quantitative level. However, most theories to date have not been applied to the accompanying pattern of response latencies ... Data from three experiments ... rule out three of the four representational mechanisms. The data support the notion that serial order is represented by *a primacy gradient that is accompanied by suppression of recalled items* [italics mine].

Electrophysiological experiments of Averbeck et al. (2002) studied macaque monkeys performing arm movement sequences that copy geometrical shapes. The data curves at time zero in the four graphs in Figure 3.10a(a) exhibit the primacy gradients of four lists that were stored in dorsolateral prefrontal cortex. These curves also show that the most active cells are read-out earliest and self-inhibit to permit read-out of the entire list in the stored order. The data were simulated (Figure 3.10a(b)) by an Item-and-Order working memory in the LIST PARSE laminar cortical model (Figure 3.9; Grossberg and Pearson, 2008).

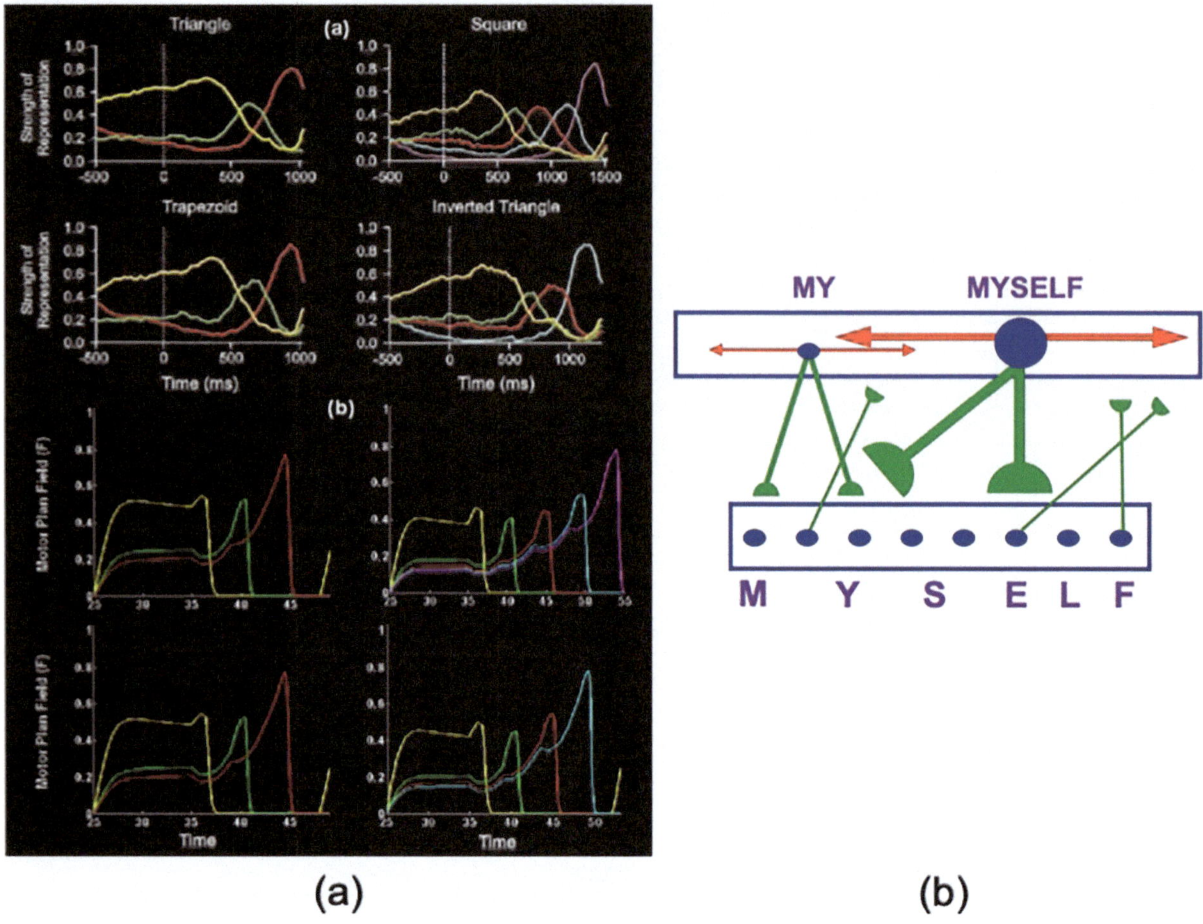

FIGURE 3.10 (left panel) (a) Neurophysiological data that conform to Item-and-Order working memory properties were recorded in a series of sequential copying experiments in monkeys (Adapted with permission from Averbeck et al. (2002)). Each of the four figures shows a primacy gradient in working memory whose most active cell is performed first as its activity self-inhibits, followed by the next most active cell, and so on. When only the last item remains, it has the highest activity because it was freed from inhibition by earlier items. The LIST PARSE laminar cortical model of working memory and list chunking simulates these data. (right panel) Circuit illustrating how Item-and-Order stored working memory item chunks (M, Y, S, E, L, F) activate list chunks (such as MY and MYSELF) in a Masking Field network. Masking Field cells respond selectively to lists of item chunks of variable length. Reprinted with the author's permission from Grossberg (2020b).

These properties also occur in the Item-Order-Rank, or IOR, generalization of Item-and-Order working memories, that can temporarily store sequences of pitches that may include repeated notes, as in the first four phrases of Bach's Partita #1; see Figure 3.1 and Section 3.14.

Figure 3.10a shows that LIST PARSE successfully models a *motor* working memory in the *dorsolateral* prefrontal cortex that can quantitatively simulate neurophysiological data about sequential recall of stored motor sequences. LIST PARSE was also shown in Grossberg and Pearson (2008) to model a *linguistic* working memory in the *ventrolateral* prefrontal cortex that quantitatively simulates psychophysical data about immediate serial recall, and immediate, delayed, and continuous distractor free recall, among other data properties.

3.6 Bowed gradients, grouping, and chunking

Not all sequences of items that are stored in a working memory can be recalled in their correct temporal order. Only primacy gradients have this property. How and when primacy gradients occur thus clarifies how music is typically performed in the correct temporal order.

Free recall tasks illustrate how a sequence of items can be performed in an incorrect order. During free recall, a list is recalled in whatever order comes to mind after hearing it just once (e.g., Murdock, 1962). If the stored list is too long, a *bowed* serial position curve is often observed. Here, items at the beginning and the end of the list are recalled earliest, and with the highest recall probability.

Grossberg (1978a, 1978b) noted that these free recall properties have a natural explanation if the pattern of cell activities that stores the list items in working memory is also bowed, with the first and last item chunks having the largest activities, while those in the middle having less activity. Then, the item chunk with the largest activity is read out first, whether at the list's beginning or end, and self-inhibits its item representation to prevent preservative performance of the same item (Figure 3.6). Then the next largest item chunk is read out, and so on in the order of stored relative activity.

As in free recall data, the model recalls items at the beginning and end of the list with greater probability because larger activities can better survive perturbations due to internal cellular noise and attentional fluctuations. *Transpositions* of recall order by items close together in the list are explained by the fact that they have similar stored activities, so their relative size, and thus temporal order, can more easily be reversed by internal noise or attentional fluctuations.

These facts about when primacy or bowed gradients are stored in working memory constrain strategies for storing all sorts of sequences in working memory, including lyrics and pitches, so that they can be learned and performed in the correct order. To understand this issue better, I will provide answers to the following questions:

- What are the longest lists that a working memory can store in the correct temporal order?
- Why can only relatively short lists be stored with the correct temporal order?

In an Item-and-Order working memory, these questions become:

- What is the longest primacy gradient that the working memory can store?
- Why is it so short?

3.7 Memory span and musical groupings: The magical numbers four and seven

What is the longest primacy gradient that can be stored? The answer to this question constrains all strategies for learning to correctly perform skilled sequences, whether during speech production, dance movements, spatial navigation, or musical performance. The upper bound during free recall has been called the Magical Number Seven, or *immediate memory span*, of 7 ± 2 items by George A. Miller (Miller, 1956).

The explanation in Grossberg (1978a) of the immediate memory span distinguished it from the then new concept of *transient memory span*. The *transient* memory span was predicted to be the longest list for which a primacy gradient may be stored in short-term memory solely as the result of bottom-up inputs, and without the benefit of read-out from active list chunks of their learned expectations into working memory. The *immediate* memory span, in contrast, was predicted to arise from the combined effect of bottom-up inputs and top-down read-out from learned expectations.

My article Grossberg (1978a) proved that read-out of top-down expectations can only increase the maximal primacy gradient that can be stored, thereby predicting that the immediate memory span exceeds the transient memory span. Given the known estimated immediate memory span of approximately seven items, the transient memory span was estimated to be approximately four items. There should thus also be a Magical Number Four when top-down effects are removed. This prediction was experimentally supported by data of Nelson Cowan (Cowan, 2001), who demonstrated a working memory capacity of 4 ± 1 items when influences of long-term memory (LTM) and grouping effects were minimized in his experimental design.

The magical numbers four and seven for storage of items in working memory shed mechanistic light on the maximum length of musical groupings that can readily be performed in the correct order. In particular, these constraints clarify how grouping small numbers of notes together in Bach's Partita Number 1 (Figure 3.1) facilitates how this exquisite piece of music is stored, learned, and performed in the correct order.

3.8 LTM Invariance Principle: How working memory supports list chunk learning

Why is the transient memory span so short? An intuitively plausible answer to this question is that the length of the transient memory span is well-suited to the length of item sequences that can be learned and performed in the correct temporal order. In particular, temporary storage of sequences in working memory is useful only if it can support stable learning of list chunks, and read-out during performance by those list chunks of the sequences in working memory that they code. In the case of music, list chunks are learned from both lyrics working memories and pitches working memories. List chunks that encode rhythms are also learned, as I will explain below.

After I could clearly articulate this problem about sequence learning, in Grossberg (1978a, 1978b) I *derived* Item-and-Order working memory circuits from hypotheses that ensure their ability to support learning and stable memory of list chunks. When this insight is applied to music, it clarifies that musical phrases are short both to store them in working memory for possible immediate performance in the correct temporal order, as well as to chunk them via learning for future performance in that order.

I derived Item-and-Order working memories from two simple postulates that enable their list chunks to be learned and stably remembered: the *LTM Invariance Principle* and the *Normalization Rule*. I used these postulates in 1978 to derive mathematical equations for Item-and-Order working memories, and to mathematically prove how they generate primacy and bowed gradients.

The LTM Invariance Principle prevents storage of longer lists of events in working memory (such as MYSELF) from causing catastrophic forgetting of previously learned list chunks of shorter lists that form parts of MYSELF (such as MY, SELF, and ELF). In particular, suppose that bottom-up inputs activate the working memory representation of a familiar list chunk, such as the word MY. Then storing in working memory the remaining portion SELF of the novel word MYSELF will not cause forgetting of the learned weights that activate the list chunk of MY. When applied to music, the LTM Invariance Principle enables larger groupings of notes to be learned without forgetting previously learned smaller groupings.

Incremental refinements of these models have been made over the years (e.g., Bradski, Carpenter, and Grossberg, 1992, 1994), eventually leading to models of how the layered circuits in the prefrontal cortex compute IOR working memories, among other properties needed to achieve higher-order properties of biological intelligence (Grossberg, 2013a, 2018, 2021; Grossberg and Pearson, 2008; Silver et al., 2011).

3.9 Stable list chunking exploits classical laws for STM, MTM, and LTM

Stable list chunks can be learned because, as new inputs are stored in working memory, the *relative sizes*, or ratios, of previously stored working memory *cell activities*, or short-term memory (STM) traces, are preserved (Figures 1.20 and 1.21), even if the newly arriving inputs may change their total activities. As a result, the relative activities of previously learned *adaptive weights*, or long-term memory (LTM) traces, are also preserved (Figure 1.22). This happens because the bottom-up signals in the axons from the working memory to the list chunks are *multiplied* by the LTM traces before the net signals activate list chunks.

This is a good place to provide the mathematical equations that define STM, MTM, and LTM traces for readers who may find this additional precision helpful. Readers can skip the remainder of this section without reducing their understanding of the subsequent text.

The picture above the equation in Figure 1.20 provides a summary of the main processes that are useful to know about: First, cell bodies and their STM traces $x_i(t)$ at the ith cell; second, signals $f(x_i(t))B_{ij}$ down the pathways, or axons, with constant path strength B_{ij} to the jth cell body with activity $x_j(t)$; and third, the adaptive weight, or LTM trace, $z_{ij}(t)$ in the hemidisk-shaped synaptic knob abutting the jth cell. The equation governing STM in Figure 1.20 is said to obey the Additive Model because it adds all of its inputs and signals. The signals themselves are products of several terms, and hence nonlinear.

Each cell in the Additive Model that is displayed in Figure 1.20 obeys a differential equation for the rate of change, denoted by dx_i/dt, of the ith STM trace x_i. The net rate of change is the sum of four terms on the right-hand side of the equation: a passive decay term—A_ix_i (note the minus sign)—which ensures that the activity decays to zero if there are no external influences; a total excitatory signal that is a sum of signals $f_j(x_j(t))B_{ji}$ times LTM traces $z_{ji}(t)$, a total inhibitory signal that is a sum of signals $g_j(x_j(t))C_{ji}$ with components like the excitatory signal times LTM traces $Z_{ji}(t)$, and an external input I_i.

The Shunting Model in Figure 1.21 generalizes the Additive Model to the more realistic situation in which the cell activities have fixed upper and lower bounds that are imposed by the multiplicative terms in red. The excitatory upper bounds are B_i/C_i, because the corresponding term in red equals zero when x_i equals that value, thereby shutting off the effect of the excitatory signals. Likewise, the inhibitory lower bounds are –Fi/Ei, because the corresponding term in red equals zero when x_i equals that value, thereby shutting off the effect of the inhibitory signals.

These bounding terms in red occur in all the *membrane equations* that define the activities of brain cells. Because the terms in red *multiply* the excitatory and inhibitory signals, they are *mass action*, or *shunting*, terms that realize *automatic gain control* of the cell responses to the signals that they shunt. Automatic gain control enables cell activities to respond more quickly to larger total excitatory and inhibitory inputs, until they start to approach their fixed upper and lower bounds.

Figure 1.22 summarizes a typical equation for Medium-Term Memory, or MTM, which describes how chemical transmitters y_{ki} in the synaptic knobs get released via term $-Lf_k(x_k)y_{ki}$. The minus sign (–) in this term causes y_{ki} to decrease, by mass action, at a rate proportional to the product of the signal size $f_k(x_k)$ and the amount of available transmitter y_{ki}.

MTM traces, as well as LTM traces, multiply the signals in their axons, as shown in the terms $f_j(x_j)D_{ji}y_{ji}z_{ji}$ in Figure 1.21. That is why, as transmitters are released, and thereby the amount of available transmitter for release is reduced, one often says that the net signals *habituate* in an activity-dependent way. In this sense, persistent activation of an axon can "tire it out." The MTM equation in Figure 1.22 omits various finer processes like transmitter production, mobilization, and release. See Grossberg (2021) for further details.

Also shown in Figure 1.22 is perhaps the simplest law for learning and memory, as embodied by Long Term Memory, or LTM traces. This law describes a process of *gated steepest descent* (Grossberg 1978a, 1978b). The *gate* in this equation is defined by signal $f_k(x_k)$, which is positive only if the activity x_k of the kth cell is positive. If signal $f_k(x_k)$ equals zero, then the rate of change dz_{ki}/dt of the LTM trace z_{ki} is also zero, so no learning occurs. That is why $f_k(x_k)$ is called a *stimulus sampling* signal: If it equals zero, no stimuli that can drive learning are "sampled."

Learning by *steepest descent* can occur only when sampling signal $f_k(x_k)$ is positive. Then the LTM trace z_{ki} approaches, or tracks, the sampled signal $h_i(x_i)$ until it equals it, at which time dz_{ki}/dt equals zero, so no further learning occurs.

Cells that send out signals along all the axons that they emit can be sampling cells. Alternatively, cells that receive signals from all the axons that they receive can also be sampling cells. In the former case, the LTM traces at the ends of all the axons that are *emitted* by a sampling cell can learn a *spatial pattern* of activity across the sampled cells. Such a spatial pattern can, for example, learn a *top-down expectation*. This network is called an *outstar* (Figure 3.11). In the latter case, the LTM traces at the ends of all the axons that are *received* by a sampling cell can also learn a spatial pattern of activity across the sampled cells. Such a spatial pattern tunes a *bottom-up adaptive filter* that can selectively activate a recognition category. This network is called an *instar* (Figure 3.11). Outstars and instars are found throughout our brains because our brains are self-organizing adaptive processors of both bottom-up and top-down signal patterns.

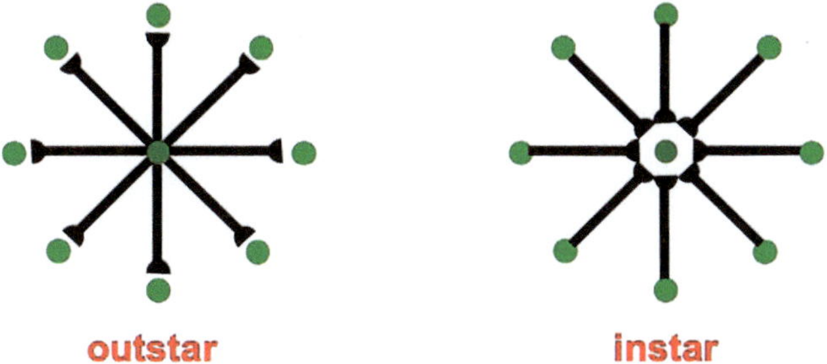

FIGURE 3.11 An outstar (left image) can learn an arbitrary spatial pattern of activity across its sampled cells. An instar (right image) can act as an adaptive filter that inputs to a recognition category. An instar can learn to selectively respond to an arbitrary spatial pattern of activity across the network that activates it. The duality of the outstar and instar networks is evident when they are drawn as above. Reprinted with the author's permission from Grossberg (2021).

3.10 Stable list chunking uses classical laws for adaptive filtering and competition

In the *conscious ARTWORD*, or cARTWORD, model of conscious speech perception (Figure 3.12; Grossberg and Kazerounian, 2011), the LTM traces are computed in synaptic knobs at the ends of bottom-up axons, in the abutting postsynaptic membranes, or both. These synaptic knobs are represented by black hemidisks in Figure 3.12. The white squares abutting the hemidisks in Figure 3.12 denote that their MTM traces can habituate, and thereby modulate the read-out of the LTM traces with which they are multiplied, as in the shunting equation in Figure 1.21. A habituated synapse releases less chemical transmitter than an unhabituated one in response to an input signal of fixed size. Habituation of a previously active pathway helps the network to choose different, as yet unhabituated pathways, to represent novel inputs, thereby preventing previous cell activations from perseverating for too long.

We can now better understand why SELF does not recode a previously learned category for MY when MYSELF is presented through time: The bottom-up adaptive signals from multiple axons are added up at each recipient list chunk via an instar (Figure 3.11). The total input to such a list chunk is a *pattern*, or vector, of activities multiplied by a *pattern*, or vector, of LTM traces. By preserving relative activities, the relative sizes of these total inputs to the category cells do not change through time; thus, nor do the corresponding LTM patterns that track these activities when learning occurs at their category cells. This is the property that explains why SELF does not degrade the learned chunk for MY when MYSELF is presented through time. The bottom-up LTM-gated pathways from the working memory to the list chunking level constitute the *adaptive filter* pathways depicted in Figure 3.12.

The same argument holds if the words MY, SELF, and MYSELF are replaced by the words in lyrics or the pitches in a melody. Larger groupings of musical elements can hereby be learned without forcing forgetting of previously learned subgroupings of them.

The Normalization Rule means that the *total* activity across all active cells in a working memory has a maximal value that is approximately independent of the number of stored items. Thus, storing more items in working memory forces each item to be stored with less activity. As shown in Figure 3.13, storage through time of more items in working memory converts a primacy gradient (left image) into a bowed gradient (right image). As additional items are presented through time, normalization forces the stored item activities to become smaller (see the

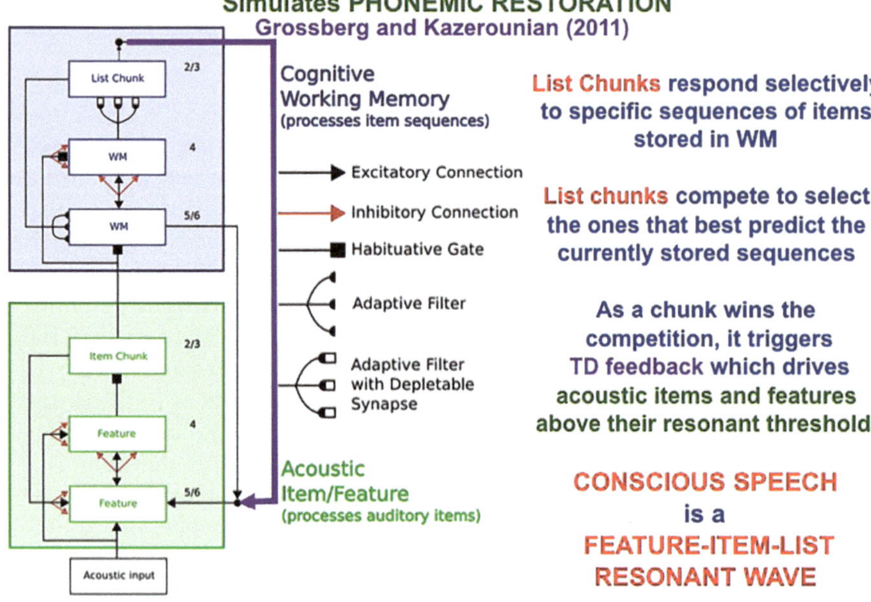

FIGURE 3.12 The conscious ARTWORD, or cARTWORD, laminar cortical speech model simulates how future context can disambiguate noisy past speech sounds in such a way that the completed percept is consciously heard to proceed from past to future as a feature item list resonant wave propagates through time. Reprinted with the author's permission from Grossberg (2021).

Bottom-up inputs from all the feature-selective cells at a given processing level can activate multiple cells at the next level that code list chunks. These cells compete to choose a winning cell, or small set of cells, that receives the largest inputs (Figures 3.10b and 3.14). The winning cells code list chunks that have the most support from their bottom-up inputs in the current context. Winning cells drive learning whereby their abutting LTM traces track the bottom-up input patterns that they filter. This adaptive tuning process enables the winning cells to fire more selectively to the input patterns that activated them, leading to the name *competitive learning* for this kind of category learning network (Grossberg, 1976a, 1976b, 1978a; Rumelhart and Zipser, 1985; Willshaw and Malsburg, 1976).

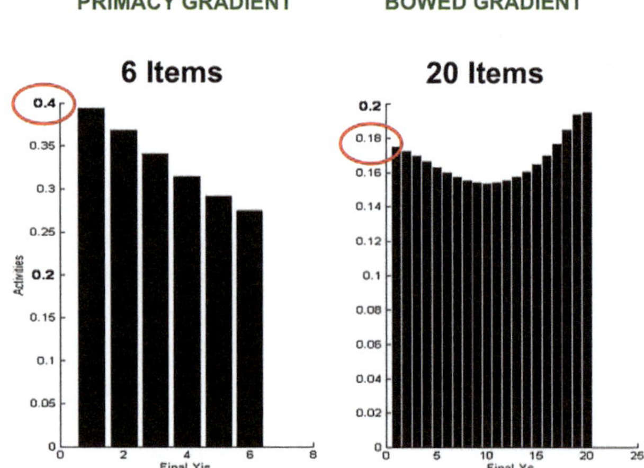

FIGURE 3.13 Simulation of a primacy gradient for a short list (left image) being transformed into a bowed gradient for a longer list (right image). Activities of cells that store the longer list are smaller due to the Normalization Rule, which follows from the shunting inhibition in the working memory network. Reprinted with the author's permission from Grossberg (2021).

numbers inside the closed red ellipses). Normalization mechanizes the *limited capacity* of working memory (Baddeley, 1986; Cowan, 2001; Grossberg, 1978a, 1978b; Miller, 1956; Posner, 1980).

3.11 Masking Fields can learn and perform musical groupings of variable length

The example of MY and MYSELF illustrates that list chunks can selectively represent *lists of variable length* in order to learn language, music, or motor skills like dancing, playing the piano, and navigating routes in space. The category cells that occur in Masking Field networks can be used for these tasks, because they can learn and store such variable-length list chunks (Figures 3.10b and 3.14). To accomplish this, a Masking Field consists of a multiple scale, self-similar, recurrent shunting on-center off-surround network (Cohen and Grossberg, 1986, 1987), properties that I clarify in the next paragraph:

> Masking field cells develop with multiple sizes, or scales, due to activity-dependent growth during development. During development, item chunks (e.g., in Figure 3.10b that categorize the letters M, Y, S, E, L, and F in the word MYSELF) are endogenously active during a critical period, leading to the growth of bottom-up connections from the item chunk level to the list chunk level, where the list chunk for MYSELF, as well as for MY and SELF, will develop. These connections

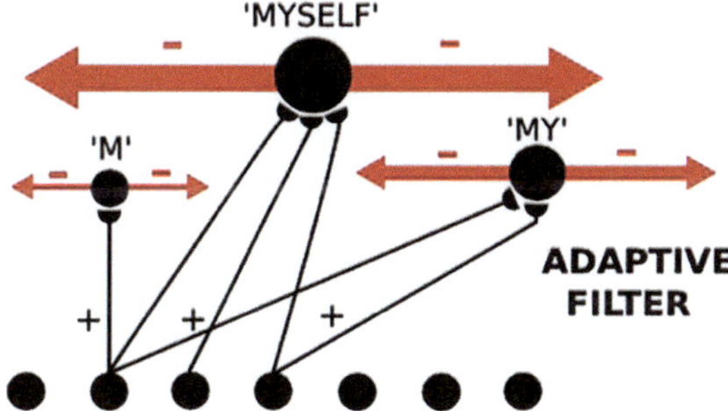

FIGURE 3.14 A Masking Field working memory is a multi-scale self-similar recurrent shunting on- center off-surround network. It can learn list chunks that respond selectively to lists of item chunks of variable length that are stored in an item working memory at the previous processing stage. Chunks that code for longer lists (e.g., MY versus MYSELF) are larger, and give rise to stronger recurrent inhibitory neurons (red arrows). Reprinted with the author's permission from Grossberg (2021).

> grow accordingly to a probabilistic law whereby variable numbers of connections contact list chunks across the network. List chunks that receive bottom-up inputs from more item chunks can code longer lists. They also receive larger total inputs, on average, through time because they receive endogenous activity simultaneously from more item chunks.

During the network's development, this input activity triggers self-similar cell growth whereby both the list chunk cell bodies and their connections grow proportionally. This growth continues until the total activity density at each cell is reduced to a threshold intensity. The net result is a Masking Field wherein longer lists are coded by larger cells with stronger recurrent inhibitory interneurons within the list chunk level (red connections in Figures 3.10b and 3.14), and stronger top-down excitatory priming pathways to the item chunk level (top-down green connections in Figure 3.10b).

Masking Field nodes are list chunks in the second and third processing levels in Figure 3.6. When representing language or lyrics, the first level can represent letters, the second level words, and the third level sequences of words. As noted above, because the second and the third levels are also Item-Order-Rank working memories, the words coded at the second level can include repeated letters, as in the words "repeated" and "letters," and the sequences coded at the third level can include repeated words, as in the phrase "our true love was true." The same is true for pitches. Then the first level can code musical notes or chords, the second level can code short pitch phrases that may contain repeated chords, and the third level can code sequences of pitch phases in a melody.

3.12 Universal design of linguistic, motor, spatial, and musical working memories

If all linguistic, motor, spatial, and musical working memories obey the LTM Invariance Principle and the Normalization Rule, then they should all share a similar design. Both psychological and neurobiological data support this prediction. Models that explain and simulate linguistic, motor, and spatial working memory data include the laminar cortical LIST PARSE model (Figure 3.9, right panel; Grossberg and Pearson, 2008) that uses a prefrontal *linguistic* working memory to explain and quantitatively simulate psychophysical data about immediate serial recall, and immediate, delayed, and continuous distractor free recall. Note the cortical layers 5/6, 4, and 2/3 in the Cognitive Working Memory in this figure. LIST PARSE also describes a prefrontal *motor* working memory that quantitatively simulates neurophysiological data about sequential recall of stored motor sequences (Figure 3.10, left and right panels).

The lisTELOS neural architecture (Figure 3.15; Silver et al., 2011) incorporates LIST PARSE as its prefrontal *spatial* Item-Order-Rank working memory. The lisTELOS model quantitatively simulates psychological and neurophysiological data about the learning and planned performance of saccadic eye movement sequences. To accomplish this, the model simulates how several parietal and prefrontal cortical areas interact with each other and with three basic ganglia gating circuits. The simulated cortical areas include the Supplementary Motor Area, or SMA, and pre-SMA, whose damage degrades performance of stored sequences in working memory (Kennerley, Sakai,

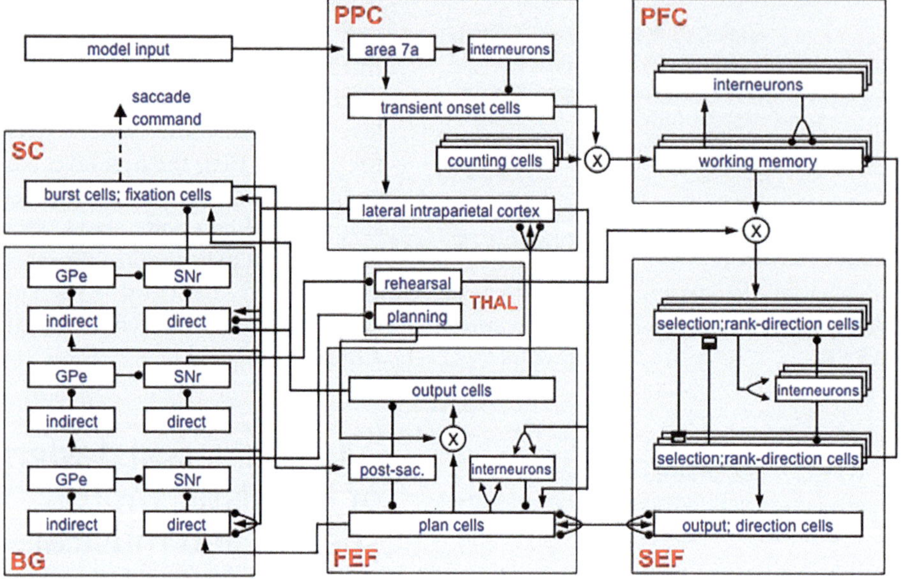

FIGURE 3.15 The lisTELOS architecture explains and simulates how sequences of saccadic eye movement commands can be stored in a spatial working memory and recalled. Multiple brain regions are needed to coordinate these processes, notably three different basal ganglia (BG) loops to regulate saccade storage, choice, and performance, and the supplementary eye fields (SEF) to choose the next saccadic command from a stored sequence. Because all working memories use a similar network design, this model can be used as a prototype for storing and recalling many other kinds of cognitive, spatial, and motor information. See the text for details. Reprinted with the author's permission from Grossberg (2021).

and Rushworth, 2004; Nachev, Kennard, and Husain, 2008; Shima and Taji, 2000; Zatorre, Chen, and Penhune, 2007).

Due to the shared circuit design of all linguistic, motor, and spatial working memories, the lisTELOS modeling results apply to how the prefrontal cortex interacts with multiple cortical areas to carry out any sequentially organized behaviors, including musical behaviors.

3.13 Working memories are recurrent shunting on-center off-surround networks

The LTM Invariance Principle and Normalization Rule are realized by a type of circuit that occurs ubiquitously throughout our brains. It is a recurrent on-center off-surround network of cells that obey the membrane equations of neurophysiology, otherwise called shunting dynamics (Figure 3.7). I mathematically proved how these networks process ratios (LTM Invariance Principle) and conserve total activity (Normalization Rule) many years ago in Grossberg (1973); also see the reviews in Grossberg (1978a, 1980b, 2013b, 2021).

In such a network, excitatory feedback signals due to recurrent on-center, or excitatory, interactions (green vertical curved arrows in Figure 3.7) help to store an evolving spatial pattern of activities in response to a sequence of inputs, while recurrent off-surround, or inhibitory, shunting interactions balance the on-center to store input relative activities (horizontal red arrows in Figure 3.7), thereby generating the desired properties of contrast normalization and conservation of total activity.

A rehearsal wave from the basal ganglia (Figure 3.7; see also Section 3.3) reads-out the highest stored activity first, and self-inhibitory feedback prevents its perseverative performance, while the network gradually renormalizes its activity through time.

The effects of recurrent inhibition can be seen in the data and simulation summarized in Figure 3.10a. After the next-to-last item is performed, the cell activity that stores the last item is no longer inhibited by other cells, so it becomes more active than previously active cells.

3.14 Recurrent shunting networks also control beat and gamma oscillations

Remarkably, the oscillatory circuits that control musical beat (see Section 5) are also recurrent on-center off-surround networks with cells that obey shunting dynamics. What distinguishes them from working memories are the following kinds of differences:

> First, the beat oscillator does not include self-inhibitory interneurons to prevent cyclic performance, although they may include self-inhibitory interneurons as part of a recurrent shunting off-surround.
>
> Second, the beat oscillator is driven by a sufficiently large arousal, or GO, signal that converts it from a network that responds phasically to external inputs, into one that tonically, or persistently, oscillates. This distinction between phasic and tonic dynamics is also illustrated by the fact that the same corticospinal

circuitry can regulate standing, which is phasic, and walking, which is tonic. Sufficiently aroused working memories can also oscillate, albeit with faster alpha, beta, gamma, and theta oscillations, during perceptual and cognitive processing (Grossberg, 2017a, 2017c). This kind of oscillator can thus support a range of oscillatory periods, depending upon parameter choices. I will provide more detail about how this oscillator works in subsequent sections.

3.15 Item-Order-Rank: Numerical hypercolumns store lists with item repetitions

How do working memories store lists with repeated items, as in the list ABACBD? This competence is needed to store lyrics with repeated words, and melodies with repeated pitches. For this to occur, a working memory must be able to selectively store items in a way that is sensitive to their list position, or *rank*. Such a working memory is called and Item-Order-Rank, or IOR, working memory.

Cognitive data demonstrating sensitivity to rank include spoonerisms, during which words or syllables in similar positions, but in different words, are interchanged; for example, “hissed my mystery lesson” (Henson, 1998). Neurophysiological data from cells in prefrontal cortex also exhibit rank sensitivity (Averbeck et al., 2003; Barone and Joseph, 1989; Funahashi, Inoue, and Kubota, 1997; Inoue and Mikami, 2006: Kermadi and Joseph, 1995; Ninokura, Mushiaske, and Tanji, 2004).

How is rank information incorporated into an Item-Order-Rank working memory model that can store item repeats at arbitrary list positions? I proposed with my PhD student Lance Pearson (Grossberg and Pearson, 2008) that an Item-Order-Rank working memory in prefrontal cortex derives its rank selectivity via a parietal-to-prefrontal projection from the analog number map that exists in parietal cortex. This prediction built upon a model of the parietal number map called the Spatial Number Network, or SpaN, model (Figure 3.16), that I developed with my PhD student, Dmitri Repin Grossberg and Repin, 2003). SpaN simulates how the parietal number map may control the ability of animals and humans to estimate and compare small numerical quantities without requiring that they count these quantities with numbers (Naccache and Dehaene, 2001; Rickard et al., 2000). Such numerical estimation is used in many animals during foraging to decide which way to turn to acquire the best sources of food.

Figure 3.16a summarizes how SpaN works. Each event in a sequence activates a transient detector that generates a brief burst of activity in response to each input (see the series of rectangular inputs denoted by xz). These bursts are added up by an accumulator (see the series of steps denoted by y). This stored total input is broadcast uniformly across the *spatial number map*. Cells at different positions in the number map have thresholds and sensitivities that increase from left to right across the map (see the sigmoid, or S-shaped, signal functions). As a result, small inputs selectively activate cells near the left of the map, whereas larger inputs activate positions to the right. Increasingly large inputs activate a series of unimodal response profiles centered at positionally displaced positions toward the right.

Figure 3.16b shows the simulated unimodal response profiles, as well as their close fit to numerosity data collected during animal conditioning experiments. The response profiles of SpaN model parietal neurons are also matched by neurophysiological data of Nieder and Miller (2004a).

Nieder and Miller (2004b) also reported a projection in vivo from the parietal number map to the prefrontal cortex. In the SpaN model, this parietal–prefrontal projection uses the ordered number map in the parietal cortex to embed numerical *hypercolumns* within the prefrontal working memory. Figure 3.17 describes how the parietal number map projects to hypercolumns in the prefrontal cortex (see red activity profile and red pathways). Each item in the list is stored in a different position in its hypercolumn if it is repeated more than once. Each item’s hypercolumn representation is denoted by a circle in Figure 3.17, and each of its four hypercolumn positions is denoted by a pie-shaped region within this circle and is numbered from 1 to 4.

For example, consider how a numerical hypercolumn can store and perform in its correct order the short list ABAC: Item A is stored in positions 1 and 2 within its hypercolumn, item B is stored in position 1 within its hypercolumn, and item C is stored in position 1 within its hypercolumn. A primacy gradient of activity can store the temporal order of a short list, whether or not it has repeated items. In response to the list ABAC, for example, each of the successively activated map positions would have a progressively smaller activity stored in the primacy gradient.

One of the parsimonious properties of the model is that it can use a primacy gradient to control storage and ordered read-out from working memory for lists with repeats and without them. In either case, an Item-Order-Rank working memory can still have self-excitatory feedback from each cell population to itself, and a broad off-surround that inhibits all other populations equally across the hypercolumns. When self-inhibitory feedback inhibits the last-performed item, the next item is performed, as usual, until the entire list ABAC is performed.

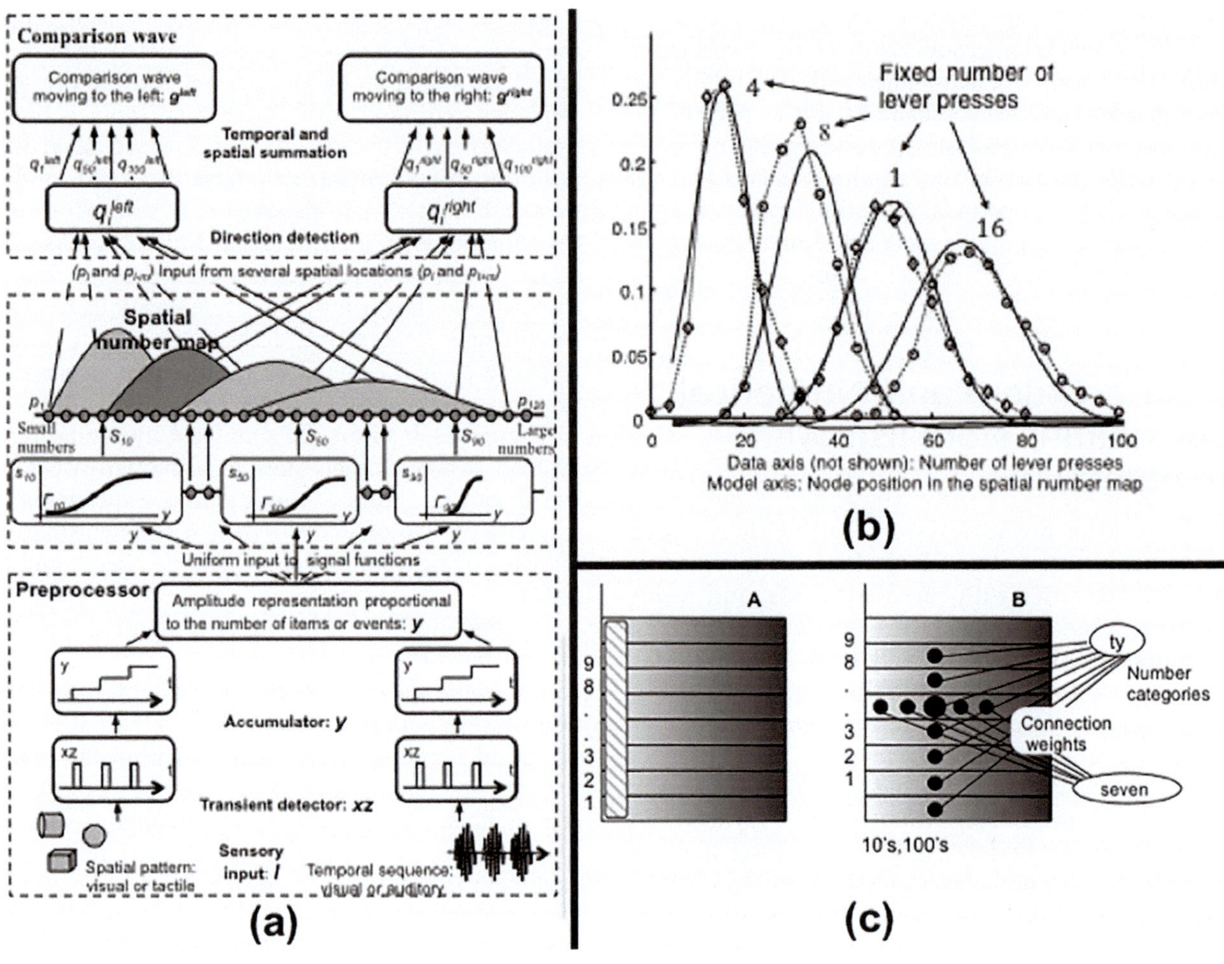

FIGURE 3.16 (a) The SpaN model simulates how spatial representations of numerical quantities are generated in the parietal cortex. See text for details. (b) Behavior numerosity data and SpaN model simulation of it. The responses to increasingly large numerical inputs activate different, but contiguous, positions in the analog number map, with larger numbers generating responses further to the right on the map. The larger variance of responses to larger numbers is called the Weber Law. All these properties obtain in both the data and the model simulation. (c) The Extended SPaN, or ESpaN, model extends SpaN to be able to learn and perform place value numbers. Interactions from the What cortical stream to the Where cortical stream accomplish this extension. In particular, previously learned phonetic categories in the What stream become associated with corresponding locations of the SpaN spatial number map in the Where stream. (cA) The striped area shows the location of the primary number map. This is extended into parallel strips, all of whose cells also respond to the inputs to the primacy number map. (cB) An example of where the association for *seven-ty* is learned in this strip map. The size of the solid circles encodes weight magnitude; the strongest association for *seventy* arises at the spatial location where both the associations for categories *seven* and *ty* are present. Reprinted with the author's permission from Grossberg and Repin (2003).

Such a cortical circuit can perform *any* list with repeated items in the correct order, whether of letters, lyrics, or pitches, just so long as it is short enough to be stored in working memory by a primacy gradient.

The numerical hypercolumns are represented by segmented circles in Figure 3.16 for convenience. They are more realistically represented by strips in a cortical map. Taken together, the totality of these strips across a cortical region provides our first example of a cortical design that I call a *strip map*. I will now summarize some surprisingly varied functions of strip maps that are all useful for understanding music.

3.16 Keeping time by counting: Place-value numbers

Sometimes we keep time in music by counting using numbers. Grossberg and Repin (2003) showed how number names may get associated with the spatial number map (Figure 3.16c). First, number names get learned in the What, or ventral, cortical stream as part of language development and learning (Figure 1.32). Then these number names get linked via associative learning with a corresponding spatial numerical map representation in the Where, or dorsal, cortical stream (Figure 1.32).

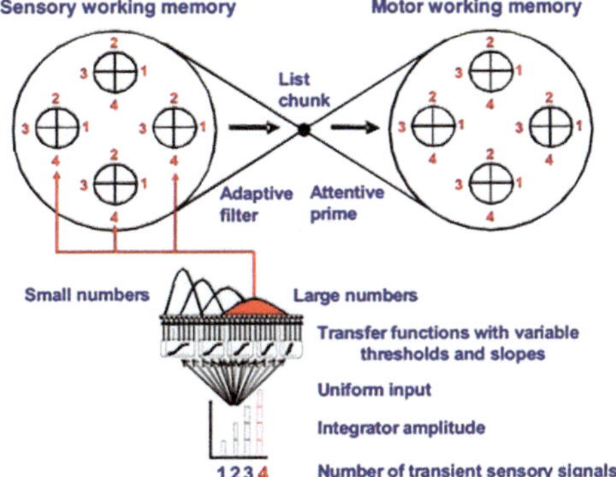

FIGURE 3.17 Grossberg and Pearson (2008) proposed the LIST PARSE model for encoding a conjunction of item, order, and sequence position, or rank, in sensory and motor working memory. Cells in the sensory and motor working memories need a second input, beyond their sensory and motor inputs, that codes rank information in order to fire. The model proposes that number maps in parietal and frontal cortex provide this positional information. The circles with numbers represent cortical hypercolumns, each coding a different sensory or motor event, with positions (for illustration) 1, 2, 3, and 4. The sensory working memory supports learning of list chunks. The list chunks learn to attentively prime the Item-Order-Rank motor working memory during reactive performance of a sequence of actions. During planned performance, cells in the motor working memory fire their motor commands when they receive a list chunk priming signal and the correct positional, or rank, input from the corresponding number map. The lower part of the figure illustrates how this number map works. In it, transient inputs in response to each sensory event are integrated into a signal whose size is proportional to the total number of sensory inputs that have occurred in the sequence. This integrated signal generates a uniform input to all the cells in the parietal number map. The signal functions with variable thresholds and slopes in the number map cause distinct populations of cells to get activated as a larger number of transients is stored. These distinct populations provide a spatial representation of the numbers. The number map cells, in turn, broadcast their positional information to the sensory working memory. A similar scheme occurs in the motor working memory. Reprinted with the author's permission from Grossberg and Pearson (2008).

Grossberg and Repin (2003) explained and simulated how this associative What-to-Where map can learn representations of larger numbers, called *place-value numbers*, such as "twenty," "thirty," or "one hundred." To accomplish this extension, the primary spatial number map (shown as a vertical striped bar in Figure 3.16c(A)) is extended to a set of horizontal strips, each of which extends the representation of its number in the primary map. Then associative learning from the number names to the number map takes place (Figure 3.16c(B)), as illustrated by the number seventy. These horizontal strips constitute another example of a strip map, a general brain design that will be seen below to have multiple functions in music.

3.17 Strip maps: A cortical design with multiple uses in music

Both place-value numbers and working memories with repeated items are coded in the cerebral cortex using strip maps. Indeed, strip maps occur throughout our brains. The orientation columns within cortical area V1 are the most famous example of strip maps (Figure 3.18). Here each strip, or hypercolumn, represents a map position whose different locations in the strip respond selectively to objects with different orientations at the position that it codes. The entire cortical map contains multiple strips that together are sensitive to all visible positions (Hubel and Wiesel, 1962, 1963).

In general, a strip map represents one feature throughout its extent (e.g., position), as well as another feature in an ordered array of positions throughout the strip (e.g., orientation). In addition to strip maps that represent orientation columns, place-value numbers, and IOR cognitive working memories that can code repeated items, strip maps also occur in our neural models of auditory streaming and speaker-normalized speech (Ames and Grossberg, 2008; Grossberg et al., 2004).

All of these strip maps are relevant to music. For example, auditory streams can separate and track the

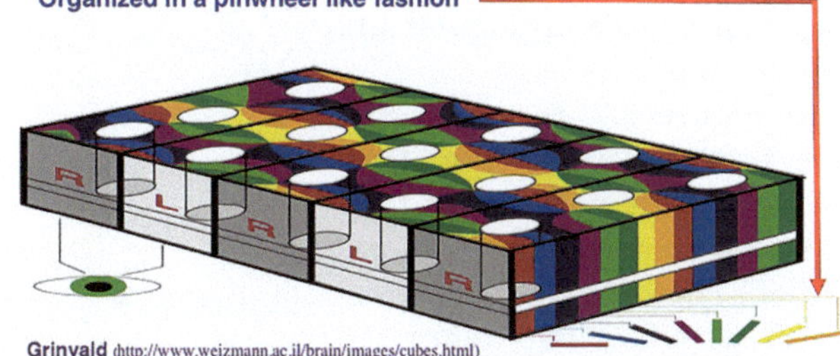

FIGURE 3.18 The hypercolumns that occur in LIST PARSE are a special case of strip maps that also occur in the ARTSTREAM and NormNet models. All of these are variants of a cortical design that also creates ocular dominance columns in the visual cortex. Reprinted with the author's permission from Grossberg (2021).

instruments that we hear during a string quartet. Place-value numbers can be used to identify a piece of music, such as the BWV (Bach Works Catalogue) number of a particular piece of music by Bach, or the page number of a particular composition in a book of music, or the fingerings for a pianist to use when playing a piece of music (Figure 3.1). The exposition in the next section will show how specialized strip maps can also represent musical lyrics, pitches, and rhythms.

4 Storing, Learning, and Performing Musical Lyrics, Pitches, and Rhythms

4.1 Storing and performing lyrics during resonant learning with its list chunk

This section will propose how a phrase of lyrics or melody can be stored in working memory and performed in the correct order with a regular rhythm whose delays can be learned by counting the beats between notes. I will propose how the counting process for one word or note delays the performance of the next word or note for the correct duration.

This explanation builds upon the fact that an Item-Order-Rank working memory can use numerical hypercolumns to store words or notes that may occur in multiple positions within a short enough phrase; for example, "our true love was true." To be learned and performed in the correct order, a phrase of lyrics or notes must be short enough to be stored as a primacy gradient (Figures 3.1, 3.7, and 3.10a).

In Figure 3.7, the rehearsal wave acts at a processing stage that occurs *after* the items that are stored by a working memory gradient compete to choose the largest activity. The rehearsal wave allows this winning activity to be read out for performance, while it begins to self-inhibit its own working memory activity using a specific inhibitory feedback pathway. Figure 3.19 uses the above foundation to begin to explain how the *lyrics* of a song can be performed with a learned and possibly variable rhythm, as illustrated by the songs described above. The same mechanisms and circuits can be used in a parallel architecture to explain how the *melody* of a song can be performed along with the lyrics.

Figure 3.19a repeats the circuit in Figure 3.7 using a notation that will be convenient for representing a larger cognitive architecture in the auditory cortex whose rhythmic performance is regulated by volitional control from the basal ganglia and cerebellum. In Figure 3.19a, excitatory connections are depicted by green arrows and inhibitory connections are depicted by red connections that end in red disks. The two cells that are represented by black disks at the top of Figure 3.19a also interact via a recurrent shunting on-center off-surround network and can thus store an activity pattern in working memory. The blue vertical bars of unequal height illustrate an activity pattern that is currently stored in working memory by this network. Everything that is written below generalizes to networks with an arbitrary finite number of cells, not just two.

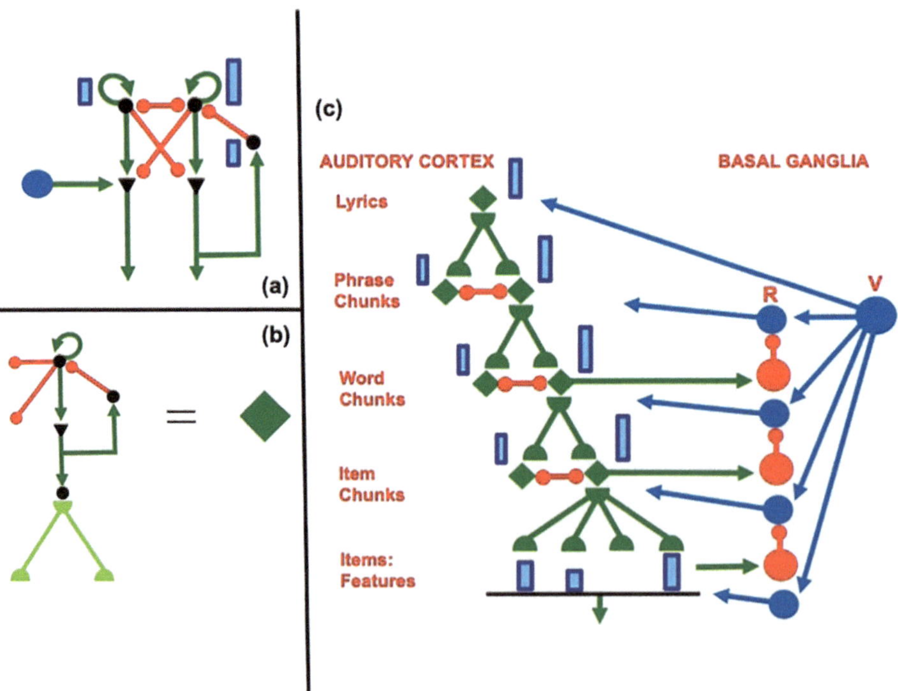

FIGURE 3.19 (a) When a rehearsal wave R (blue disk) turns on, the item that is stored in working memory can be rehearsed while it self-inhibits its working memory representation. Relative activity amplitudes are represented by the sizes of vertical blue rectangles. Triangular cells are polyvalent. (b) Filled diamond summarizes key stages in choosing items to be rehearsed and associating them with bottom-up adaptive filter and learned top-down expectations. (c) Recursive read-out, under volitional control, from the hierarchy of processing stages that represents the lyrics of a song. Green represents excitatory connections. Red represents inhibitory connections. Blue disks represent volitional gain control signals. See the text for details. Reprinted with the author's permission from Grossberg (2023).

The working memory network in Figure 3.19a outputs via a nonrecurrent, or feedforward, on-center off-surround network to a pair of cells that are denoted by black triangles. This feedforward competitive network chooses the larger activity that is stored in working memory for further processing, while inhibiting the smaller activity. The competitive network allows only one cell at a time to receive a positive input at the next processing stage; namely, the cell beneath the one that has been storing the larger activity in working memory.

The black triangles denote *polyvalent* cells that can fire only when they receive simultaneous inputs from specific and nonspecific input sources. The specific input comes from the working memory network. It is "specific" because it corresponds to a particular chunk that is being stored in working memory. The nonspecific input comes from the blue cell (population). Activating this blue cell occurs when rehearsal is desired. When activated, the blue cell releases a rehearsal wave across the entire network. In the figure, this rehearsal wave is represented by the horizontal green arrow. Only one rehearsal wave output pathway is shown. In fact, the blue cell sends equal excitatory signals to all of the polyvalent cells whenever it fires, in keeping with it being a source of nonspecific arousal.

A polyvalent cell that receives both a specific and a nonspecific input can fire. It can then send a signal along its output pathway, which is denoted by a downward-facing green arrow. This signal also activates a recurrent specific inhibitory interneuron, which shuts off the working memory cell that activated it. When this working memory cell is silenced, so too are its feedforward inhibitory signals that competitively silence outputs from the rest of the working memory network. Because the rehearsal wave is brief, all additional outputs from the working memory network are prevented.

The circuit on the left hand side of Figure 3.19b depicts a single cell in the working memory network and all of its output connections. This unit occurs multiple times in larger cognitive architectures. It includes bidirectional adaptive pathways at the next processing stage. These pathways are drawn in light, rather than dark, green because the rest of the circuit with which these adaptive pathways interact is not shown. Both the bottom-up and top-down pathways are adaptive and are thus denoted by light green hemidisks. This notation will also be used in larger architectures.

In order to facilitate drawing such architectures, the circuit in Figure 3.19b is denoted in various other figures as the filled green diamond on the right hand side of Figure 3.19b. Figure 3.19c shows how this circuit is repeated in one such architecture, which I will discuss below.

4.2 Learning lyrics in a hierarchical cortical architecture: Recursive read-in

Previous articles have modeled how sequences of items that are stored in working memory can be used to drive the learning of list chunks at the next processing stage (Bradski, Carpenter, and Grossberg, 1994; Grossberg and Kazerounian, 2011, 2016; Kazerounian and Grossberg, 2014). A volitional gain control source initiates this storage process. This gain control process can act iteratively to support learning of a hierarchy of ever-more-complex sequential representations. Figure 3.6 summarizes a three-level processing hierarchy that can learn phrases and sentences with repeated items. The volitional gain control process that regulates such storage and learning was omitted from this figure. The top level in Figure 3.6 corresponds to the Phrase Chunks level in Figure 3.19c.

4.3 Rehearsing lyrics using hierarchical cortical architecture: Recursive read-out

Figure 3.19c incorporates the network components in Figures 3.19a and 3.19b into a hierarchical cortical architecture that can represent working memory storage and fluent read-out of the lyrics of an entire song using multiple cortical processing areas.

The top-most list chunk in this figure is a Lyrics Chunk. When its top-down learned expectation pathways are activated by a rehearsal wave R, it reads out Phrase Chunks that are stored in working memory as a primacy gradient (for sufficiently short songs). When the top-down learned expectation pathway of the most active Phrase Chunk are activated by a rehearsal wave R, it reads out Word Chunks that are stored in working memory as a primacy gradient of the words in that phrase. Similarly, when the top-down learned expectation pathway of the most active Word Chunk is activated by a rehearsal wave R, it reads out the Item Chunks which spell out that word, again in a primacy gradient. Each Item Chunk, in turn, can read out the distributed pattern of features that it codes. This description clarifies that the recursive read-out of primacy gradients can regulate performance of lyrics that are much longer than any one primacy gradient can accomplish.

As shown in Figure 3.19c, each rehearsal wave node R is quickly inhibited by activation of the processing level whose chunk it activates. These inhibitory signals are activated by inhibitory gain control signals from the red disks that alternate with the blue rehearsal wave sources R. For example, activating a Phrase Chunk sends excitatory signals to its Word Chunks. The most active Word Chunk is chosen by the circuit in Figures 3.19a and 3.19b. When that Word Chunk also receives a rehearsal wave from the rehearsal node R at its level, it can fire. In addition to sending top-down excitatory priming signals to the Item Chunk level, it also activates the inhibitory gain control node at its level, which inhibits the rehearsal node R at the Phrase Chunk level and thus terminates read-out of the next-most-active Phrase Chunk. This process of brief activation occurs at each level, terminated by an inhibitory gain control signal at the level just below.

When all the Items of an Item Chunk are rehearsed, the Item level can no longer activate an inhibitory gain control signal. Then a rehearsal wave can activate the currently most active Item Chunk, whose Items can be rehearsed in the same way. This process continues until all the Item Chunks of a Word Chunk are rehearsed. Then the next Word Chunk can be read out, and the process repeats itself until all the Word Chunks of a Phrase Chunk are rehearsed. And so on, up the hierarchy until all the lyrics are performed.

BRAIN SOURCES OF VOLITIONAL SIGNALS

The volitional rehearsal processes V and R in Figure 3.19c are regulated by the basal ganglia, whose substantia nigra pars reticulata (SNr) opens gates that release perceptual, cognitive, and motor processes (Alexander and Crutcher, 1990; Alexander, DeLong, and Strick, 1996; Chapin et al., 2010; Fujiok, Zendel, and Ross, 2010; Grahn and Brett, 2007, 2009; Grahn and Rowe, 2009, 2013; Grossberg, 2016; Hikosaka and Wurtz, 1983, 1989; Kung et al., 2013; Mink, 1996; Mink and Thach, 1993). Whether the nonspecific input is excitatory, as in the simplified circuits of Figure 3.19, or an inhibitory signal whose activation disinhibits a tonically inhibitory gating signal, as occurs throughout the basal ganglia (Alexander and Crutcher, 1990; Alexander, DeLong, and Strick, 1996), is an important detail that is modeled by Brown, Bullock, and Grossberg (2004) and Silver et al. (2011) in large-scale neural architectures wherein the basal ganglia regulate learning and performance of observable behaviors (Figure 3.15). Activating R is controlled by the basal ganglia *direct pathway*, whereas inhibiting R is controlled by the basal ganglia *indirect pathway*.

POLYVALENT CELLS: FROM INVERTEBRATE COMMAND CELLS TO HUMAN BRAINS

Polyvalent cells that fire in response to converging specific and nonspecific inputs have been a design feature in brains for many thousands, if not millions, of years, ranging from command cells in invertebrates, like the ones that control swimming by crayfish swimmerets, and the pattern generators that control songs in birds (Figure 3.20), to cells that regulate reinforcement learning, motivated attention, and decision-making in humans (Figure 1.30). How rewards and punishments interact with polyvalent cells to modulate our learning of concepts, values, and actions has guided a lot of my neural modeling work for over 50 years; for example, Grossberg (1970, 1971a, 1971b, 1972a, 1972b, 1974, 1978a, 1982, 1984, 2021).

4.4 Auditory streaming, speaker normalization, and learning what lyrics mean

Before discussing how lyrics can be learned and performed at different rhythms and beats, I will summarize in greater detail how our neural models of how auditory streaming and speaker normalization work in order to mechanistically clarify how lyrics, or indeed all auditory communications, including language, can be learned by listening to a speaker whose sounds are uttered with different acoustic frequencies than our own voice produces. I will review some basic concepts to make the addition of illustrative mechanistic details self-contained.

Speaker normalization enables us to understand speech spoken by children, women, and men in different frequency ranges than our own. In particular, when babies babble sounds, they hear their own babbled sounds and use them to learn an associative map between their heard sounds and the motor commands that generated them. This kind of interaction is called a *circular reaction* by Piaget (1963). Babies use this associative map to learn how to imitate adult language sounds and thus to begin learning language.

Speaker normalization makes this possible (Figure 3.5). Without speaker normalization, the sounds that are produced by women and men could not activate the associative map that a baby learns, since it would be restricted to the sound frequencies that a baby can babble and the motor commands that made them. This restriction is overcome by normalizing the sounds that the baby babbles *before* the processing stage where the associative map is learned. Then the associative map can be learned between the normalized babbled sounds and the baby's motor commands that caused them. Sounds that are uttered by women and men in different frequency ranges are also normalized before they reach the learned associative map. As a result, their sounds can also activate the associative map to enable the baby to begin to imitate and learn the language utterances of other individuals. These language utterances include the lyrics of songs.

Before a baby or adult can process the sounds from any acoustic source, such as a speaker or singer, *auditory*

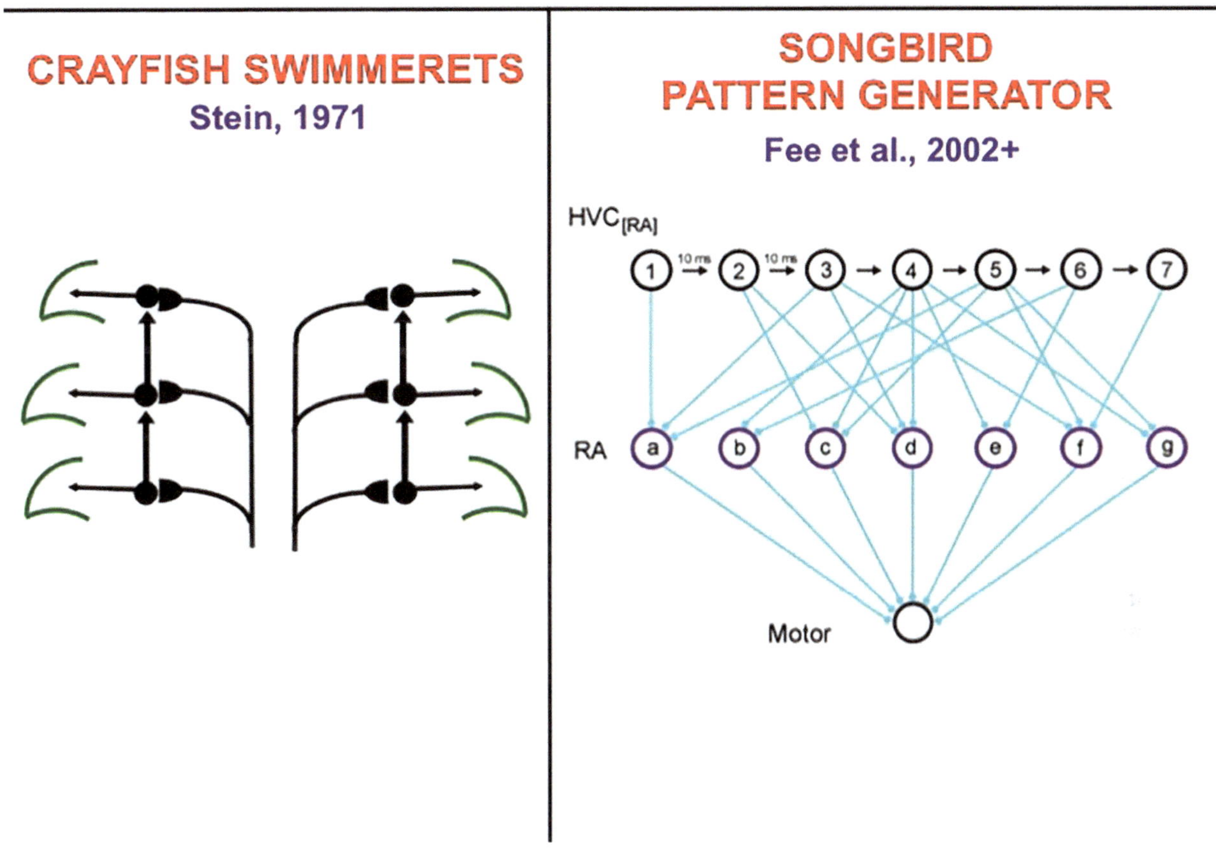

FIGURE 3.20 (left column) An early embodiment of nonspecific arousal was as a command cell in such primitive animals as crayfish. (right column) The songbird pattern generator is also an avalanche. This kind of circuit raises the question of how the connections self-organize through development and learning. Reprinted with the author's permission from Grossberg (2021).

streaming must first occur (Figure 3.2). Auditory streaming is also called *auditory scene analysis* (Bregman, 1990). It enables the sounds from multiple acoustic sources, including both voices and instruments, to be separated and tracked through time, including through intervals of noise or overlapping frequencies from multiple acoustic sources. Only after acoustic sources like voices are separated can they be individually normalized.

The above facts raise the following question: How can brain evolution be smart enough to discover the processes of auditory streaming, and the speaker normalization that follows it? Our modeling work has led to the parsimonious proposal that auditory streaming and speaker normalization both seem to use homologous neural circuits. The ARTSTREAM model of auditory streaming (Figure 3.2; Grossberg et al., 2004) simulates how auditory streams are separated using strip maps, asymmetric competitive circuits, and ART category learning circuits. The circuits within the NormNet model of speaker normalization (Figure 3.21; Ames and Grossberg, 2008) replicate, in specialized form, the auditory streaming circuits whose outputs they process. NormNet hereby clarifies how auditory streaming supports speech normalization.

This proposal offers a parsimonious solution to the speaker normalization problem because only after acoustic sources, including speakers, have been separated from one another by streaming can they be speaker-normalized. Likewise, it is only after visual figure-ground separation occurs that invariant visual object categories can be learned (see Chapter 2 and Figure 2.18 to Figure 2.22).

NormNet uses strip maps (Figure 3.21a) to simulate the transformation from speaker-dependent to speaker-normalized language representations (Figure 3.21b). After speech is transformed to become speaker-invariant (that is, normalized) and rate-invariant, it is in a form where our brains can learn and recognize language *meanings* from multiple speakers—without having to relearn them for each speaker and speech rate. Speaker-dependent speaker utterances are converted into speaker-normalized spectral representations before they are learned as categories using ART circuits, which can quickly learn and stably remember them using both bottom-up and

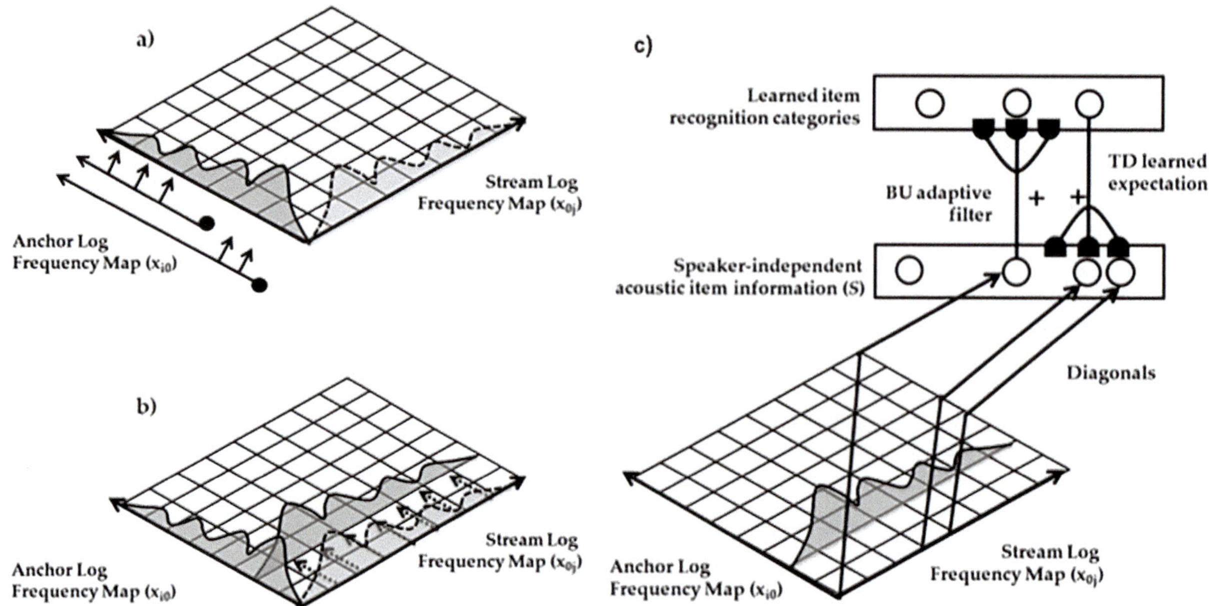

FIGURE 3.21 The NormNet model shows how speaker normalization can be achieved using specializations of the same mechanisms that create auditory streams. See the text for details. Reprinted with the author's permission from Ames and Grossberg (2008).

top-down adaptive pathways (Figure 3.21c). These speech item categories are then input to, and stored by, an Item-Order-Rank Speaker-Independent Working Memory (Figure 3.5) whose primacy gradients can be categorized into item, word, and phrase list chunks (Figure 3.19c).

NormNet was tested by simulating synthesized steady-state vowels from the Peterson and Barney (1952) vowel database and achieved accuracy rates similar to those achieved by human listeners. Vowels are the speech sounds that are most similar to musical pitches. Much further modeling work needs to be done to apply these streaming and normalization models to process conversational speech, and the great variety of musical lyrics and melodies.

4.5 Changing musical key: Speaker normalization and relative pitch

A basic issue in music is how to play or sing the same piece of music in a different key. I propose that a striking homology exists between normalizing speech and changing musical key. Just as speaker normalization uses a variant of the circuit that controls auditory streaming, changing a musical key uses a variant of the circuit that controls speaker normalization. Both processes shift all frequencies by a given amount, whether for purposes of speaking language or performing music. In Figure 3.21, this frequency is called the *anchor frequency*. Moreover, auditory streaming, speaker normalization, and changing musical key all use strip maps.

By normalizing speech, one can recognize a language utterance independent of its absolute pitches. By changing musical key, one can recognize a melodic utterance independent of its absolute pitches. Speaker normalization and changing musical key thus seem to use homologous circuits that each specialize mechanisms of auditory streaming. This homologous mechanism also makes it possible for the meanings of words sung in a different key to be understood. In both processes, one can still hear the speech or music frequencies as they are uttered, while also recognizing their invariant meaning (Figure 3.5), via parallel cortical streams.

4.6 Rehearsing a phrase of pitches in a melody at the correct rhythm

Having summarized how the *order information* of both lyrics and pitches may be stored in their respective working memories, the next issue is how they are performed at a desired *rhythm*. This issue will be discussed for the case of lyrics, leading to the ability to perform the words of a song at a learned rhythm. The same kind of analysis applies to the pitches with which the song is sung.

As I noted in Section 1.7, a phrase of lyrics or melody can be stored in a working memory and performed with a rhythm whose variable delays can be regulated by counting the number of beats between notes, while inhibiting the performance of the next word or note for the correct duration until the count for the last word or note is complete. Figure 3.16 summarizes part of the circuit in the parietal cortex that controls counting. The production of counts, whether vocally or subvocally, requires that this circuit interacts with prefrontal cortical circuits that regulate sequential performance of any list of items (Section 3) and with motor cortical circuits that read out the counting behaviors. As noted in Section 4.3, all of these processes are regulated by the substantia nigra pars reticulata (SNr) of the basal ganglia, which opens gates that release perceptual, cognitive, and motor processes.

The SNr also prevents the performance of behaviors whose gates are not open while counting is occurring. This happens using competitive interactions among the gating pathways. In particular, read-out of the next word in a lyric is prevented while counting the number of beats during which the current word is being performed.

4.7 Adaptively timed classical conditioning: Cerebellar spectral timing

How are temporally discrete counts converted into a learned duration that is continuously read out during skilled performance while a given word is being performed? This process is controlled by adaptively timed learning within the cerebellum, which is known to be important in learning the adaptively timed behaviors during musical performance (Doyon, Penhune, and Ungerleider, 2003; Ivry et al., 2003; Penhune and Doyon, 2005; Ramnani and Passingham, 2001; Sakai, Ramnani, and Passingham, 2002; Thach, 1998; Zatorre, Chen, and Penhune, 2007).

Adaptively timed learning by the cerebellum builds upon the Spectral Timing model, wherein a spectrum of cells, each of which reacts at a different rate, can together, as a population, learn to adaptively time movements that occur over a time span of hundreds of milliseconds or several seconds (Figure 3.22b). My colleagues and I have shown that spectral timing circuits can also successfully model adaptively timed learning by the hippocampus (Grossberg and Merrill, 1992, 1996; Grossberg and Schmajuk, 1989) and the basal ganglia (Brown, Bullock, and Grossberg, 1999, 2004), which contribute to dance movements in different ways, the former by regulating timed movements through space, and the latter by coordinating when perceptual, cognitive, emotional, and movement gates open and close.

The Spectral Timing model was progressively developed until it could quantitatively simulate experimental facts about the biochemical processes that span times as

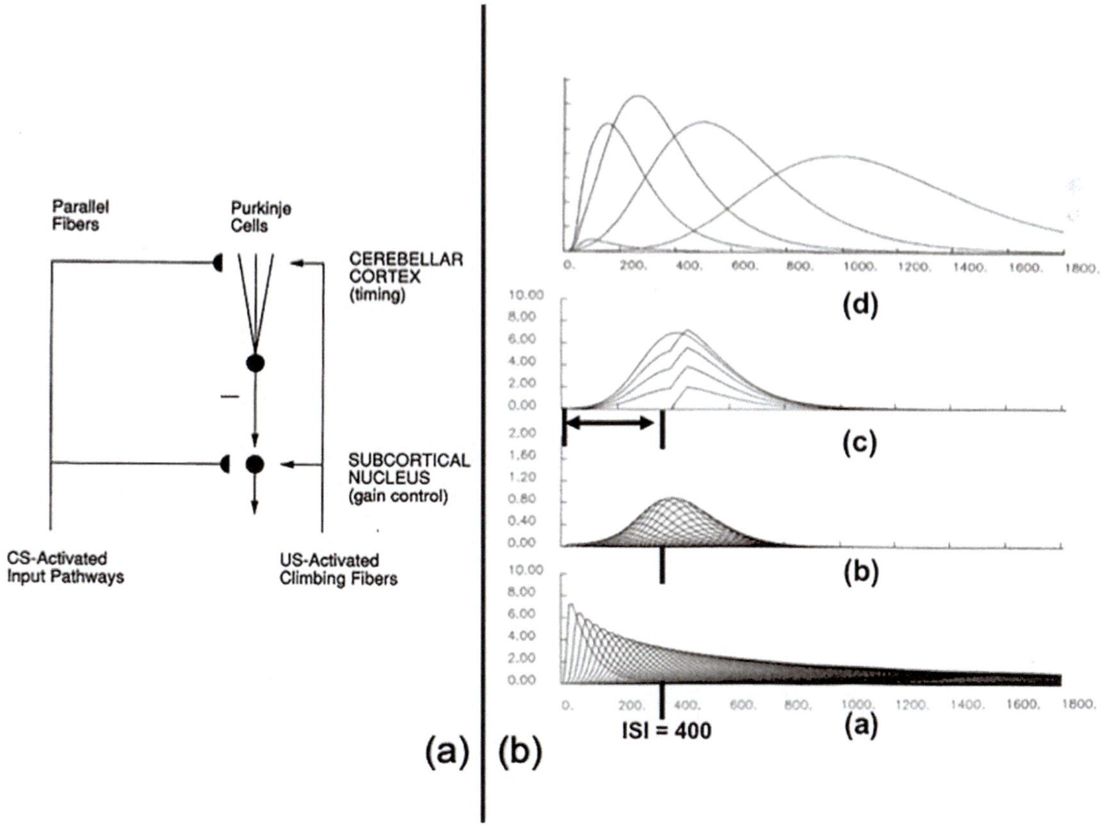

FIGURE 3.22 (a) Schematic of how learning in the cerebellum occurs. Reprinted from the author's permission from Grossberg and Merrill (1996). (b) Summary of a computer simulation showing how the cerebellum learns adaptively timed responses over a duration of hundreds of milliseconds using the population response of a tunable spectrum of cells whose individual peak responses occur at different times during this time interval. Reprinted with the author's permission from Grossberg and Schmajuk (1989). See the text for details.

long as hundreds of milliseconds, or even a few seconds, even though neuronal STM traces respond much more quickly. My PhD student John Fiala and I, with help from our colleague Daniel Bullock (Fiala, Grossberg, and Bullock, 1996), have shown how the metabotropic glutamate receptor, or mGluR, system accomplishes this feat, with the spectrum specified by a Calcium gradient (Figure 3.23). We used this mGluR model to quantitatively simulate behavioral, neuroanatomical, neurophysiological, biophysical, and biochemical data about adaptively timed learning by the cerebellum. The spectral timing circuits in basal ganglia, cerebellum, and hippocampus all share variations of the same circuit design, and all of these brain regions contribute to our understanding of music, as I noted above.

It seems that adaptive timing using mGluR is an ancient discovery in brains and even non-neural tissues. Even in non-neural tissues, these timed spectra are due to quantal release of Ca2+ gradients from inositol 1,4,5-trisphosphate (IP3)-sensitive intracellular stores, as in Figure 3.23. One non-neural example of a timed spectrum occurs in HeLa cancer cells, which are a cell type in an immortal cell line that came from cervical cancer cells of Henrietta Lacks on February 8, 1951 (Masters, 2002). The importance of HeLa cells in medical research cannot be overstated. For example: "Over the past several decades, this cell line has contributed to many medical breakthroughs,

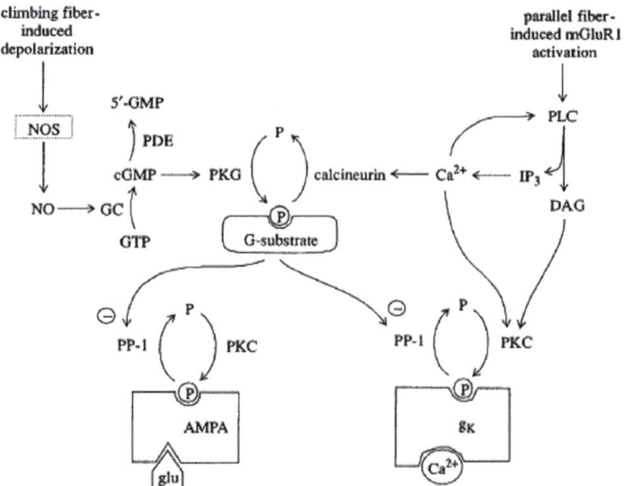

FIGURE 3.23 Cerebellar biochemistry that supports the hypothesis of how the metabotropic glutamate receptor (mGluR). supports adaptively timed conditioning at cerebellar Purkinje cells. AMPA, Amino-3-hydroxy-5-methyl4-isoxazole propionic acid-sensitive glutamate receptor; cGMP, cyclic guanosine monophosphate; DAG, diacylglycerol; glu, glutamate; GC, guanylyl cyclase; gK, Ca2+-dependent K+ channel protein; GTP, guanosine triphosphate; IP 3'inositol,4,5-trisphosphate; NO, nitric oxide; NOS, nitric oxide synthase; P, phosphate; PLC, phospholipase C; PKC, protein kinase C; PKG, cGMP-dependent protein kinase; PP-I, protein phosphatase-i. Reprinted with the author's permission from Fiala, Grossberg, and Bullock (1996).

from research on the effects of zero gravity in outer space and the development of polio and COVID-19 vaccines, to the study of leukemia, the AIDS virus and cancer worldwide." (https://www.hopkinsmedicine.org/henrietta-lacks/importance-of-hela-cells) Other mGluR examples include the "puffs" in the oocytes, or developing eggs, of *Xenopus*, a type of aquatic frog (Yao, Choi, and Parker, 1995) and the "sparks" in cardiac myocytes, or mature contractile muscle cells (Cannell, Cheng, and Lederer, 1995; López-López et al., 1995).

The role of mGluR in adaptively timed learning raises the question of what can go wrong if the mGluR system does not function normally? My PhD student Devika Kishnan and I explained how a breakdown in mGluR functioning explains data about abnormal timing in autistic individuals, as well as individuals with the related Fragile X syndrome (Grossberg 2021; Grossberg and Kishnan, 2018).

Figure 3.22a summarizes the anatomy within the cerebellum whereby a learned sensory or cognitive representation (labeled CS-Activated Input Pathways) sends trainable signals to both the cerebellar cortex and its subcortical nuclei. In this figure, the sensory representation selectively responds to a conditioned stimulus, or CS, during a classical conditioning experiment. Classical conditioning is an ancient kind of learning that occurs across multiple species, ranging from invertebrates like *Aplysia Californica* through rabbits to humans (Buonomano, Baxter, and Byrne, 1990; Buonomano and Mauk, 1994, Carew, Walters, and Kandel, 1981; Kamin, 1968; Millenson, Kehoe, and Gormezano, 1977; Smith, 1968; Woodruff-Pak, Papka, and Ivry, 1996).

During classical conditioning, the CS may be chosen to be any perceived signal with no prior learned associations, such as a brief tone or light. The CS is then paired on multiple learning trials with an unconditioned stimulus, or US. During classical conditioning, the US may be a positive reward or negative punishment, such as food or shock. In either case, classical conditioning is a special case of *reinforcement learning*, with the reward or punishment acting like a positive or negative *reinforcer*.

The US occurs after a delay, or interstimulus interval (ISI), of between 50 to several hundred milliseconds after the onset of the CS. After sufficiently many trials, the CS can learn to activate some of the consequences that the US originally caused. Figure 3.22b shows a computer simulation in which the spectrum is trained with an ISI of 400 milliseconds. When all the spectrum outputs are summed, the resulting output signal generate its peak activation at a delay of 400 milliseconds. I will now explain how this happens in greater detail.

The trainable CS-activated signals to the cerebellar cortex are carried by parallel fibers that end in Purkinje cells (Figure 3.22a; Eccles, Ito, and Szentagothai, 1967;

Fiala, Grossberg, and Bullock, 1996). Learning that occurs at the parallel fiber/Purkinje cell synapses is called Long-Term Depression, or LTD, because pairing at Purkinje cells of a CS-activated signal along a parallel fiber with a teaching signal along a US-activated climbing fiber causes the adaptive weight at the end of the active parallel fiber to *decrease*, or be depressed (Ito and Kano, 1982). Because Purkinje cells are tonically active, a smaller learned signal from a parallel fiber will activate target Purkinje cells less. These Purkinje cells will consequently inhibit their target subcortical nuclear cells less, thereby *dis*inhibiting them and releasing a timed output signal from the subcortical nucleus to the cells elsewhere in the brain to which they project.

Figure 3.22b provides a more detailed insight into properties of adaptively timed learning at the Purkinje cell synapses. I call this kind of learning *spectrally timed* learning, or *spectral timing* for short (Fiala, Grossberg, and Bullock, 1996; Grossberg and Merrill, 1996), because a signal down the parallel fiber has multiple branches that activate a spectrum of cells, each of which reacts at a different rate (Figure 3.22b(a)). Learning occurs at the synapses of each cell's output pathway. More learning occurs in cells that are more active at times when the CS and US signals are paired. In the computer simulation results shown in Figure 3.22b, the most learning occurs at the interstimulus interval, or ISI, between the CS and US, which is 400 milliseconds. These learned weights multiply their signals in the spectrum, leading to a spectrum of learned signals that are largest at the ISI (Figure 3.22b(b)). When all of these signals are added up at a fixed ISI, unimodal output signals are created that peak at the ISI, such as those shown in Figure 3.22b(c) during the first five learning trials.

When the ISI is varied, the learning curves at multiple ISIs look like the curves in Figure 3.22b(d). These curves become broader as the ISI increases. This property is called the Weber Law (Smith, 1968). Note also that the envelope of all the curves has an inverted-U shape. The Weber Law and its envelope are a signature of spectral timing wherever it occurs in the brain.

4.8 Learning to perform timed lyrics using cerebellar spectral timing

This background enables us to mechanistically understand how spectral timing helps to time the performance of musical lyrics: When counting signals that occur regularly in time activate training inputs via climbing fibers to a cerebellar spectrum, then, as Figure 3.22 shows, the Purkinje cell output will learn to remain small throughout that timed duration during which it disinhibits its target cerebellar nuclear cells (Figure 3.21a).

How is the correctly learned timed duration linked to a given word in a song's lyrics? This requires an interaction between multiple brain regions, including the thalamus, cerebral cortex, cerebellum, and basal ganglia (Bostan, Dum, and Strick, 2013; Bostan and Strick, 2010; Gao et al., 2018; Grossberg and Paine, 2000; Hintzen, Pelzer, and Tittgemeyer, 2018). Figure 3.24a shows how a Phrase Chunk, which can code for a learned word, can accomplish this. Such a Phrase Chunk can read out its series of words into an Item-Order-Rank working memory. It can also, in parallel, read out an array of pathways that activate a band of parallel fibers in the cerebellum (Figure 3.24a). Before reaching the cerebellum, each pathway inputs to a polyvalent cell (black triangle) that also receives an excitatory input from a Word Chunk in working memory when it generates an output signal to be rehearsed. Convergent signals from the Phrase Chunk and the Word Chunk at this polyvalent cell can fire it, so that its output signal to the cerebellum only samples counting inputs when a particular word in a prescribed phrase of lyrics in a song is being rehearsed. In this way, for example, when the word "true" in the phrase "my true love is true" is read out of working memory, it can learn to be performed with two different durations at its two locations in the phrase.

Figure 3.24b shows that the disinhibited cerebellar nuclear cell inputs to a polyvalent cell (blue triangle) also receives a volitional signal V from the basal ganglia. When both inputs are on, the polyvalent cell can fire and support rehearsal of the word using the LIST PARSE working memory and list chunking circuit in Figure 3.9. LIST PARSE regulates timed read out of a word at variable speeds by using acceleration and deceleration signals to control a smooth transition to the next sound in the word as performance of the previous sound is almost completed. The execution of each sound is, in turn, regulated by a Vector Integration to Endpoint, or VITE, motor trajectory controller. This happens in the following way.

The Vector Integration to Endpoint, or VITE, model and its extensions (e.g., Bullock and Grossberg, 1988, 1991; Bullock, Grossberg, and Guenther, 1993) were originally used to explain and quantitatively simulate psychophysical and neurobiological data about how our arms reach for objects. VITE dynamics were later shown to play a role in controlling other types of movements than reaching movements, including movements to play musical instruments, and movements of the speech articulators that are used to sing (Bohland, Bullock, and Guenther, 2010; Guenther, 1995; Jacobs and Bullock, 1998).

The individual movements that are controlled by LIST PARSE in Figure 3.9 (right panel) use a VITE trajectory generator, whose circuit is summarized at the top of the figure. In such a circuit, a *present position vector* P_i computes the current position of a limb, and a *target position vector* T_i computes the desired final position of the limb. Vector P_i is subtracted from T_i (dashed red line

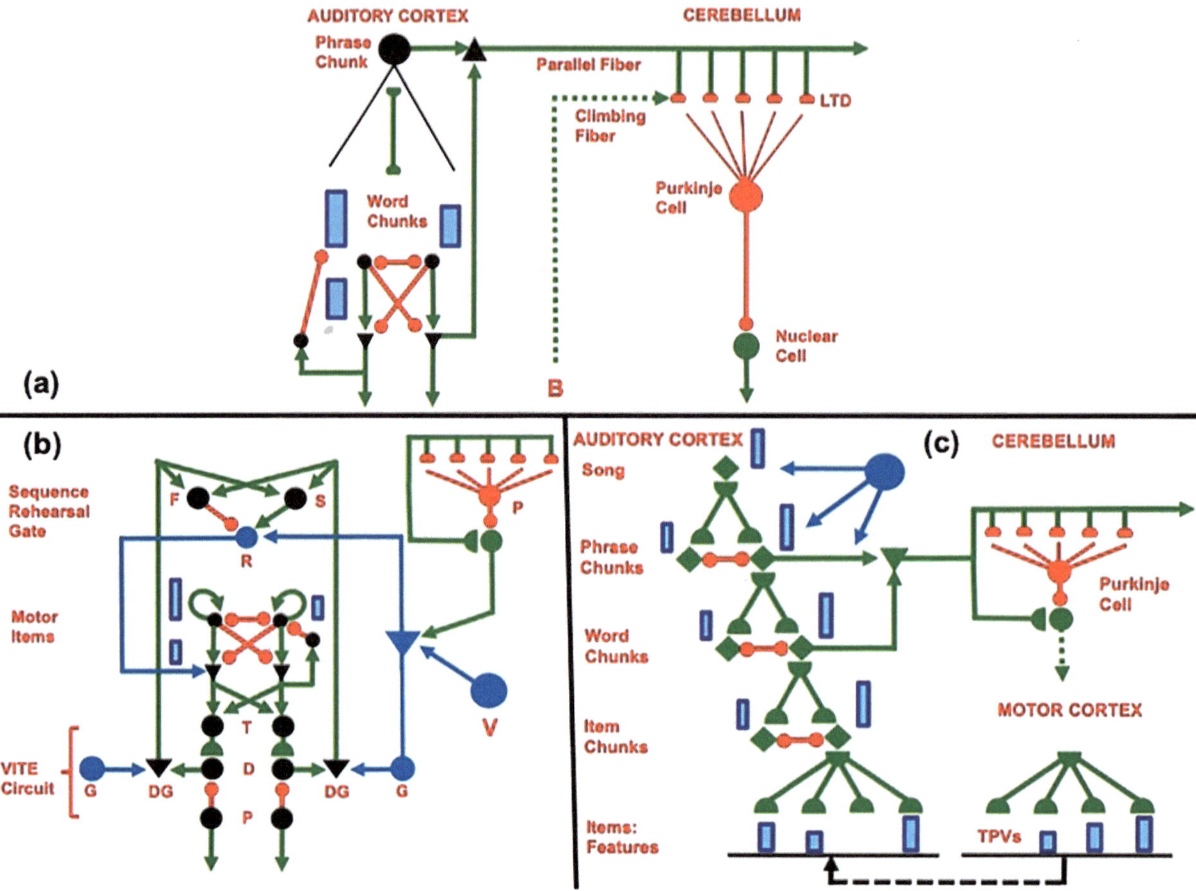

FIGURE 3.24 (a) Sequences of Word Chunks that are stored in working memory in the auditory cortex interact via adaptive bottom-up and top-down pathways in an ART circuit to learn Phrase Chunks. Each Phrase Chunk emits parallel pathways to polyvalent cells. Each polyvalent cell also receives a specific input from a particular Word Chunk. This polyvalent cell thus responds selectively to its Word Chunk when it is currently being performed as part of a prescribed phrase of lyrics. The polyvalent cell activates parallel fibers to the cerebellum where they learn an adaptively timed duration to perform that word in a lyric. Climbing fibers are teaching signals that drive adaptively timed learning using Long-Term Depression, or LTD, in (parallel fiber)—(Purkinje cell) synapses. During such a trained duration, Purkinje cell outputs are depressed and their target subcortical nuclear cells are disinhibited. (b) Circuitry whereby cerebellar adaptive timing regulates the learned durations between successive words in a lyric. (c) Circuitry whereby cerebellar adaptive timing regulates timed performance by the hierarchy of processing stages in the auditory cortex that represent a song's lyrics. See the text for details. Reprinted with the author's permission from Grossberg (2023).

between them) to compute a *difference vector* D_i that codes the direction and distance of the desired straight movement. Then D_i is multiplied by a volitional GO signal G to determine an *outflow movement speed vector* D_iG that is integrated through time by P_i until P_i equals T_i, at which time D_i equals zero. The desired target position has then been reached, so movement stops. Increasing (decreasing) the volitional signal G increases (decreases) movement speed without changing its final endpoint.

LIST PARSE can interact with a VITE trajectory generator to smoothly perform a sequence of straight movements at variable speeds, such as a skilled sequence of arm movements, dance, or performance of the sounds in a lyric or melody. Given that this can be achieved at variable speeds by varying the size of a volitional signal V, how do the circuits in Figure 3.9 know when an ongoing movement is almost complete, so that the next movement can begin to be rehearsed from working memory in a smooth way?

In Figure 3.9, the rehearsal signal R that controls read-out of successive movement commands from motor working memory is regulated by a difference signal (B—A) that is activated by cells that slowly (B) and more quickly (A) time-average outflow velocity signals DG. Signal (B—A) is computed at R by combining an excitatory signal from B with an inhibitory signal from A.

Due to the bell shape through time of the velocity vector DG for each straight movement, (B—A) is sensitive to whether DG is increasing, and thus accelerating in its growth, or decreasing, and thus decelerating. In particular, due to the bell shape of DG, the net signal (B—A) is initially negative, thereby keeping R off, but (B—A) becomes positive towards the end of the movement, thereby activating R and initiating a smoothly interpolated rehearsal of the next movement in the performance of the word. As indicated in Figure 3.24c, the next word in the phrase can then be rehearsed with its learned timing, until the entire musical phrase has been performed with the rhythm of the song.

The rate at which such movement trajectories may be completed in response to a GO signal G of a given size helps to determine the "resonance" frequency of performing a sequence of such movements. The use of a shared GO signal in many movement control circuits clarifies how preferred performance rates during walking and running are linked to rates of musical performance. Various studies have suggested that "people can synchronize their walking movements with music over a broad range of tempi, but that this synchronization is most optimal in a rather narrow range around 120 BPM [beats per minute]. This finding can be connected with previous findings indicating that most music has a tempo in this range" (Styns et al., 2007, p. 784).

5 Basal Ganglia Control of Periodic Dynamics, including Beats

A great deal has been written about the psychology and neurobiology of musical beat in both normal, or typical, individuals and clinical patients, including how it engages sensory, cognitive, emotional, and motor systems in both musicians and non-musicians (e.g., Beier and Ferriera, 2018; Buhmann et al., 2017; Jones, 1976; Lagrois, Palmer, and Peretz, 2019; Lenc et al. 2018; London, 2004; Rajendran et al., 2017; Slater et al., 2018; Tal et al., 2017; Toiviainen et al., 2019). Several models have been proposed to clarify the underlying mechanisms (e.g., Large, 2010; Large, Herrera, and Velasco, 2015; Lerdahl and Jackendoff, 1983). It has also been shown that interactions exist across rhythmic circuits that oscillate with very different frequencies, such as circadian and motor circuits (Ivanov et al., 2007).

I will not review this extensive literature here. Rather, I will discuss how the same kind of recurrent shunting on-center off-surround network that has been used earlier in this chapter to explain and simulate multiple types of data can also create beat-like oscillations whose frequency can be regulated by volitional signals. These oscillations can periodically move our fingers and legs, among other body parts, and enable them to "move with the beat."

5.1 Finger and gait oscillations: Shunting competition with slow inhibition

The circuit in Figure 3.25a is defined by a recurrent shunting on-center off-surround circuit. It can oscillate in response to external volitional inputs I_1 and I_2 because the inhibitory interneurons (black disks) in its off-surround vary more slowly through time than the self-excitatory cells (white disks) in its on-center. As a result, the recurrent on-center can rapidly amplify the activity of its cell, followed by slow inhibition that shuts it down, after which another cell can get activated. My colleagues Christopher Pribe, Michael Cohen, and I have used this kind of oscillator to simulate oscillatory movements of pairs of corresponding fingers in the two hands (Grossberg, Pribe, and Cohen, 1997).

In particular, the circuit in Figure 3.25a was used to simulate data of Yamanishi, Kawato, and Suzuki (1980). These authors studied a bimanual finger tapping task whereby subjects were required to start from a stable posture before periodically tapping keys in time to visual cues that occur at an increasing frequency. The model circuit accurately simulates the resulting behavioral data (Figure 3.25b), including the fact that anti-phase oscillations at low frequencies switch to in-phase oscillations at high frequencies, in-phase oscillations occur at both low and high frequencies, phase fluctuations occur at the anti-phase/in-phase transition, a "seagull effect" of larger errors occurs at intermediate phases, and oscillations slip toward in-phase and anti-phase when driven at intermediate phases. Also, when one of the periodic inputs I_1 or I_2 is missing, then the recurrent interactions continue to oscillate, as reported in data about the so-called missing beat phenomenon (Tal et al., 2017).

Figure 3.26a describes a variant of the recurrent shunting on-center off-surround network in Figure 3.25a. This network simulates a circuit in the spinal cord that controls the locomotion of quadruped animals (Pribe, Grossberg, and Cohen, 1997). Note in Figure 3.26a the separate, but interacting, recurrent networks that control fore limbs and hind limbs. Opening an appropriate gate in the SNr disinhibits a GO signal. Increasing this GO signal causes the circuit to generate a sequence of different movement gaits, as well as the gait transitions, that are familiar in different quadrupeds. Figure 3.26b shows the gaits and gait transitions that are familiar in cats; namely, walk-trot-pace-gallop. The model has been used to also simulate the gaits and gait transitions in humans (walk-run) and elephants (amble-walk).

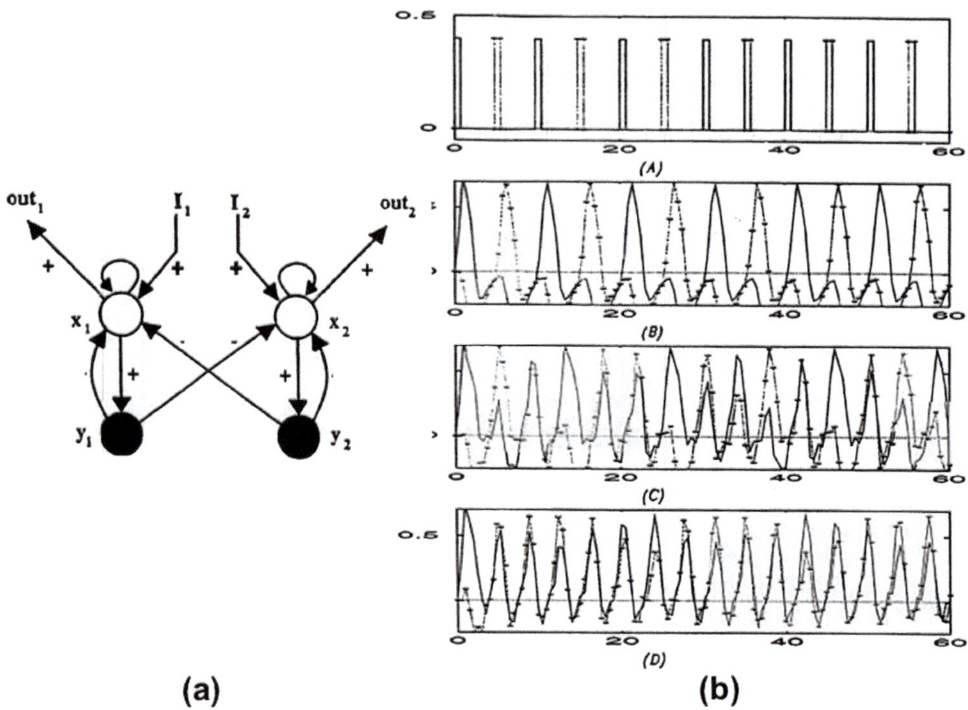

FIGURE 3.25 A recurrent shunting on-center off-surround network models data about synchronization of bimanual tapping. Bifurcation from anti-phase to in-phase oscillation occurs in response to anti-phase inputs of increasing frequency. Reprinted with the author's permission from Grossberg, Pribe, and Cohen (1997).

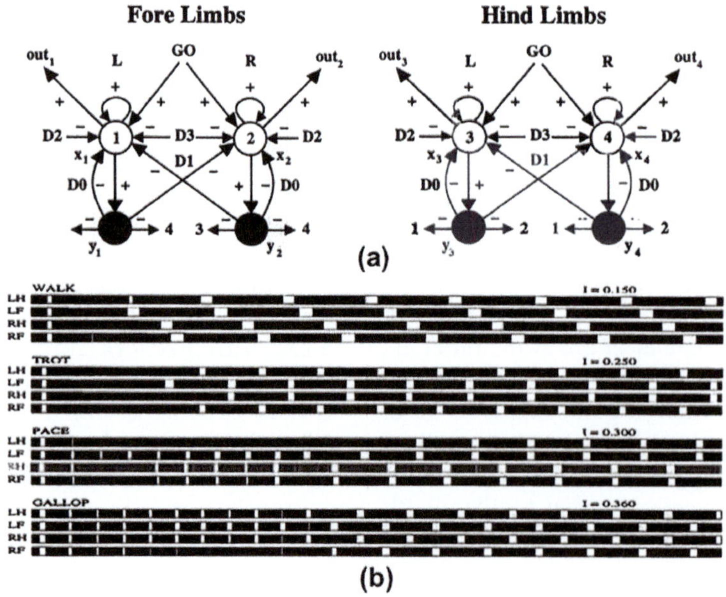

FIGURE 3.26 A pair of interacting recurrent shunting on-center (+ signs) off-surround (– signs) networks model the central pattern generator, or CPG, that controls quadruped movement gaits. Movement gait transitions are activated by an increasing volitional GO signal. Cells that emit excitatory signals are denoted by open circles. Inhibitory interneurons that emit inhibitory signals are denoted by closed disks. (a) Self-inhibitory feedback is labeled by the parameter D0, inhibition between forelimbs and between hindlimbs is labeled by D1, inhibition between matched forelimbs and hindlimbs is labeled by D2, and inhibition between crossed forelimbs and hindlimbs is labeled by D3. (b) Computer simulation of how increasing the GO signal, when it is combined with GO-modulated modulation of the inhibitory coefficients, triggers an ordered series of gaits (walk, trot, pace, and gallop). Reprinted with the author's permission from Pribe, Grossberg, and Cohen, 1997).

5.2 Similar design for beat and movement circuits explains how they synchronize

Recurrent shunting on-center off-surround circuits, such as those in Figures 3.25 and 3.26, can also be used to simulate musical beats. This shared design between musical beat and movement circuits, all coordinated with shared GO signals, clarifies how musical beat circuits can synchronize with motor circuits that enable us to move with the beat (Burger et al, 2014; Phillips-Silver and Trainor, 2007; Schaefer and Overy, 2015; Sowiński and Bella, 2013; Trainor, 2007; Wallin, Merker, and Brown, 2000; Zatorre, Chen, and Penhune, 2007). The observation that music can induce movement goes back at least to Aristotle who asserted that movement "follows" sound (Helmholtz, 1954; Phillips-Silver and Trainor, 2007).

5.3 Oscillators that can carry a beat

This section summarizes some technical details, including mathematical equations for a recurrent shunting on-center off-surround network that can carry a beat. As always, readers who prefer sticking to qualitative explanations can skip this section.

ELLIAS–GROSSBERG OSCILLATOR

Perhaps the simplest shunting competitive network that can be used as a central pattern generator, or CPG, for carrying a beat is the Ellias–Grossberg model that I developed with my PhD student, Samuel Ellias (Ellias and Grossberg, 1975). Here, a single cell with activity x excites itself and sends excitatory signals to, and receives inhibitory feedback from, the activity y of a slow inhibitory interneuron:

$$dx/dt = -Ax + (B - x)[f(x) + I] - Dxf(y) \quad (1)$$

$$dy/dt = E(x - y). \quad (2)$$

Our article Ellias and Grossberg (1975) published theorems that characterize how this system oscillates when the GO input I in equation (1) is sufficiently large. The oscillation frequency increases with I until it reaches a critical value, after which it decreases. The range where increasing I also increases oscillation frequency clarifies how an individual can speed up his/her periodic movements, including movements to keep up with a faster beat.

To quantitatively explain data about oscillatory movements, equations (1) and (2) were generalized in three ways:

> In equation (1), inhibition drives the excitatory potential to 0, which is also the passive equilibrium point. More generally, large inhibitory inputs can *hyperpolarize* a cell, thereby driving its potential to negative values.
>
> Also in equation (1), the excitatory feedback signal $f(x)$ uses the same signal function as the inhibitory feedback signal $f(y)$. In general, this is not true, so a different signal function $g(y)$ was chosen to define inhibitory feedback.
>
> Finally, in equation (2), there is no shunting inhibitory term, so we included an inhibitory shunting term $(C + x)$.

The simplest system that includes these three generalizations is:

$$dx/dt = -Ax + (B - x)[f(x) + I] - (C + x)Dg(y) \quad (3)$$

$$dy/dt = E[(1 - y)[x]^+ - y]. \quad (4)$$

In equation (3), term $(C + x)$ allows x to become hyperpolarized to the value $-C$, and the term $g(y)$ defines an inhibitory feedback signal that may differ from the excitatory feedback signal $f(y)$.

In equation (4), the shunting term $(1 - y)$ defines a maximal possible value of 1 for y, and the threshold function $[x]^+ = \max(x,0)$ equals x when x is non-negative, and 0 when x is negative.

This system can, in turn, be generalized to the following system of shunting recurrent competitive equations which, in different specialized anatomies, quantitatively simulates oscillations that occur when humans coordinate finger movements across both hands (Figure 3.25; Grossberg, Pribe, and Cohen, 1997) and quadruped animals move with one of several possible gaits (Figure 3.26; Pribe, Grossberg, and Cohen, 1997):

$$dx_i/dt = -Ax_i + (B - x_i)[f(x_i) + I_i] - (C + x_i)\sum D_{ij}\, g(y_j) \quad (5)$$

$$dy_i/dt = E[(1 - y_i)[x_i]^+ - y_i]. \quad (6)$$

FINGER OSCILLATIONS: FROM ANTI-PHASE TO IN-PHASE SYNCHRONY AND THE MISSING BEAT

As noted above, Yamanishi, Kawato, and Suzuki (1980) described a bimanual finger-tapping task whereby subjects were required to tap keys in time to visual cues. The timing of the cues was varied across ten relative phases. Subjects' fingers tended to slip from intermediate phase relationships toward being in-phase or anti-phase. The in-phase and anti-phase relationships exhibited less variability than intermediate phase relationships, leading to a "seagull" effect to describe the standard deviation of observed movements as a function of relative phase. Yamanishi, Kawato, and Suzuki (1980) defined a formal phase oscillator to describe these effects. Grossberg, Pribe, and Cohen (1997) showed how these data properties emerge from the dynamical system (5) and (6) when the inputs I_i mimic the timing of the experimental inputs.

Scott Kelso (Kelso, 1981) described a related task in which a metronome signaled when the fingers should move. Both in-phase and anti-phase movements were possible at lower frequencies, but the movements spontaneously switched to in-phase movements at higher frequencies (Kelso, 1984). These properties were also simulated by equations (5) and (6). Figure 2.25a describes the CPG oscillator that couples the two fingers when $i = 1, 2$. Figure 3.25b shows a computer simulation of CPG dynamics wherein anti-phase inputs (Figure 3.25b(A)) to equation (5) first lead to anti-phase finger movements, as simulated in (Figure 3.25b(B)). These anti-phase finger movements spontaneously switch to in-phase movements as the anti-phase inputs occur more rapidly, as simulated in (Figure 3.25b(C)) and (Figure 3.25b(D)).

These modeling results do not incorporate influences that occur in our limbs when they are attached to our bodies and subjected to perceptual, cognitive, and motor feedback factors that include different delays in processing visual or auditory periodic stimuli, different delays of sensory and motor afferent pathways, temporal dynamics of motor cortical movement commands, dynamics and kinematics of opponent motor effectors, effects of afferent feedback from movements on their controllers, anticipatory properties of prefrontal cortical circuits, and so on. Bruno Repp has particularly well documented such subtleties during finger movement synchronization (e.g., Repp, 2005, 2006a, 2006b; Repp and Su, 2013). Such additional influences may help to explain why musicians typically do not keep time to a strict periodic beat (Hennig et al., 2011). Equations (5) and (6) model a core oscillator that these various additional inputs can be used to modulate in future studies.

QUADRUPED GAIT OSCILLATIONS AND MOTOR SEQUENCE PERFORMANCE AT VARIABLE SPEEDS

A main difference between finger synchronization studies and quadruped gaits is that during finger synchronization studies, *external* inputs, like visual cues, drive finger synchronization (Figure 3.25). In contrast, during quadruped gaits, an *internal* volitional GO signal drives leg movements (Figure 3.26a), whose increasing size triggers different gaits as the legs oscillate faster (Figure 3.26b). Quadruped movements also need to cope with the force of gravity to maintain bodily balance despite movement-generated forces through time (Brown, 1911). One important challenge is ensuring that one leg sufficiently stabilizes the body while the other leg is being launched toward the body's next target position, and to do so at variable speeds. How does one foot know how far along the other foot has gone, and to do it at variable speeds? The circuit in Figure 3.9 clarifies how it may happen using information about leg movement acceleration and deceleration. Embedding such oscillators in humanoid robots will require that the basic oscillatory signals be appropriately modulated by force-related factors.

Although models as simple as the VITE model do not include the relevant processes, more complicated models do. In particular, the VITE circuit cannot maintain a desired trajectory under variable force and loading conditions. It needs to interact with suitably designed cerebellar and spinal cord circuits to accomplish this feat. The spinal cord Factorization-of-LEngth-and-TEnsion, or FLETE, model that my colleague Daniel Bullock and I developed together (Bullock and Grossberg, 1989, 1991) shows how movement trajectories that are computed by cortical circuit models like VITE can preserve accurate movement trajectories of an arm under conditions of variable speed and force when they interact with FLETE in an expanded VITE-FLETE model. FLETE explains and simulates many data about how identified cells and circuits in the spinal cord interact with the motor cortex and parietal cortex to achieve these goals. These cortico-spinal model circuits were additionally joined with cerebellar circuits (Figure 3.27) to show how cerebellar learning can maintain accurate control of a realistic multi-joint arm model with antagonistic muscles, as described in an article published with our PhD student Jose (Pepe) Contreras-Vidal (Contreras-Vidal, Grossberg, and Bullock, 1997).

6 Models of Emotional, Cognitive-Emotional, and Expectation Violations in Music

Brain networks that control emotional, cognitive-emotional, and expectation violation processes (e.g., Egermann, Pearce, Wiggins, and McAdams, 2013) also play a major role in the enjoyment and understanding of music. Knowing how emotions interact with visual and auditory perception and cognition via cognitive-emotional interactions can greatly clarify how this happens. My colleagues and I have been incrementally developing such a unifying and principled theory of cognitive-emotional brain dynamics for over fifty years (e.g., Dranias, Grossberg, and Bullock, 2008; Fiala, Grossberg, and Bullock, 1996; Franklin and Grossberg, 2017; Grossberg, 1971, 1972a,

CORTICO-SPINO-CEREBELLAR MOVEMENT CIRCUIT

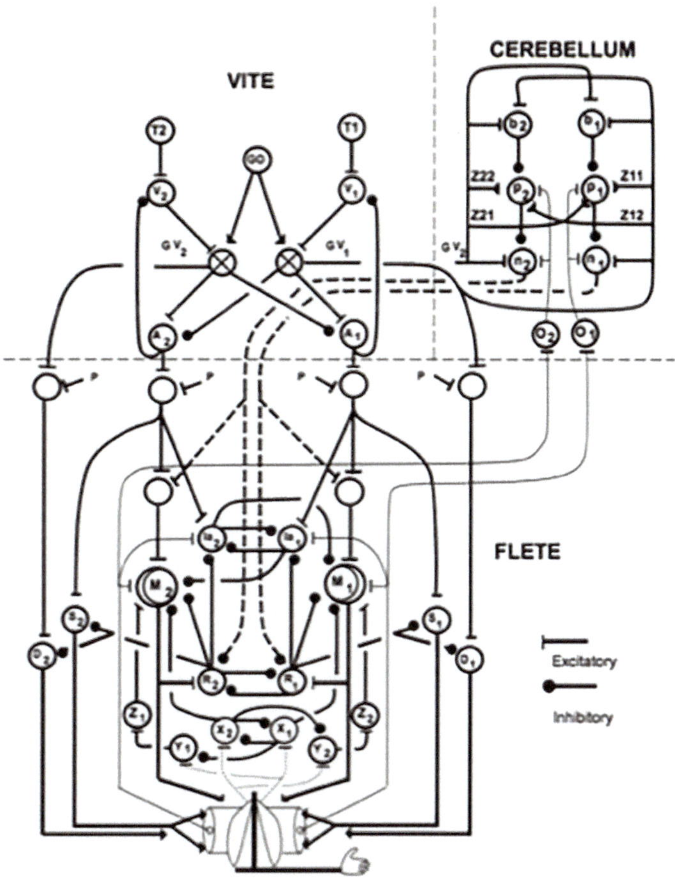

FIGURE 3.27 The combined VITE, FLETE, cerebellar, and multi-joint opponent muscle model for trajectory formation in the presence of variable forces and obstacles. Reprinted with the author's permission from Contreras-Vidal, Grossberg, and Bullock (1997).

1972b, 1975, 1982, 1984, 2017a, 2017c, 2018, 2019, 2021; Grossberg, Bullock, and Dranias, 2008; Grossberg and Levine, 1987; Grossberg and Merrill, 1992, 1996; Grossberg and Schmajuk, 1987, 1989)).

I already mentioned our simplest cognitive-emotional model, called CogEM (Cognitive-Emotional-Motor), in Chapter 1 (see Figure 1.30). Just as the VITE model was extended to a VITE-FLETE-cerebellum model, CogEM has been extended to the MOTIVATOR (Matching Objects To Internal VAlues Triggers Option Revaluations) model that I developed with my PhD student Mark Dranias, with help from Daniel Bullock (Dranias, Grossberg, and Bullock, 2008; Grossberg, Bullock, and Dranias, 2008). As Figure 3.28 illustrates, MOTIVATOR models multiple perceptual and cognitive cortical areas that compute object categories and object-value categories, while they interact with subcortical areas like amygdala and hypothalamus that compute value categories, all of them coordinated by a reward expectation filter in the basal ganglia that regulates all of these regions by opening and closing gates to enable their expression.

The MOTIVATOR model can be used to clarify emotional reactions to musical experiences, in much the same way that it has been used to explain emotional reactions to perceptual and cognitive experiences. The model simulates how active emotions can trigger incentive motivational signals that select and sustain attention upon emotionally valued information, and thereby enable such information to control temporally coordinated actions to realize valued goals. In the case of music, such sustained attention helps to maintain performance of a piece of music, modulated by the changing emotions felt through time by the performer. A self-contained review of CogEM and MOTIVATOR is provided in Grossberg (2021). I will summarize some highlights of CogEM in the next few paragraphs.

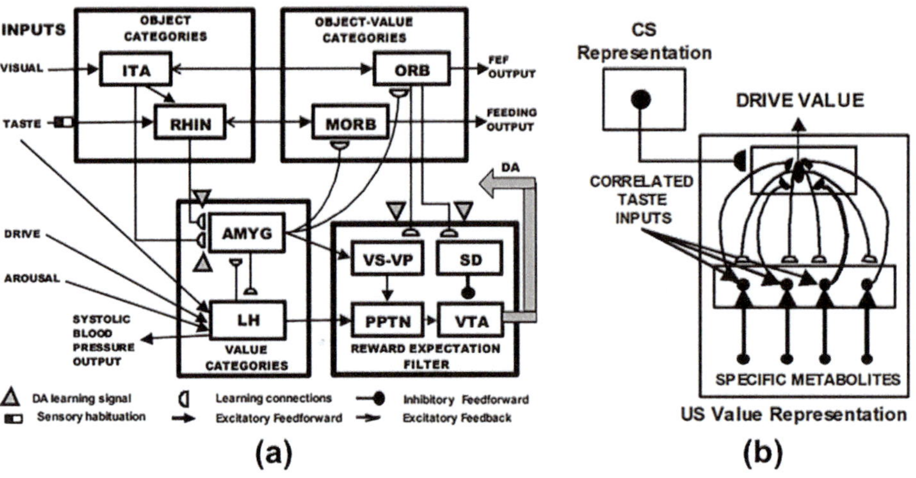

FIGURE 3.28 (a) The MOTIVATOR neural model generalizes CogEM by also including the basal ganglia. It can hereby explain and simulate complementary functions of the amygdala and basal ganglia (SNc) during conditioning and learned performance. The basal ganglia generate Now Print signals in response to *unexpected* rewards. These signals modulate learning of new associations in many brain regions. The amygdala supports motivated attention to trigger actions that are *expected* to occur in response to conditioned or unconditioned stimuli. Object Categories represent visual or gustatory inputs in anterior inferotemporal (ITA) and rhinal (RHIN) cortices, respectively. Value Categories represent the value of anticipated outcomes on the basis of hunger and satiety inputs, in amygdala (AMYG) and lateral hypothalamus (LH). Object-Value Categories resolve the value of competing perceptual stimuli in medial (MORB) and lateral (ORB) orbitofrontal cortex. The Reward Expectation Filter detects the omission or delivery of rewards using a circuit that spans ventral striatum (VS), ventral pallidum (VP), striosomal delay (SD) cells in the ventral striatum, the pedunculopontine nucleus (PPTN), and midbrain dopaminergic neurons of the substantia nigra pars compacta/ventral tegmental area (SNc/VTA). The circuit that processes CS-related visual information (ITA, AMYG, ORB) operates in parallel with a circuit that processes US-related visual and gustatory information (RHIN, AMYG, MORB). (b) Reciprocal adaptive connections between lateral hypothalamus and amygdala enable amygdala cells to become learned value categories. The bottom region represents hypothalamic cells, which receive converging taste and metabolite inputs whereby they become taste-drive cells. Bottom-up signals from activity patterns across these cells activate competing value category, or US Value Representations, in the amygdala. A winning value category learns to respond selectively to specific combinations of taste-drive activity patterns and sends adaptive top-down priming signals back to the taste-drive cells that activated it. CS-activated conditioned reinforcer signals are also associatively linked to value categories. Adaptive connections end in (approximate) hemidiscs. Reprinted with the author's permission from Grossberg (2021).

To ensure our survival in an ever-changing world, our brains have evolved to realize mental capabilities that support behavioral success. These mental capabilities range from perception through learning, cognition, and emotion to actions like navigation, looking, and reaching to realize valued goals. When all these processes interact properly together, they define the human condition and enable civilizations to emerge and succeed.

Learning enables us to learn language and to have thoughts about all of our experiences. But without emotion, we do not know what we want and value, and cannot effectively make decisions that can flexibly adjust to changing environmental constraints to have a high probability of realizing valued goals. Indeed, when cognition and emotion fail to interact properly, serious mental disorders can occur, including autism and schizophrenia. My neural models have proposed what changes in normal brain processes can cause autistic and schizophrenic behavioral symptoms (Grossberg, 1999, 2000, 2003, 2012, 2017b; Grossberg and Kishnan, 2018; Grossberg and Pepe, 1970; Grossberg and Seidman, 2006; Vladusich et al., 2010). The Cognitive-Emotion-Motor, or CogEM, model provides mechanistic insights about how these brain processes work. CogEM's discovery and development were guided and tested by principled and unifying explanations of data from many psychological and neurobiological experiments about normal or typical humans and animals.

Figure 1.30 illustrates the simplest version of the CogEM macrocircuit. It shows how *thoughts*, in the form of learned categories that learn from *external* perceptual cues (CS) in sensory cortices, notably the temporal lobe, and *feelings* that are activated by *internal* homeostatic cues in subcortical regions like the amygdala, are combined in the orbitofrontal cortex. When the polyvalent cells of the orbitofrontal cortex simultaneously receive these perceptual and homeostatic cues, these cells can fire and send feedback signals to the sensory cortices to select and focus motivated attention upon visual objects and other events that are currently valued, while inhibiting motivationally irrelevant events. As valued objects and events are chosen, they also generate learned outputs to acquire valued goals.

Activation of the feedback circuit that is defined by temporal-amygdala-orbitofrontal interactions can create a *cognitive-emotional resonance* that maintains motivated attention upon a motivationally salient object category or event. If this resonance is sustained long enough, it can also support conscious feelings about this object or event, including about music that is currently being played, heard, or imagined, and the musicians who may currently be playing it.

I reviewed in Section 2 of Chapter 1 how Adaptive Resonance Theory, or ART, supports a *feature-category resonance* to support conscious recognition of learned categories and the critical features that control their learning and prediction. In the case of music, the conscious recognition can be of the music itself, as well as of the conductor, soloist, and other familiar artists who are playing it. I also noted that a *surface-shroud resonance* supports conscious seeing of attended objects, and regulates looking and action in response to them. In the case of music, these consciously seen and attended objects can again be of the conductor, soloist, and other familiar artists who are playing it.

In Sections 1 and 2 of this chapter, I reviewed how a *stream-shroud resonance* supports conscious hearing of attended auditory communications, and a *spectral-pitch-and-timbre* resonance supports conscious hearing of pitches and timbres during auditory communication, including music.

When all of these kinds of conscious resonances interact synchronously—including the feature-category, surface-shroud, stream-shroud, and cognitive-emotional resonances—all of our senses combine to enable us to enjoy, understand, and even perform music that we love.

Disconfirmation of expectations in a piece of music may influence emotional reactions to it (e.g., Abdallah and Plumbley, 2009; Egermann et al., 2013; Hanslick, 1854). Our neural network models clarify how this happens. Indeed, ART explains how learned expectations dynamically stabilize all kinds of cognitive and emotional learning (Figures 2.20–2.26 in Chapter 2 and Sections 2 and 3 in this chapter), including how music may be learned rapidly without incurring the risk of catastrophically forgetting musical phrases that have already been learned. Disconfirmation of these expectations, whether during musical or non-musical experiences, leads to an ART reset (Figures 2.21–2.23 in Chapter 2) that influences ongoing cognitive and emotional events, including antagonistic rebounds that reset currently active emotions, and enable a shift of attention to focus on currently important environmental events, whether in the forest primeval or the music that we hear in the concert hall.

In contrast to models like ART and CogEM, recent AI algorithms like ChatGPT, where GPT stands for Generative Pre-trained Transformer, do not include mechanisms to guide their thoughts to realize valued goals. They literally do not know what they are talking about. They have no goals, values, or a concept of meaning.

Chapter 4 presents an alternative model to ChatGPT that does embody these qualities, This alternative model does not have to be "pre-trained," as in the P of GPT. Rather, it shows how children can learn perceptual and affective meanings from their parents or other teachers, and do this learning incrementally in real time. I have called that model ChatSOME, where SOME stands for Self-Organized-MEaning.

Chapter 5 contrasts properties of ART and related biological neural network models with Deep Learning and ChatGPT. This contrast includes a list of seventeen foundational problems of Deep Learning that demonstrate what a weak learning model it is, and one that has no resemblance to the kind of brain dynamics that support human learning and intelligence.

In particular, Deep Learning is *untrustworthy* (because it is not explainable) and *unreliable* (because it can experience catastrophic forgetting). The biological neural networks that are reviewed in Chapter 5 include various of the models that are discussed in Chapters 2–4, but from the perspective that all these biologically derived models are trustworthy and reliable. Chapter 5 notes that our biological models for adaptive sensory-motor control are also trustworthy and reliable, and reviews key properties of them that enable humans to learn to speak, dance, and sing.

Language and Meaning

How children learn to understand language meanings

"The child … projects the whole of his verbal thought into things."
Jean Piaget

"The most complex type of behavior that I know: the logical and orderly arrangement of thought and action."
Karl Lashley

"The more I think about language, the more it amazes me that people ever understand each other at all."
Kurt Gödel

"Language does not just describe reality. Language creates the reality it describes."
Desmond Tutu

"A language is an exact reflection of the character and growth of its speakers."
Cesar Chavez

"Poetry is what gets lost in translation."
Robert Frost

"Speak the speech, I pray you, as I pronounc'd it to you, trippingly on the tongue."
William Shakespeare

Your Creative Brain and AI. Stephen Grossberg, Oxford University Press. © Oxford University Press (2026).
DOI: 10.1093/9780198965367.003.0004

1 Introduction

Our conscious experiences of seeing and hearing have permitted humans to create visual arts and music, as Chapters 2 and 3 have discussed. But there would be no art or music without languages to support the development of human symbol-making and civilizations, and to allow us and our ancestors to ascribe meanings to life experiences that find their way into our artistic creations.

Language is an ability that is unique to humans, and is key to the human condition. The integrated system of words, semantics, syntax, grammar, and the like, in human languages, which most children learn to use effortlessly, enable us to effectively communicate with each other about the open-ended possibilities of life, including changing environmental circumstances and unexpected events.

Indeed, as summarized by Pagel (2017):

> "Human language is distinct from all other known animal forms of communication in being *compositional*. Human language allows speakers to express thoughts in sentences comprising subjects, verbs and objects—such as 'I kicked the ball'—and recognizing past, present and future tenses. Compositionality gives human language an endless capacity for generating new sentences as speakers combine and recombine sets of words into their subject, verb and object roles ... Human language is also *referential*, meaning speakers use it to exchange specific information with each other about people or objects and their locations or actions."

Language accordingly plays a crucial role in supporting the visual arts and music. In the visual arts, language enables us to describe the colors and shapes that are embodied in works of art, the meanings that we ascribe to the resulting creations, as well as the methods and tools that are used to create them. In music, language plays a crucial role not only by enabling societies to understand the culture in which the music is created but also to describe and remember the lyrics, melodies, rhythms, and emotions that a piece of music expresses.

In all these cases, a key theme is how languages acquire *meaning* through learning experiences during which language utterances are reliably associated with the perceptual, affective, and motoric events that they signify. This chapter describes a biological neural network model that can be used to explain how children and adults learn to understand language meanings about the perceptual, affective, and motoric events that they consciously experience. This kind of learning often occurs when a child, or adult, interacts with a teacher to learn language meanings about events that they experience together. Multiple types of self-organizing brain processes are involved in learning language meanings, including many processes that are also needed to understand art and music. These brain processes control conscious visual and auditory perception, joint attention, object learning and conscious recognition, cognitive working memory, cognitive planning, emotion, cognitive-emotional interactions, volition, and goal-oriented actions.

This chapter explains how all these brain processes interact to enable learning of language meanings to occur. It also contrasts these human capabilities with currently popular Artificial Intelligence, or AI, models like ChatGPT which, unlike humans, have no concept of goals, values, or meaning. These models literally do not know what they are talking about. As I noted in Chapter 3, GPT stands for Generative Pre-trained Transformer.

To contrast the current model—which, because it is a neural network model, is also part of AI—with ChatGPT and related models, I call it the ChatSOME model, where SOME abbreviates Self-Organizing Meaning.

The community of AI practitioners who have viewed ChatGPT as a way to understand human intelligence has, at least so far, achieved at best limited success. One way out of this impasse is for them to learn about, and to invest enough of their vast resources to further develop, known biological neural network models about how our brains learn to understand the world and its multi-faceted meanings. I like to call the product of that hopeful result Natural Intelligence, or NI.

My own long-standing interest in NI is illustrated by the title of my book called *Neural Networks and Natural Intelligence*, which was published in 1988 by MIT Press. Scientists and AI practitioners who want to contribute to the development of NI can find over 500 downloadable archival articles that provide a blueprint for developing multiple facets of NI at sites.bu.edu/steveg/. These articles provide the technical details needed to flesh out the self-contained and non-technical overview and synthesis of neural models that explain the main brain processes that give rise to our conscious minds that I provided in Grossberg (2021).

2 Toward Understanding How Children and Adults Learn Language Meanings

I will now provide a functional and mechanistic analysis of key brain processes that enable a child to learn language utterances and their meaning. Learning the meaning of language utterances allows children and adults to describe and understand their perceptual, and emotional, experiences in the world, and to generate appropriate actions based upon this understanding. Such learning typically begins when a baby who knows no language interacts with someone who does, often a parent or other caregiver.

Multiple brain regions interact to support the learning of language utterances and their meanings. My explanation accordingly builds upon biological neural network models of how our brains make our minds that have been getting steadily developed during the past half century. These models, which were not derived to explain how children and adults learn language meanings, provide principled and unifying explanations of hundreds of psychological and neurobiological experiments about other faculties of human intelligence. Many model predictions have also been confirmed by subsequent experiments over the years. This chapter will review enough of this background to make it self-contained.

The importance of how language learning and its meaning occur for understanding the human condition is reflected by the large number of articles and books that have been written about it; for example, Aziz-Aadeh, 2013; Bergen, 2012; Boroditsky, 2000; Boroditsky and Ramscar, 2002; Boulenger, Hauk, and Pulvemuller, 2009; Bruner, 1985; Carey, 2022; Casasanto and Dijkstra, 2010; Chatterjee, 2008, 2010; Colston, 2019; Dove, 2009; Fernardino et al., 2013; Fischer and Zwaan, 2008; Gallese and Lakoff, 2005; Gibbs, 1994, 2006; Gleitman and Gleitman, 2022; Glenberg, 2010; Glenberg and Kaschak, 2010; Hauk and Pulvemuller, 2004; Kable et al., 2005; Kable, Lease-Spellmeyer, and Chatterjee, 2002; Kemmerer et al., 2008; Mahon and Caramazza, 2008; Pecher and Zwaan, Eds., 2005; Raposo et al., 2009; Richardson et al., 2003; Saygin et al., 2010; Tronick, 2007; Wallentin et al., 2005, 2011; Watson et al, 2013; Zwaan and Taylor, 2006). Although such contributions include many facts about language learning and meaning, they do not provide a unifying mechanistic neural explanation of these data, or of the organizational principles that shape these brain mechanisms.

This chapter models how brain processes of language learning and meaning occur in *real time*. Such a "real time" exposition explains how these processes unfold moment-by-moment across multiple brain regions as the learning individual interacts with the world. No available experimental method can provide a mechanistic bridge between such interactive brain dynamics and the behavioral properties that they cause. That is because these behaviors are *emergent properties* of these brain interactions. A rigorous mechanistic theory is needed to bridge this explanatory gap.

Indeed, a major fact about *all* the cortical areas that support intelligent faculties like attention, learning, vision, speech, language, and cognition is that they share the same *laminar* organization. Chapters 1 and 2 have already discussed aspects of what I have called the paradigm of *laminar computing*. All the cells in these cortical areas are organized into layers, with characteristic interactions within and between these layers (see Figures 1.33–1.36). These shared laminar circuits are the "little grey cells" that Agatha Christie's redoubtable detective, Hercule Poirot, used to solve his mysteries. Although all parts of neocortex share the same canonical laminar organization, specializations of this canonical circuit design carry out all the higher processes of human intelligence. An explanation of how this works was outlined in the previous chapters, and explained in detail in Grossberg (2021). Here I focus on those concepts, and the corresponding brain interactions that enable children and adults to learn perceptual, affective, and motoric language meanings.

2.1 Visual and auditory circular reactions

I will now discuss how a baby first learns, during visual and auditory *circular reactions*, how to look at, or point to, objects of interest, as well as how to produce, store, and learn simple speech sounds that create a foundation for learning to imitate the language utterances of its teachers.

Before a baby or child can learn language from an adult, it must first be able to pay attention to, and learn to recognize, that adult's behaviors, notably to pay attention to and recognize an adult caregiver's face when he or she is speaking. To accomplish this feat, several additional learning processes need to occur.

Learning to recognize an adult caregiver's face includes the ability to recognize multiple views of the face as experienced from multiple positions, orientations, and distances. In other words, a young child effortlessly learns to solve the *invariant pattern recognition problem*, a problem that is just as important for human development and social success as it is in engineering, technology, and AI. I discussed this problem in Sections 2.17–2.19 of Chapter 2 as part of my analysis of how humans consciously see and recognize visual art. This invariant learning process includes learning of multiple views of a mother's face while a baby looks up at its mother as it suckles milk from her breast or a bottle.

Chapter 2 explained how the learning of invariant object representations interacts with the learning of individual views of the object. In addition to enabling the child to recognize its mother's face from multiple perspectives, this ability enables the child to also learn that specific views of her face correlate with her looking, pointing, or otherwise acting on objects in a given direction. These invariant and view-specific representations reciprocally interact with each other via learned connections so that a child can invariantly recognize its mother's face even as it uses her currently perceived view to predict her ongoing actions.

Section 3 will review how learning of invariant object categories of objects like a face requires interactions between *spatial attention* in the dorsal, or Where, cortical processing stream with *object attention* in the ventral, or What, cortical processing stream, notably between cortical areas PPC (posterior parietal cortex) and IT (inferotemporal cortex), respectively (Figures 1.32, 1.37, and 2.17). Figure 4.1 shows, in addition, that there are analogous Dorsal and Ventral streams for the processing of auditory information, including speech and language. More generally, all the learning that is important for a child's understanding and survival require interactions between multiple brain regions.

Another important example of inter-region interactions during learning occurs when the cognitive representations for learned object recognition interact reciprocally via learned connections with emotional representations. These cognitive-emotional interactions, which include brain regions like the prefrontal/temporal cortices and amygdala/hypothalamus, amplify the cognitive representations of currently valued objects in a scene (see Figure 1.30). These cognitive-emotional interactions will be discussed in Section 4 in terms of their relevance for understanding language meanings.

Competitive interactions among object representations in the temporal and prefrontal cortices can choose the currently most valued objects in a scene that may be cluttered with multiple other objects. I will explain in Section 4 how these motivationally amplified representations successfully compete for spatial attention to attract an observer's gaze, such as when a child orients to look at its mother's face. The child can then begin to learn how current views of her face predict where she is looking, pointing, or otherwise acting.

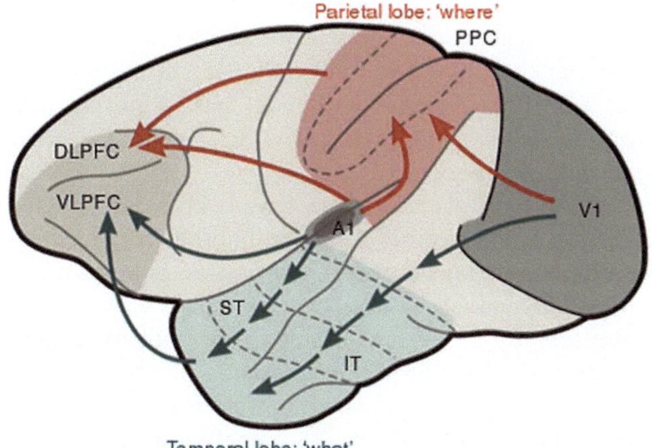

FIGURE 4.1 A What ventral cortical stream and Where/How dorsal cortical stream have been described for audition, no less than for vision. Reprinted with permission from Figure 1 in Rauschecker and Scott (2009).

A prerequisite for learning to understand and produce simple language utterances is for a *sequence* of speech sounds, or items, to be stored temporarily in a *working memory* that is designed to support learning and stable memory of stored phonemes, syllables, words, and sentences. Such working memories occur in multiple regions of the prefrontal cortex. Reciprocal interactions occur between sequences of stored working memory items and the learned speech categories, or *list chunks*, that categorize the sequences.

Bottom-up interactions from the working memory to the list chunk level enable the list chunks to be learned. Top-down interactions from an active list chunk to the working memory can dynamically stabilize the learning of the item sequence that it codes. I will supplement the discussion of working memories in Section 1 of Chapter 3 with further explanations in Section 13 in this chapter of how and why working memories and their learned list chunks have the designs that they do.

During language recall, list chunks read out into working memory their learned item sequences. *Volitional signals*, or GO signals, from the basal ganglia enable a sequence that is stored in working memory to be performed in the correct order and at a desired speed (see Figures 3.9, 3.12, 3.13, 3.15, 3.17, and 3.19 in Chapter 3). This role for the basal ganglia will be briefly reviewed in Section 14 in order to enable a self-contained reading of each chapter. Also discussed there is how more complex rhythmic performances, as during the singing of musical lyrics and melodies, can be achieved using *spectral timing circuits* in the cerebellum and basal ganglia, a topic that was also discussed in Chapter 3; see Figure 3.22.

When these capabilities interact in real time, a child can learn that specific language phrases and sentences strongly correlate with specific visual objects and events that the child is simultaneously watching a teacher use or perform. In this way, a child can learn that a phrase like "mommy walks" correlates with a simultaneous percept of the described action. The phrase hereby acquires meaning by representing this perceptual experience.

Learning of auditory language utterances occurs in the ventral, or What, cortical processing stream for object attention and categorization, including the temporal and prefrontal cortices (Sections 13 and 19; Kimppa et al., 2015; Rauschecker and Scott, 2009). Learning to visually recognize mommy also occurs in the What stream, including the inferotemporal cortices (Tanaka, 1996; Tanaka et al., 1991).

In contrast, the perceptual representation of mommy walking includes representations in the dorsal, or Where cortical stream for spatial attention and action, including cortical areas MT and MST. Learning language meanings thus often require What-to-Where cortical stream interactions that link language utterances to the perceptual experiences that they describe. Sections 4 and 5 will discuss how learning between these What and Where stream interactions occurs.

Mommy can also activate affective representations, which enable a baby or child to have feelings about mommy. As noted above, midbrain regions like the amygdala and hypothalamus help to generate such feelings when they interact reciprocally with perceptual and cognitive representations that represent mommy. Section 4 will discuss how interactions between cognitive and emotional representations are learned and how they focus motivated attention upon a valued caregiver like mommy.

Language meaning is thus embodied in the interactions between a language utterance and the perceptual and affective experiences with which it is correlated, much as the word "mommy" can activate a complex set of learned associations that prime perceptual, cognitive, emotional, and action representations that have been experienced during previous interactions with her. When the word "mommy" is internally heard as a child thinks about mommy, activation of this language representation can also prime multiple perceptual and affective representations. Volitional GO signals from the basal ganglia (Section 14) can then activate internal visual experiences such as visual imagery and felt emotions, thereby enabling a child or adult to understand the perceptual and affective meaning of "mommy" without needing to observe her.

The OpenAI project ChatGPT (GPT = Generative Pre-trained Transformer) has shown some ability to provide answers expressed in language about many topics, as well as frequent failures to do so. ChatGPT has acquired this ability after having been fed immense language databases by human operators, as well as a probabilistic algorithm to predict what comes next in its currently active linguistic context (http://chat.openai.com/; Newport, 2023).

In contrast to the biological neural network models that are described below, ChatGPT cannot do fast incremental learning with self-stabilizing memories about novel situations that include unexpected events. It therefore cannot rapidly learn about the rare events that are often the basis of great discoveries, such as being able to learn the first breakout of a new disease in an environment where lots of people are sick with other diseases, or to make a great scientific discovery based on just a few key facts combined with a deep experimental intuition and a creative imagination. In fact, ChatGPT cannot learn anything after its parameters are frozen before it is used in applications.

ChatGPT would not seem nearly as impressive if it were not being interpreted by humans who do know the real-world meaning of the language that they use, including ChatGPT creators and users.

Perhaps most importantly, and central to the main theme of the current article, ChatGPT does not know the real-world meaning of its predictions. It literally does not know what it is talking about, if only because its language phrases have not been learned in real time along with the perceptual, affective, and motor experiences that could give them meaning.

Those parts of ChatGPT that are learned use the Deep Learning algorithm (LeCun, Bengio, and Hinton, 2015) to do so. Deep Learning has at least seventeen foundational computational problems. I will discuss these problems in Chapter 5 and show that Adaptive Resonance Theory, or ART, has none of them, nor has had any of them since I first published articles about ART in 1976 (Grossberg, 1976a, 1976b, 1980), long before Deep Learning was proposed and before the current popularity of the back propagation algorithm, upon which Deep Learning builds, was triggered by the 1986 article of Rumelhart, Hinton, and Williams (1986).

The current article describes brain processes that enable humans to learn language that expresses the meaning of perceptually and affectively experienced events in the real world. As I noted above, I call the neural model that accomplishes this ChatSOME, where the acronym SOME stands for Self-Organizing MEaning.

2.2 Building upon visual and auditory circular reactions

Early steps in language development prepare a learned scaffold upon which later learning of language meanings can build. These processes are reviewed in Grossberg (2021). As noted there, and briefly in Chapter 3, perception-cognition-emotion-action *circular reactions* occur during auditory and visual development. The pioneering Swiss clinical and developmental psychologist, Jean Piaget, has richly contributed to understanding what circular reactions are, and how they work (e.g., Piaget, 1945, 1951, 1952). Recent reviews and discussions of Piaget's work include those of Burman (2021), Carey, Zaitchik, and Bascandziev (2015), and Singer and Revenson (1997). In particular, all babies normally go through a *babbling phase* during which a *circular reaction* can be learned (Figure 4.2).

During a *visual* circular reaction, babies endogenously babble, or spontaneously generate, hand/arm movements to multiple positions around their bodies. Babbled movements endogenously sample the workspace within which a baby can reach. As their hands move in front of them, their eyes reactively look at their hands. While the baby's eyes are looking at its hands, an associative map is learned from its hand positions to the corresponding eye positions, *and* from its eye positions to hand positions. The learned maps between eye and hand in both directions is the "circular" reaction.

After map learning occurs, when a baby, child, or adult looks at a target position with its eyes, this eye position can

FROM SEEING AND REACHING
TO HEARING AND SPEAKING
CIRCULAR REACTIONS
Piaget, 1945, 1951, 1952
Homologous circuits for development and learning of
motor-equivalent REACHING and SPEAKING
DIRECT
DIVA
Bullock, Grossberg, and Guenther, 1993
Guenther, 1995

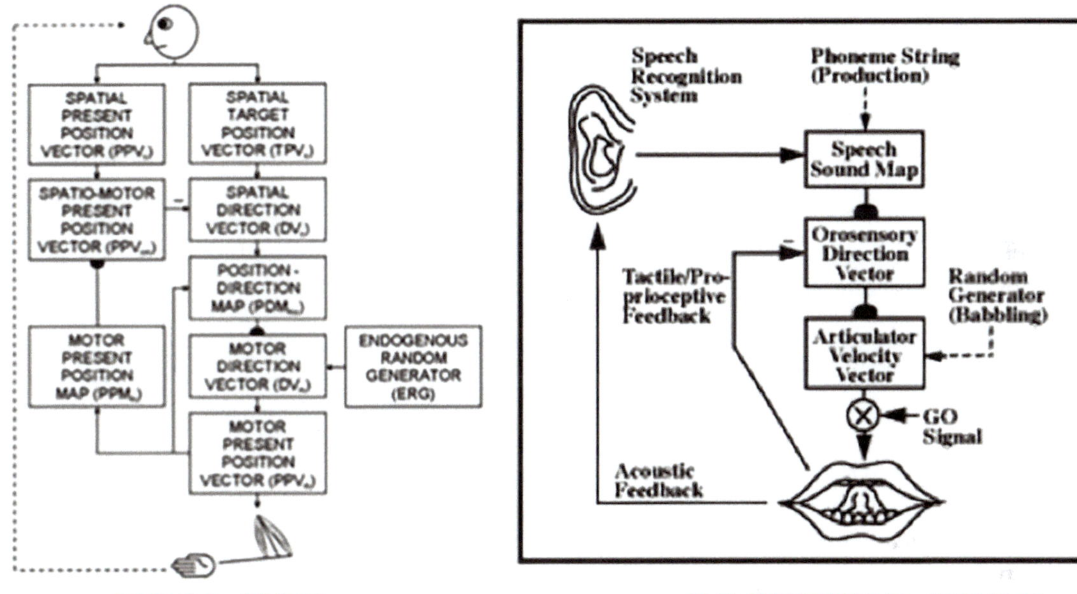

FIGURE 4.2 The DIRECT and DIVA models have homologous circuits to learn and control motor- equivalent reaching (left panel) and speaking (right panel), with tool use with the hand-arm system, and coarticulation with the speech articulator system, resulting properties. Reprinted with the author's permission from Grossberg (2021).

use the learned associative map to prime the activation of a movement command to reach the target position in space. If a volitional GO signal (see Figure 4.2, right panel), or "the will to act," is also activated by opening the correct basal ganglia gate, then the chosen target position is fully activated and enables a hand/arm movement to reach the foveated position in space, as when a baby looks at her toes and then moves her hands to hold them (Figure 4.3).

Because our bodies grow for many years as we develop from babies into children, teenagers, and adults, these maps continue updating their learned parameters to enable control of accurate movements using our current bodies and limbs. As Figure 4.4 summarizes, this kind of continually updated learning of sensory-motor maps and gains takes place in the dorsal, or Where, cortical stream, including cortical areas like the posterior parietal cortex, or PPC (Figure 4.1).

FIGURE 4.3 A baby touching its toes illustrates how actions can emerge within the reaching range due to a visual circular reaction. Image taken from the web. Westend61 GmbH/Alamy Stock Photo.

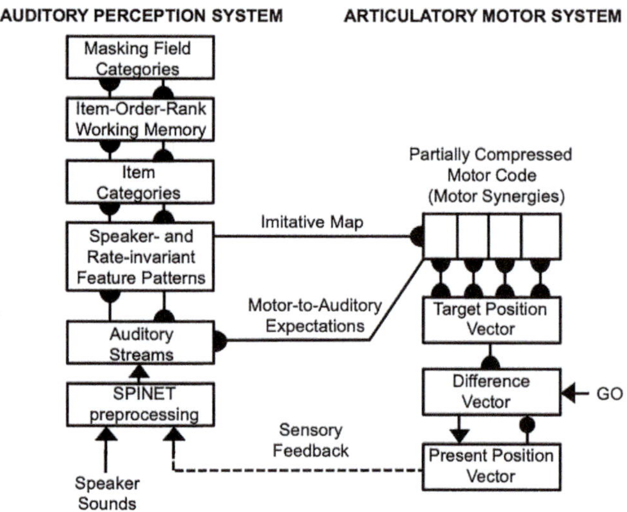

FIGURE 4.4 A model macrocircuit of interactions between the auditory perception system and the articulatory motor system that describes learning and feedback interactions within and between them during language learning and performance. Young children can hereby learn speaker-invariant and rate-invariant representations of acoustic feature patterns. These representations can be activated both by sensory feedback from the child's own babbled utterances (see the GO signal, which is endogenously active during babbling) and by external speaker sounds. Author created.

The matching and learning of spatial and action representations that takes place in the Where stream are *computationally complementary* to the matching and learning of object categories that take place in the ventral, or What, cortical stream: As Figure 1.31 summarizes, What learning of object categories is *excitatory* and occurs in a *match* state, whereas Where learning of spatial and object representations is *inhibitory* and occurs in a *mismatch* state.

In particular, as Section 2 of Chapter 2 noted (see Figures 2.20–2.22), category learning in Adaptive Resonance Theory occurs when there is a good enough *match* between an active top-down expectation and a bottom-up input pattern to trigger an *excitatory* resonance in cortical areas like the inferotemporal, or IT, cortex. Such a resonance drives *fast learning without catastrophic forgetting* of the adaptive weights in both the active bottom-up filter and top-down expectation.

In contrast, as Section 4.8 of Chapter 3 noted (see Figure 3.9), *spatially variant reaching* movements that are controlled by a Vector Integration to Endpoint, or VITE, network can occur when there is a *mismatch* between a desired target position and the present position of the arm. The movement continues, under volitional GO signal control, until the arm is where you want it to be.

Inhibitory matching in the Where cortical stream supports learning that can *continually update sensory-motor maps and gains* as our bodies continually grow and change throughout life. This kind of *mismatch learning* in the Where stream calibrates adaptive weights in the pathways that are activated by the target position vector. Learning continues until the desired target position and present position vectors that control movement use consistent parameters to do so, as modeled by the Vector Associative Map, or VAM, model (Gaudiano and Grossberg, 1991, 1992). As a result, when you have moved to where you want to be, the mismatch equals zero, so the movement stops.

Similar processes of development and learning occur in the auditory system. As illustrated by the DIVA model in Figure 4.2, an *auditory* circular reaction occurs during its own babbling phase. During such an auditory circular reaction, babies endogenously babble simple sounds that sweep out the workspace of sounds that they can create. The babies also hear the sounds that they create. When the motor commands that caused the sounds and the auditory representations of the heard sounds are simultaneously active in the baby's brain, an associative map is learned between these auditory representations and the motor commands that produced them.

After enough learning occurs, a child can use the learned *imitative map* (see Figure 4.4) to approximately imitate sounds from adult speakers. It can then incrementally learn how to speak using increasingly complicated speaker speech and language utterances, again under volitional control.

Several biological neural network models have been developed to explain how visual circular reactions enable reaching behaviors to be learned (Bullock, Cisek, and Grossberg, 1998; Bullock and Grossberg, 1988, 1991; Bullock, Grossberg, and Guenther, 1993; Gaudiano and Grossberg, 1991, 1992). Other models have been developed to explain how auditory circular reactions enable speech and language behaviors to be learned (Grossberg, 1986; Grossberg and Stone, 1986; Cohen, Grossberg, and Stork, 1988; Guenther, 1995; Guenther, Ghosh, and Tourville, 2006). Grossberg (2021) summarizes solutions of the multiple design problems that must be solved by each brain for these processes to work well, neural models that solve these problems, and lots of psychological and neurobiological data that the models explain.

Figure 4.4 describes a macrocircuit of how auditory and articulatory learning processes, and feedback interactions within and between them, enable language learning to occur. These processes enable a young child to learn speaker-invariant and rate-invariant representations of acoustic feature patterns that can be activated by both sensory feedback from the child's own babbled utterances and external speaker sounds.

Taken together, these brain processes enable babies to learn to imitate simple sentences that adult caregivers say, such as "Mommy walk," "Mommy throw ball," and so on.

2.3 Learning invariant categories of valued objects: Where-to-What stream interactions

Before a child can learn to say anything about, or to, mommy, it must be able to notice that mommy is there by paying attention to her. To explain how this happens, I will review brain processes that contribute to social cognition (Grossberg. 2021; Grossberg and Vladusich, 2010).

Before a child can pay attention to mommy, it must first learn to recognize her face using both an invariant object category representation as well as category representations of specific facial views. The ARTSCAN Search model proposes this happens. See Chapter 2, Figures 2.25 and 2.26. Such category learning can begin when baby looks up at mommy's face, say during suckling or other close encounters where mommy provides high-value primary rewards. The baby then gradually learns *view-specific* category representations of her face in the posterior inferotemporal cortex, or ITp, as well as an *invariant* category representation of it in anterior inferotemporal cortex, or ITa.

Such an invariant category can selectively fire when her face is seen from any view, position, or size on the child's retinas. The learned associations between ITp and ITa are bidirectional (see Figure 2.25), as during all object learning processes that are capable of dynamically stabilizing their learned memories, as modeled by Adaptive Resonance Theory, or ART (Section 2 of Chapter 2; Cao, Grossberg, and Markowitz, 2011; Carpenter, Grossberg, and Mehanian, 1989; Chang, Grossberg, and Cao, 2014; Fazl, Grossberg, and Mingolla, 2009; Grossberg, 1978, 2019, 2021; Grossberg, Markowitz, and Cao, 2011; Grossberg, Srinivasan, and Yazdanbakhsh, 2011, 2014; Grossberg and Wyse, 1991).

As in every ART model, the bottom-up pathways from ITp to ITa are an adaptive filter for learning the invariant category's adaptive weights. The adaptive weights in the top-down pathways from ITa to ITp learn expectations that focus attention upon the *critical features* that are used to recognize object views in ITp. The reciprocal excitatory interactions between ITp and ITa trigger a *category-to-category resonance* that dynamically stabilizes the memories of critical feature patterns that are learned by adaptive weights in both the bottom-up and top-down pathways. Recall that ART solves the *stability-plasticity dilemma* because it can support fast incremental learning of such categories without experiencing catastrophic forgetting.

As noted in Section 2.17 of Chapter 2, learning of an invariant object representation in ITa enables it to fire in response to multiple view-specific representations that are learned in ITp. This is proposed to happen as follows: The first ITp representation to be learned activates an initially uncommitted cell population in ITa that will become the invariant category representation. To become an invariant object representation, this chosen cell population in ITa must remain active while *multiple* view-specific representations in ITp are associated with it, one at a time, as the observer inspects the object.

Each view-specific category in ITp is inhibited, or reset (see Category Reset stage in Figure 2.25), when the next view-specific category is activated, learned, and associated with the emerging invariant category in ITa. Why is the ITa representation *not* reset when this happens, since it was originally activated by the first ITp representation to be reset? This is explained by how the child's *spatial attention* sustains its focus on mommy's face with a *surface-shroud resonance* that prevents reset of the ITa representation while the baby attends multiple views of mommy's face (Fazl, Grossberg, and Mingolla, 2009). An active shroud in the Posterior Parietal Cortex, or PPC, inhibits the Category Reset processing stage, also in PPC, that would otherwise have inhibited ITa (Figure 2.25). While the reset is being inhibited in this way, each ITp view-specific category that is learned can also be associated with the persistently active ITa category, thereby converting it into an *invariant* object category.

As noted in Section 2.19 of Chapter 2, an *attentional shroud* is spatial attention that fits itself to the shape of the object surface that is being attended (Tyler and Kontsevich, 1995), much as when a child focuses her attention upon mommy's face. When spatial attention shifts to another object, the category reset stage is disinhibited and learning of a new invariant category can begin.

A surface-shroud resonance is depicted in Figure 2.25 between visual cortical area V4 (lumped with V2 as V2/V4 in the figure) and the posterior parietal cortex, or PPC. A *gain field* in the lateral intraparietal area, or LIP, occurs between these cortical areas to carry out the change of coordinates from retinotopic surface coordinates of the attended object's view-specific categories to head-centered spatial attention coordinates of the invariant category.

Retinotopic coordinates register what we see when we move our eyes around the world. For example, after foveating on mommy's mouth, when a baby moves her eyes upward to look at mommy's nose, the retinotopic representation of mommy's mouth moves downward on her retinas. In contrast, a *head-centered spatial representation* of mommy's face does not move during this eye movement unless the baby also moves its head relative to mommy's face. A head-centered representation enables the baby to learn properties of mommy's face that do not change just because she moves her eyes relative to it.

Category learning between ITp and ITa goes on in the ventral, or What, cortical processing stream. Modulation of ITa learning by an attentional shroud is due to Where-to-What stream signals that occur while a surface-shroud resonance is active between the attended object's surface representation in cortical area V4 of the What stream and the shroud in the posterior parietal cortex, or PPC, of the dorsal, or Where, stream.

In summary, a surface-shroud resonance between V4 and PPC sustains spatial attention upon mommy's face during invariant category learning between ITp and ITa. Recurrent inhibitory interactions within these cortical regions choose the most predictive cell representations, focus attention upon them, and suppress outlying features.

While a surface-shroud resonance modulates invariant category learning of mommy's face in the baby's What cortical stream, it also supports *conscious seeing* of her face's surface representation in V4, and the other visual cortical regions with which V4 resonates.

Given the learning that goes on in the circuit of Figure 2.25, how does a baby use this knowledge to control actions that are directed to achieve valued goals? In particular, how does the baby find mommy's face in a room or other scene that contains other objects?

Figure 2.26 shows that the same brain regions that are used in Figure 2.25 can also be used to drive an efficient search to find, and then focus sustained attention, upon a desired target object in a cluttered scene, such as mommy's face. The ARTSCAN Search model whose microcircuit is depicted in these figures hereby embodies a solution of the famous Where's Waldo Problem (Figure 4.5; Chang, Grossberg, and Cao, 2014).

In the classical Where's Waldo Problem, Waldo typically wears a shirt with horizontal red and white stripes, glasses, and a red hat. He is standing in the lower right of Figure 4.5. Although a baby searching for her mommy's face will typically not have to search through a crowded beach scene, there will usually be other objects around than mommy when the baby searches for her. Supportive psychophysical and neurobiological data for our proposed solution of the Where's Waldo Problem are reviewed in Grossberg (2021).

FIGURE 4.5 The Where's Waldo Problem concerns how humans can detect a particular individual in a cluttered scene. In this scene, Waldo is standing in the lower-right corner of the picture wearing a shirt with horizontal red and white stripes, glasses, and a red hat.

3 Cognitive-Emotional Interactions Focus Motivated Attention on a Valued Caregiver's Face

A baby can learn its mommy's face while she is engaged in actions that reward the baby, notably feeding it with her breast or a bottle. The milk, warmth, comfort, happiness, and so on that are experienced during feeding are all positively rewarding. As category learning occurs, categories that persist over time, notably the emerging invariant face category, are bidirectionally associated with positive emotional centers, also called *drive representations* or *value categories*, that are activated in the baby's brain by mommy's rewarding activities. These drive representations are in the amygdala/hypothalamic system.

I briefly mentioned our neural model of *cognitive-emotional resonances*, called the Cognitive-Emotional-Motor, or CogEM, model in text near Figure 1.30 in Chapter 1 and its MOTIVATOR model generalization in Section 6 of Chapter 3; see Figure 3.28. Figure 4.6 provides a somewhat more detailed version of the CogEM macrocircuit. My colleagues and I have incrementally developed this model over the years to achieve an ever-broadening interdisciplinary explanatory range (Chang, Grossberg, and Cao, 2014; Dranias, Grossberg, and Bullock, 2008; Fiala,

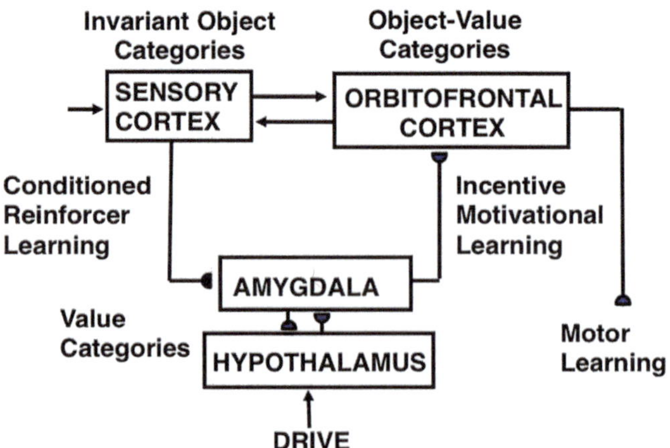

FIGURE 4.6 Macrocircuit of the functional processing stages and their anatomical interpretations within the Cognitive-Emotional-Motor, or CogEM, model of cognitive-emotional interactions. Reprinted with the author's permission from Grossberg (2021).

Grossberg, and Bullock, 1996; Franklin and Grossberg, 2017; Grossberg, 1971a, 1971b, 1972a, 1972b, 1974, 1975, 1978, 1982, 1984a, 1984b, 2018, 2019; Grossberg, Bullock, and Dranias, 2008; Grossberg and Levine, 1987; Grossberg and Schmajuk, 1987, 1989).

Figure 4.6 summarizes the positive feedback loop in CogEM that links, via learned connections, invariant object categories that are learned in sensory cortices—such as an invariant category of mommy's face—with value categories, or drive representations, in the amydala/hypothalamus that express the feelings that mommy's face elicits in a baby's brain.

When activity in this feedback loop is sustained for a long enough time with sufficient activity, it generates a *cognitive-emotional resonance* between mommy's attended invariant face category and these feelings, while it maintains motivated attention upon her face. The object-value representation in the orbitofrontal cortex that brings these sensory and affective signals together can also read out commands for actions that are compatible with them, as well as feedback signals to the sensory cortices to support sustained attention on mommy's face while moving toward her. The orbitofrontal cortex is an interface where object and value information converge, and is thus an important part of choosing a valued object upon which to attend, and acting to acquire it.

In particular, after category learning and cognitive-emotional learning occur, when the baby and its mommy are in different spatial locations, the baby's attention is drawn to whatever familiar view of mommy's face is seen. If an unfamiliar view is similar to a familiar one, then the most similar familiar view category is activated most.

Whereas invariant category learning requires a Where-to-What interaction between cortical streams, *orienting* to a valued familiar face requires a What-to-Where stream interaction (Figure 2.26). Such an interaction is needed because the motivationally amplified *invariant* face category in ITa of the What cortical stream, being positionally invariant, cannot directly control orienting to a particular position in space, such as where mommy is currently located. These positional representations are computed in the Where cortical stream.

Because the invariant ITa category of mommy's face is amplified by cognitive-emotional feedback, it will win the competition with other object categories in the scene, which will be suppressed. The chosen invariant ITa category can then *prime* all the ITp view-specific categories of mommy's face (Figure 2.25). Bottom-up inputs from a current view of mommy's face (e.g., the green arrow from the object's boundary representation in Figure 2.25) determines which ITp category will fire.

Such an ITp category is both view-specific *and* positionally-specific. It can therefore trigger reciprocal excitatory interactions with the surface representation in cortical area V2/V4 of the object view that it categorizes (Figure 2.25). When this surface representation is amplified, it triggers a surface-shroud resonance that most strongly activates the object's position via spatial attention in PPC, while inhibiting other positions via recurrent inhibition across PPC. PPC can then command looking, reaching, and other actions directly toward the position of mommy's face.

4 Joint Attention: How Looking at a Valued Face Triggers Orienting to Where It Is Looking in Space

How does a baby learn to associate an attended view of mommy's face with the position in space where she is looking or pointing (Deák, Flom, and Pick, 2000; Emery et al., 1997; Frischen, Bayliss, and Tipper, 2007; Materna, Dicke, and Their, 2008; Tomasello and Farrar, 1986)? For starters, as mommy moves her arm, and perhaps her body too, to point to something for the baby to look at, spatial attention can flow from mommy's attended face representation along her arm to her hand.

4.1 G-waves: Long-range apparent motion tracks attended moving objects

As I noted in Chapter 3 in the text around Figure 3.3, I have called a flow of spatial attention a G-wave, or Gauss-wave, because it describes how attention smoothly flows with a Gaussian profile along successive positions on mommy's body, starting with an attended initial position, and ending at a moving target position. Figure 4.7 provides more information about the steps leading to the formation of a G-wave.

I introduced G-waves to model psychophysical data about *long-range apparent motion* (Francis and Grossberg, 1996; Grossberg, 1988, 2014; Grossberg and Rudd, 1989, 1992). The simplest example of long-range apparent motion occurs when two flashes occur at different positions at successive times. Within a range of spatial separations and temporal delays, observers perceive a smooth motion from the first flash to the second. This visual illusion is called *beta motion* (Exner, 1875; Kolers, 1972).

This type of long-range apparent motion is not just a laboratory curiosity, however. Many motion percepts that

(a) TEMPORAL PROFILE OF SINGLE FLASH

Suppose that a single flash quickly turns on to maximum activity, stays there for a short time, and then shuts off

It causes an increase in activity, followed by an exponential decay of activity

The corresponding Gaussian profile waxes and wanes through time

Since the peak position of the Gaussian does not change through time, nothing moves

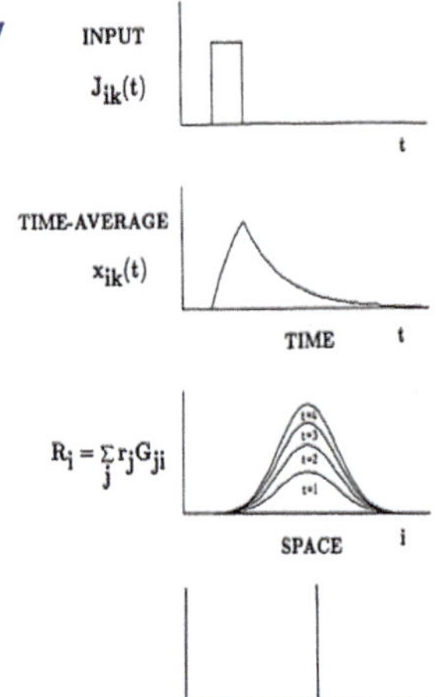

(b) TEMPORAL PROFILE OF TWO FLASHES

If two flashes occur in rapid succession, the waning of the activity due to the first flash may overlap in time with the waxing of the activity due to the second flash

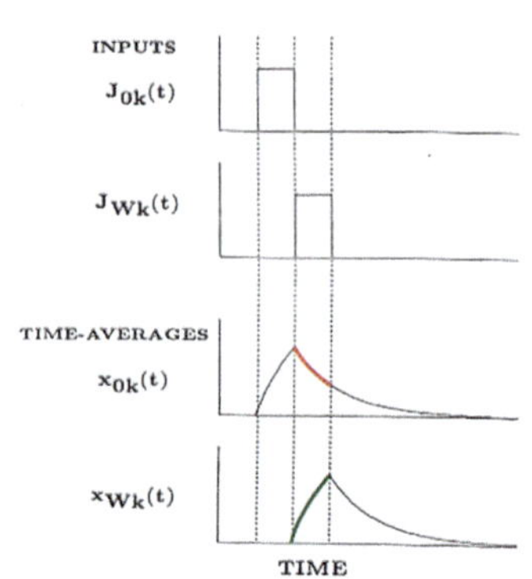

(c) TRAVELING WAVE (G-WAVE): LONG-RANGE MOTION

If the Gaussian activity profiles of two flashes overlap sufficiently in space and time, then the sum of Gaussians produced by the waning of the first flash added to the Gaussian produced by the waxing of the second flash can produce a single-peaked traveling wave from the position of the first flash to that of the second flash

The wave is then processed through a WTA choice network.

The resulting continuous motion percept is both long-range and sharp

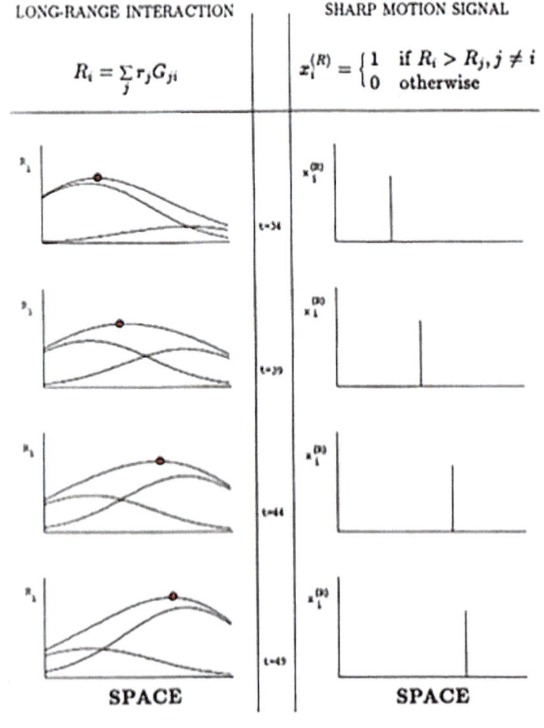

FIGURE 4.7 Dynamics of long-range apparent motion: (a) As a flash waxes and wanes through time, so too do the activities of the cells in its Gaussian receptive field. Because the maximum of each Gaussian occurs at the same position, nothing is perceived to move. (b) If two flashes occur in succession, then the cell activation that is caused by the first one can be waning, while the activation due to the second one is waxing. (c) The sum of the waning Gaussian activity profile due to the first flash and the waxing Gaussian activity profile due to the second flash has a maximum that moves like a traveling wave from the first to the second flash. Reprinted with the author's permission from Grossberg (2021).

are important for survival are long-range apparent motion percepts in naturalistic settings. For example, long-range apparent motion is used to continuously track a prey or predator as it darts behind successive bushes or trees at variable speeds. G-waves can interpolate a continuous trajectory at variable speeds between the successive "flashes" of the prey or predator when it emerges from behind an occlusion. In particular, G-waves can flow *behind* occluding objects to maintain attention on the moving target (Grossberg, 1998, 2021).

4.2 Learning to associate a view of mommy's face to where her hand is in space

In the case of mommy pointing, a G-wave can travel from mommy's currently attended face to her hand as she points at an object of interest. Attending mommy's face can occur for a longer duration than a "flash." Even if mommy's face is attended for a while, when she moves her hand, a G-wave can travel from her face to her hand (Baloch and Grossberg, 1997). As a result, the child shifts her attention to the position of mommy's hand. An association can then be learned from the view-specific category of mommy's face, which is currently stored in short-term memory, to the final position of her hand in space, which is currently stored as the position to which the child is attending. The TELOS model simulates how such an association is learned from the What stream representation of mommy's face to the Where stream representation of her current hand position (Figure 4.8; Brown, Bullock, and Grossberg, 2004).

The acronym TELOS was chosen in part because it is the ancient Greek word *telos* for goal, end, or completion of a plan. TELOS is also an acronym for TElencephalic

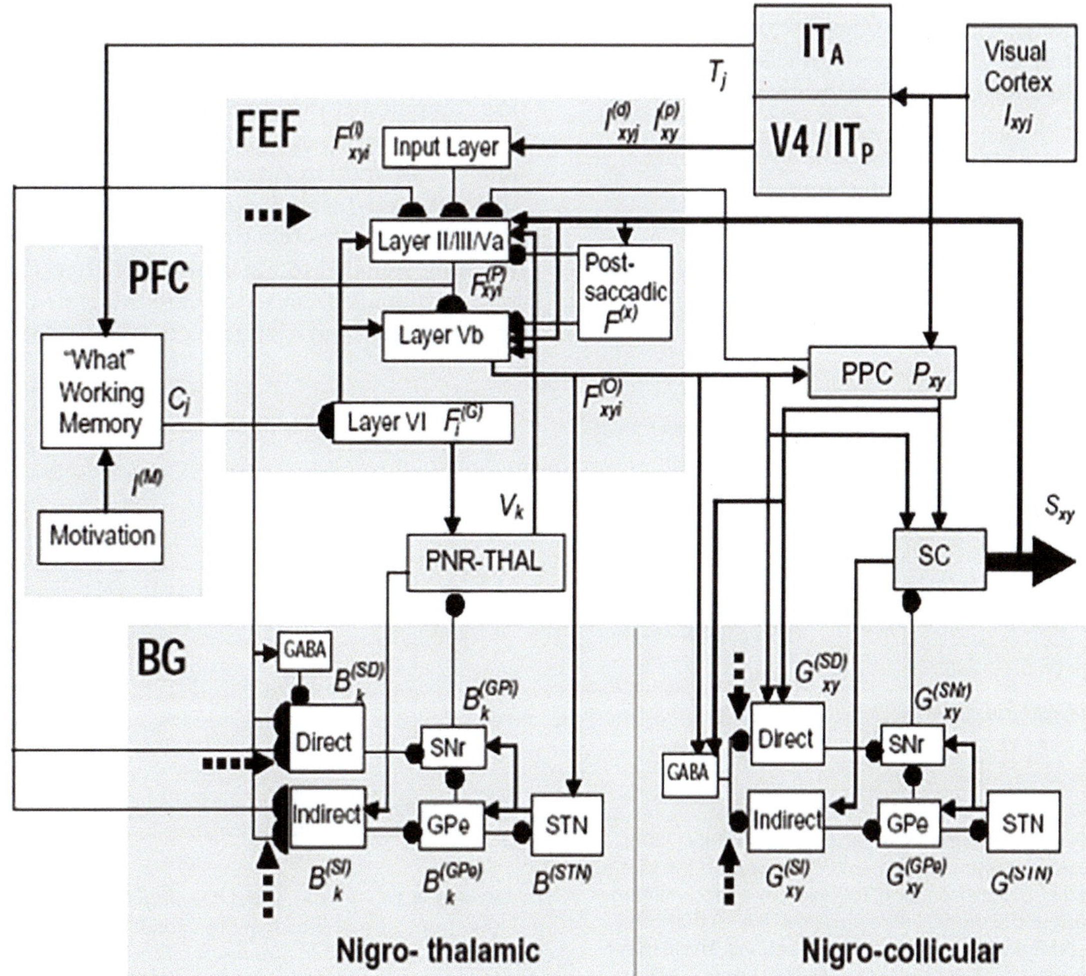

FIGURE 4.8 The lisTELOS model (see Figure 3.15) built upon key processes that were earlier modeled by the TELOS model. See the text for details. Author created.

Laminar Objective Selector. The basal ganglia and cerebral cortex together make up the Telencephalon. The basal ganglia open and close gates that select the outputs of Laminar oculomotor structures, notably the frontal eye field (FEF), in the frontal cortex and the superior colliculus (SC). Coordinated actions of these structures select the current behavioral Objective, whether a desired eye movement or maintenance of eye fixation. The lisTELOS model (Figure 3.15) includes TELOS as a part.

The view-specific category of mommy's face can, during subsequent experiences, use the learned association to predict where mommy is looking, and thus to enable a baby or child to look in the direction that mommy is looking, whether or not she is pointing there.

4.3 How the direction of object motion is converted into fixating eye movements

Considerable brain machinery is needed for the child to convert the motion of mommy's hand into an eye movement that can foveate its endpoint. The MOtion DEcision, or MODE, model, which I developed with my PhD student and postdoctoral fellow, Praveen Pilly (Grossberg and Pilly, 2008), defined and simulated on the computer the motion processing stages that compute the direction of object motion and convert it into saccadic eye movements that maintain fixation upon a moving target. Figure 4.9 summarizes the multiple processing stages in the Where stream whose interactions accomplish this.

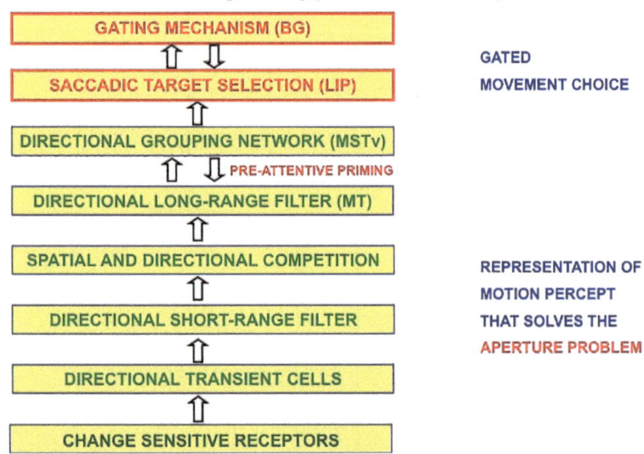

FIGURE 4.9 The MODE model of Grossberg and Pilly (2008) uses motion preprocessing stages, collectively called the Motion BCS (green letters), as its front end, followed by a saccadic target selection circuit in the model LIP region (red letters) that converts motion directions into movement directions. These movement choices are also under basal ganglia (BG) control. MT = Middle Temporal area; MSTv = ventral Middle Superior Temporal area; LIP = Lateral Intra-Parietal area Reprinted with the author's permission from Grossberg (2021).

The first problem that MODE solves is the *aperture problem* that was experimentally studied by Hans Wallach (Guilford, 1929; Wallach, 1935 [Wuerger, Shapley, and Rubin, 1996]). The MODE model's solution of the aperture problem enables our brains to transform directionally ambiguous local motion information that is received by our retinas into a representation of an object's global motion direction and speed. This is the same Hans Wallach whose work was discussed in Chapter 2 about how lightness anchoring may work (Wallach, 1948, 1976) using his highest-luminance-as-white (HLAW) rule whereby the perceptual quality "white" is assigned to the highest luminance in a scene, and gray values of less luminous surface regions are assigned relative to the white standard.

Figure 4.10 illustrates the aperture problem. Wallach (1935) noted that a moving line that is seen within a circular aperture always looks like it is moving in the direction *perpendicular* to its orientation, no matter what its real motion direction is. That is because all local motion signals are ambiguous (purple arrows).

Consider the different percept that arises when viewing a rectangular aperture in which parallel lines move at a fixed speed across the aperture. This display is often called a "barberpole" in the perceptual literature. In this situation, *feature tracking signals* (green arrows), that are computed at the boundaries of the rectangle, define a consistent direction of motion over time. In contrast, the ambiguous motion directions, that are computed at the line positions inside the rectangle (purple arrows again), keep changing through time as the lines move. The feature tracking signals are much stronger than the ambiguous motion signals because they are integrated through time in a consistent way along a rectangle boundary.

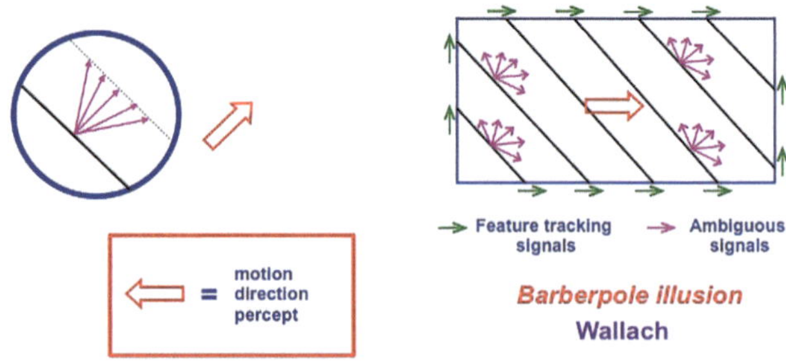

FIGURE 4.10 The perceived direction of an object is derived either from a small subset of directionally unambiguous feature tracking signals (green arrows) or by voting among directionally ambiguous signals when feature tracking signals are not available (red arrows) to determine an estimate of object motion direction and speed (outline red arrow). Reprinted with the author's permission from Grossberg (2021).

The MODE model explains how our brains choose the stronger feature tracking motion direction, while suppressing the ambiguous motion directions. The MODE model also explains how the winning motion direction propagates along the entire length of the line. The result is a percept of horizontal motion of all the lines across the aperture (red outlined arrow). This choice process is called *motion capture*. How it occurs is explained in detail in Grossberg (2021). A video of the aperture problem percept and the barberpole illusion can be seen at https://www.liverpool.ac.uk/~marcob/Trieste/aperture.html.

The MODE model also converts the direction of object *motion* into an eye movement command to *move* in that direction using a transformation from visual cortical area MSTv (ventral medial superior temporal area) to LIP (lateral intraparietal cortex). This transformation is accomplished by the two final processing stages in Figure 4.9. The two red arrows in Figure 4.11 point to the cortical areas that carry out this transformation in our brains.

When a target unpredictably changes its direction and speed of motion, saccadic eye movements are not sufficient to track the target. Instead, several interacting brain regions, including MST and LIP, help to coordinate smooth pursuit eye movements (SPEM), which maintain foveation of the target while it moves smoothly across space, and saccadic eye movements (SAC), which are triggered to restore foveation if the target unexpectedly changes its motion direction and speed. The macrocircuit of this SAC-SPEM model is shown in Figure 4.12 (Grossberg, Srihasam, and Bullock, 2012). As the figure shows, our brains need a lot of computation to solve this problem.

5 Associating Seeing Mommy with Her Name: Building upon Circular Reactions

As learning of mommy's invariant face category stabilizes, the baby can learn to associate it with an auditory production of mommy's name. This ability also requires a Piagetian *circular reaction*. As noted in Section 2, this auditory circular reaction occurs during its own babbling phase (Figure 4.4). During it, babies endogenously babble

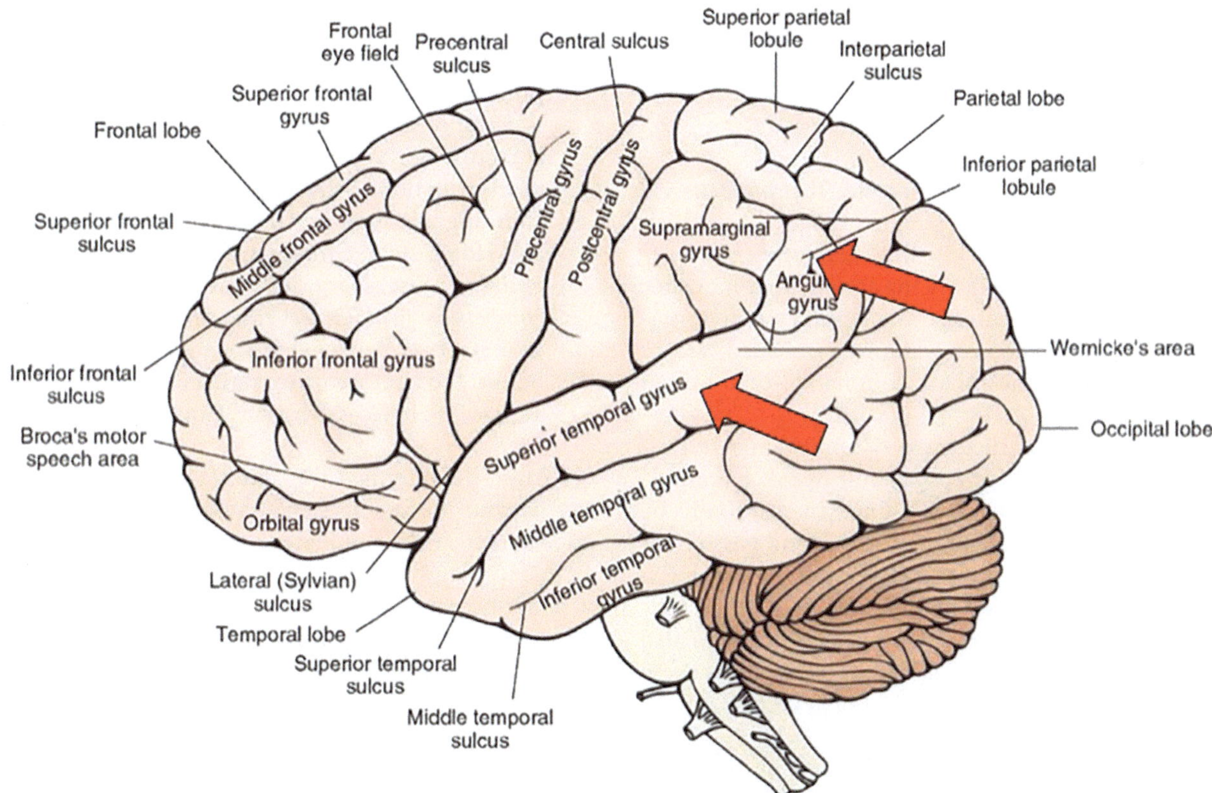

FIGURE 4.11 Anatomical locations (see red arrows) of visual cortical area MSTv (ventral medial superior temporal area) and LIP (lateral intraparietal cortex). Red arrows added by the author. Reprint with permission of Figure 66 in https://www2.tulane.edu/~h0Ward/BrLg/Cortex.html.

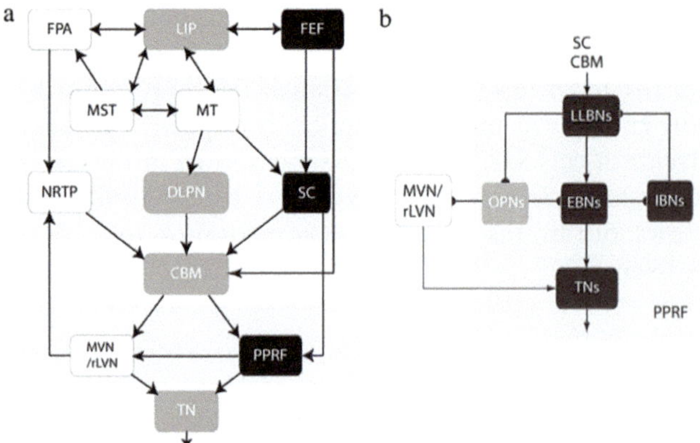

FIGURE 4.12 SAC-SPEM model macrocircuit: (a) Modeled interactions among brain regions implicated in oculomotor control. Black boxes denote areas belonging to the saccadic eye movement system (SAC), white boxes the smooth pursuit eye movement system (SPEM), and gray boxes areas that belong to both systems. LIP-Lateral Intra-Parietal area; FPA-Frontal Pursuit Area; MST-Middle Superior Temporal area; MT-Middle Temporal area; FEF-Frontal Eye Fields; NRTP-Nucleus Reticularis Tegmenti Pontis; DLPN-Dorso-Lateral Pontine Nuclei; SC-Superior Colliculus; CBM-cerebellum; MVN/rLVN-Medial and Rostro-Lateral Vestibular Nuclei; PPRF-Peri-Pontine Reticular Formation; TN-Tonic Neurons. (b) Constituents of the saccade generator in the PPRF, and projection of omnipauser neurons to the pursuit neurons of the MVN/rLVN. Arrows indicate excitatory connections, and semi-circles indicate inhibitory connections. OPN–Omni-Pauser Neurons; LLBN-Long-Lead Burst Neurons; EBN-Excitatory Burst Neurons; IBN-Inhibitory Burst Neurons; TN-Tonic Neurons. Reprinted with the author's permission from Grossberg, Srihasam, and Bullock (2012).

the simple sounds that they can create. This happens via the GO signal from the basal ganglia in Figure 4.4, which is endogenously active during this developmental critical period. The baby also hears the sounds that it creates via auditory Sensory Feedback. The auditory representations of the heard sounds can then be associated, via an Imitative Map, with the motor commands that caused these sounds.

A child uses the learned map to approximately imitate sounds that they hear from adult Speaker Sounds. Their approximations of adult sounds may initially be coarse, but with the help of adult feedback and passive listening to other speakers' utterances, the map is incrementally refined, leading eventually to adult speech and language utterances. These later speech productions are also released by activating basal ganglia GO signals, but these signals are now under the child's volitional control, rather than being endogenously active.

If the name "mommy" is then uttered by the child or another person, the child's positive feelings for her mommy can trigger cognitive-emotional interactions that strengthen the learned association between seeing mommy and saying mommy.

6 Learning Mommy's Movements: Boundaries, Surfaces, Motions, and Rebounds

Before a baby or child can learn short sentences that use mommy's name, like "mommy points" or "mommy walks," a child must first learn to recognize her movements and learn names for them. Suppose that a child sees one side of mommy as she walks. How does her brain represent any object's motion? How does it represent the motions of a complex form like a human body?

Section 4 reviewed how a G-wave can track a moving object, such as mommy's hand as she points. It can also track mommy's moving body. Before an extended object like a body can be tracked, however, the baby's brain needs to preprocess the visual signals that it receives on her retinas to create a sufficiently complete and stable representation of her body that can be tracked. Several problems need to be solved by her brain to achieve this:

> Objects are often seen in multiple lighting conditions. A baby needs to compensate for them by *discounting the illuminant* to prevent confusions of object form with ever-changing conditions of object illumination. I mentioned the discounting process in Chapter 2. Discounting the illuminant is accomplished by a recurrent on-center off-surround network whose neurons obey the membrane equations of neurophysiology, also called shunting interactions (Grossberg, 1988; Grossberg and Hong, 2006; Grossberg and Todorovic, 1988; Hong and Grossberg, 2004). Such a *recurrent competitive field* (Grossberg, 2013) occurs at multiple stages of neuronal processing, including within the photosensitive retina (Grossberg, 1987a, 1987b; Grossberg and Hong, 2006).

After discounting, the illuminant, multiple interacting cortical processing stages complete the *boundary* and *surface* representations that are used to recognize mommy. Boundaries and surfaces are incomplete when registered by a retina because retinas contain a *blind spot* where retinal photoreceptors send their signals along axons to form the optic nerve. As Figure 2.15 illustrates, the blind spot is as large as the fovea, but we never see it. Figure 2.15 also shows that the retina is covered by nourishing veins. These occlusions are also not seen (Kolb, H. Webvision: The Organization of the Retina and Visual System: http://webvision.med.utah.edu/book/part-i-foundations/ simple-anatomy-of-the-retina/).

These gaps in retinal processing do not interfere with conscious vision due to the combined effect of visual processes acting at multiple processing stages. First, our eyes jiggle rapidly in their orbits, thereby creating transient signals from even stationary objects. These transients refresh the firing of retinal neurons. Because the blind spot and retinal veins are attached to the jiggling retina, they create no transients. Their inputs therefore fade, resulting in a retinal image where light is occluded at multiple positions (Figure 4.13).

The visual cortex uses multiple processing stages to create complete enough boundary and surface representations to direct conscious looking and reaching toward valued goal objects. These stages overcome the processing uncertainties at the positions of the blind spot and retinal veins, where no visual signals are registered. I call this process *hierarchical resolution of uncertainty* (Grossberg, 1987a, 1987b, 1994, 2021; Grossberg and Mingolla, 1985a, 1985b; Grossberg and Todorovic, 1988). Completed boundary and surface representations of mommy's form are computed in the visual cortical area V4 of the What cortical stream from retina-to-LGN-to-V1-to-V2-to-V4 (Figure 2.17; Grossberg, 1994. 1997, 2016; Grossberg and Pessoa, 1998; Kelly and Grossberg, 2000).

Mommy's motion could produce streaks of *persisting* activations at cells that fired when her body moved across earlier retinal positions, much like a moving comet can leave a long tail. Such persisting streaks could seriously degrade percepts of a scene, as well as recognition of mommy's body. Remarkably, these streaks are limited by the same mechanism that helps to generate a representation of mommy's body in the first place.

EVERY LINE IS AN ILLUSION!

Boundary completion
Which boundaries to connect?

Surface filling-in
What color and brightness do we *SEE*?

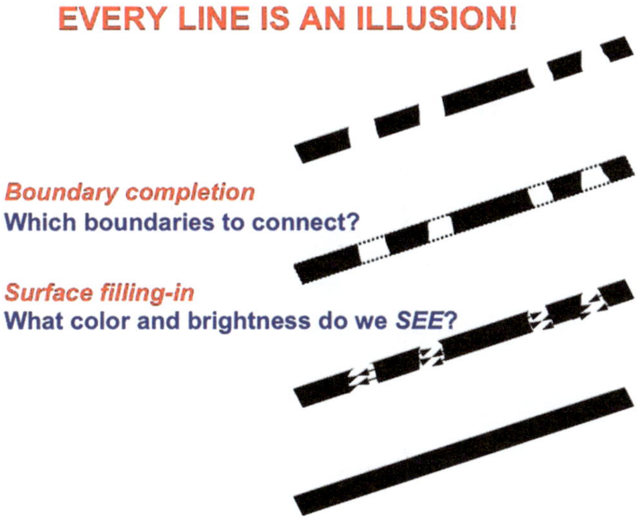

FIGURE 4.13 "Every line is an illusion" because regions of the line that are occluded by the blind spot or retinal veins (top figure: the unoccluded regions of the retinal input are a series of black colinear fragments) are completed at higher levels of brain processing by boundary completion (second figure down: dotted lines indicate boundary completion) and surface filling-in (third figure down: bidirectional arrows indicate filling-in within the completed boundaries) to yield a solid connected bar (fourth figure down). Reprinted with the author's permission from Grossberg (2021).

The primary circuit that prevents streaking is a *gated dipole* opponent process (Figure 4.14). A gated dipole ON channel selectively responds to onset of mommy's image when it crosses the receptive field of its cells. As mommy moves off the positions that define the ON channel receptive field, the gated dipole OFF channel at those positions triggers a transient *antagonistic rebound* which inhibits the ON response to mommy (the transient OFF-response in Figure 4.14), thereby limiting its persistence (Francis and Grossberg, 1996a, 1996b; Francis, Grossberg, and Mingolla, 1994; Grossberg, 2021).

How is an antagonistic rebound caused? The antagonistic rebound can occur because, in both the ON and OFF channel, there is a habituative transmitter gate (see y_1 and y_2 in Figure 4.14). A transmitter is inactivated, or habituates, in an activity-dependent way when its channel is activated, leading to transmitter release. Released transmitter activates a postsynaptic cell.

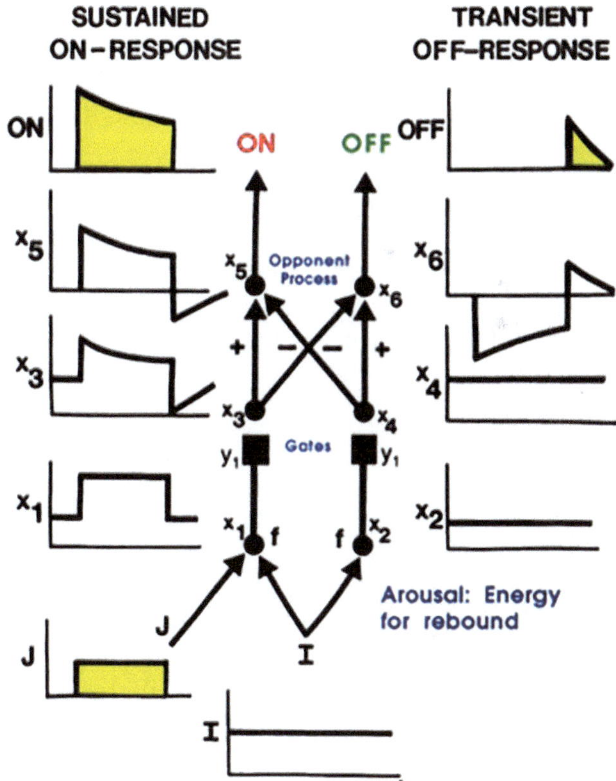

FIGURE 4.14 A gated dipole opponent process can generate a sustained habituative ON response in its ON channel in response to a steady input, J, to it, and a transient antagonistic rebound from its OFF channel in response to offset of J. The habituative transmitter gates, or medium-term memory (MTM) processing stages (filled black square synapses), are needed for a rebound to happen. See the text for details. Reprinted with the author's permission from Grossberg (2021).

Mommy's body activates the ON channel via the phasic input J in Figure 4.14. While J is on, y_1 gets progressively inactivated. Transmitters multiply the signals in their ON and OFF channels. When mommy's image, and thus J, at a given position shuts off, only equal arousal signals I input to the ON and OFF channels. The net signal to the habituated ON channel is temporarily less than the net signal to the OFF channel. A transient OFF response thus occurs, which is the antagonistic rebound that limits the persistence of mommy's representation in the visual cortex.

Opponent processes occur throughout our brains. For example, as summarized in Figure 4.15, looking at lines with different orientations that all intersect at a point for a sufficiently long time, and then shifting attention to look at a white background, can trigger an antagonistic rebound of circular lines. This is called the Mackay illusion. Looking at water flowing downward for a sufficiently long time, and then shifting attention to look at a white background, can trigger an antagonistic rebound of upward motion. This is called the waterfall illusion.

There are many other aftereffects due to antagonistic rebounds. Here is are two more: Looking at the color red for a sufficiently long time, and then shifting attention to look at a white background, can trigger an antagonistic rebound of green. Experiencing fear for a long time, whose painful inducing stimulus shuts off suddenly, can trigger an antagonistic rebound of relief.

In all these cases, a gated dipole helps to regulate their ON and OFF responses. Figure 4.15 shows that antagonistic rebounds in orientation differ by 90 degrees from their inducing stimuli, and by 180 degrees in their motion direction. These different symmetries in orientation and direction reflect the fact that *orientation* is computed in the V1-to-V2-to-V4 What cortical stream as part of visual form perception, whereas *direction* is computed in the V1-to-MT Where cortical stream as part of visual motion perception (Figure 2.17).

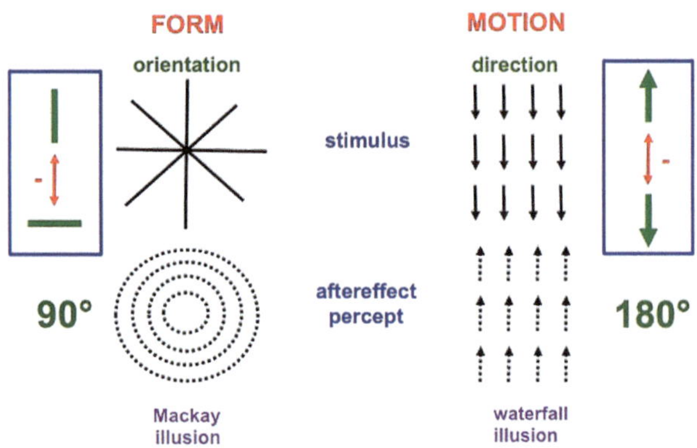

FIGURE 4.15 The MacKay and waterfall illusion aftereffects dramatically illustrate the different symmetries that occur in the orientational form stream and the directional motion stream. See the text for details. Reprinted with the author's permission from Grossberg (2021).

7 Categorizing Mommy's Movements: What-to-Where FORMOTION Interactions

As I just noted, a series of changing positions of a moving visual form like mommy's body is computed in the What cortical stream. Perceiving a series of an object's changing positions is not, however, the same thing as perceiving its motion. Indeed, object motion is computed in the Where cortical stream.

Computations of form and motion are done in separate cortical streams because object form is sensitive to the *orientation* of an object's boundaries, thereby enabling pairs of like-oriented boundaries to compute binocular disparity (Figures 2.9 and 2.10). In contrast, object motion is sensitive to an object's *direction* of motion. A good estimate of motion direction pools directional estimates from all an object's boundaries, across all their different orientations, that move in the same direction (Figure 4.16). A computation of motion direction hereby destroys the information that computes object orientation.

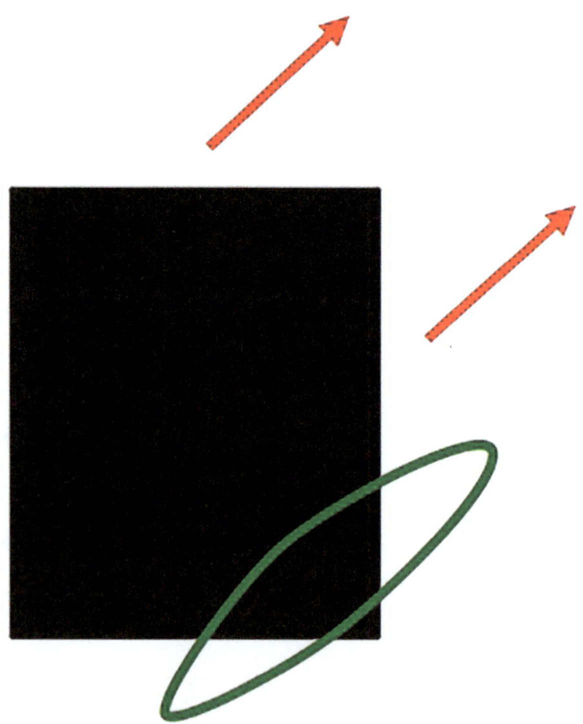

FIGURE 4.16 A directionally tuned motion-sensitive cell pools all possible contrast-sensitive sources of information that are moving in its preferred direction, and at similar motion directions. In this figure, the black rectangle is moving diagonally upward. The directional motion cell has a similar preferred motion direction. It pools diagonal motion signals from both the black-to-white vertical right-hand edge of the rectangle and from the white-to-black horizontal bottom edge of the rectangle. Reprinted with the author's permission from Grossberg (2021).

These parallel computations of object form and object motion are *computationally complementary* (Figures 1.31, 1.32, and 1.38; Grossberg, 1991). The What cortical stream uses representations of object form to learn categories whereby to recognize objects. The Where cortical stream uses representations of object motion to navigate while avoiding obstacles in cluttered scenes, and to track moving targets (Browning, Grossberg, and Mingolla, 2009a, 2009b; Elder, Grossberg, and Mingolla, 2009; Grossberg, 2021; Grossberg, Mingolla, and Pack, 1999).

The Where stream needs a complete visual representation of an object's form to successfully track it. Such a representation in the What stream is topographically mapped into the Where cortical stream, whose dynamics can then track its motion through time. The 3D FORMOTION model simulates how this happens (Berzhanskaya, Grossberg, and Mingolla, 2007). Figure 4.17 shows the processing stages of the 3D FORMOTION model. These include the stages that are needed to complete a representation of an object's form boundaries in cortical area V2 before they are topographically mapped into cortical area MT of the motion stream, where they can be tracked through time.

When a complex object like mommy walks or points, different parts of her body move with different directions and speeds. Leg movements while walking, and arm movements while pointing, are perceived relative to the *reference frame* of mommy's moving body that is tracked to keep her foveated. The MODE, or Motion DEcision, model and the SAC-SPEM model together explain how tracking happens (Figures 4.9 and 4.12).

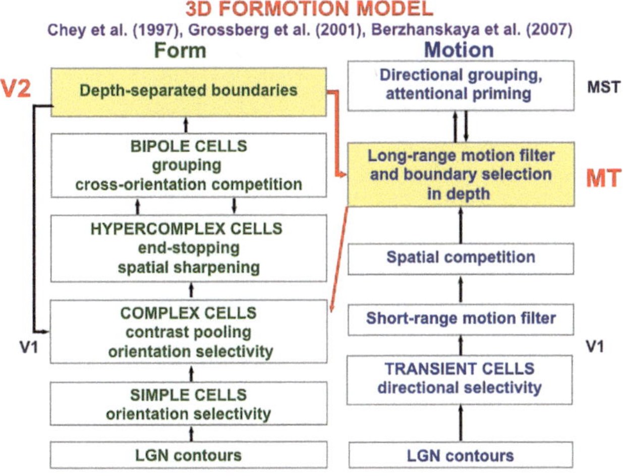

FIGURE 4.17 The 3D FORMOTION model combines mechanisms for determining the relative depth of a visual form in = cortical area V2 with mechanisms for both short-range and long-range motion filtering and grouping in cortical area MT. A formotion interaction across these form and motion processing streams, from V2 to MT, is predicted to enable the motion stream to track objects moving in depth. Reprinted with the author's permission from Grossberg (2021).

8 Peak Shift and Behavioral Contrast: Directions of Moving Parts and Wholes

It remains to explain how the movements of mommy's body define a reference frame relative to which the directions of her arm and leg movements are perceived (Grossberg, 2011; Grossberg, Leveille, and Versace, 2011). Various types of reference frames are needed, including a retinocentric frame, where an object is coded relative to the location of the activity it induces on the retina, and an object-centered reference frame (Bremner, Bryant and Mareschal, 2005; Wade and Swanston, 1996), where an object is perceived relative to other objects. For example, on a cloudy night, the moon may appear to be moving in a direction opposite to that of the clouds. In a laboratory setting, this effect can be replicated in *induced motion* experiments, wherein the motion of one object appears to cause motion in the opposite direction in another, otherwise static, object (Duncker, 1938).

Why is not the concept of object reference frame, and the properties of relative motion of object parts relative to that reference frame, too clever for evolution to discover? As I have always found throughout my work as a theorist, an elementary neural mechanism that occurs in multiple parts of the brain is used. This is another example of how evolution has exploited variations and specializations of basic designs to carry out multiple functions.

The neural property that naturally computes object frames and the relative motion of their parts is called *peak shift and behavioral contrast*. This property occurs in *all* recurrent shunting on-center off-surround networks whose cells interact via Gaussian receptive fields. As my previous discussions have illustrated in Chapter 2, such networks occur throughout our brains.

Peak shift and behavioral contrast occur when two inputs activate an on-center off-surround network whose neurons respond selectively to different *motion directions* in a *topographic map*. In particular, neurons that are sensitive to oblique-upward motion directions occur between cells that are sensitive to upward and horizontal motion directions (Figure 4.18). Likewise, cells that are sensitive to oblique-downward motion directions occur between cells that are sensitive to downward and horizontal directions.

Motion directional inputs to such a network from mommy's moving body and arm lead to a *vector decomposition* under the following conditions: Suppose that the horizontal leftward motion direction of mommy's body activates a horizontally tuned motion direction cell in the middle of its Gaussian on-center. Oblique motion directions of mommy's arm activate the Gaussian

WHAT IS THE MECHANISM OF VECTOR DECOMPOSITION?
Grossberg, Leveille, and Versace, 2011

Prediction: DIRECTIONAL PEAK SHIFT!

...specifically, a peak shift due to Gaussian lateral inhibition

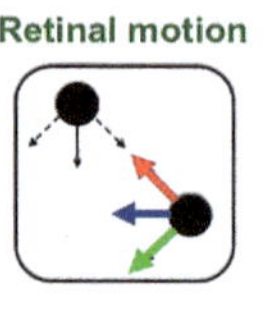

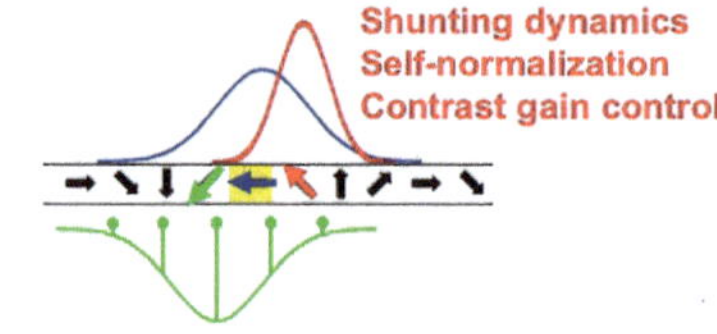

part motion
common motion
relative motion

FIGURE 4.18 (left panel) The upper-left black dot moves up and down, while the lower-right black dot moves left (blue arrow) and right. Under proper spatial separations and motion speeds, the dots are perceived to be moving back and forth (red arrow) to and from each other, as their common motion moves diagonally downward (green arrow). (right panel) This motion illusion is due to a directional peak shift (red Gaussian) in a directional hypercolumn (spatially arrayed orientationally tuned cells). The common motion direction activates an inhibitory Gaussian receptive field (green), while the horizontal part motion activates an excitatory Gaussian receptive field (blue) over the directional hypercolumn. When these inhibitory and excitatory Gaussians interact, the result is an excitatory Gaussian (red) whose maximal activity is shifted away from the inhibitory Gaussian (peak shift). Because interactions across this recurrent shunting on-center off-surround network are normalized, the red Gaussian, being narrower than the blue Gaussian, has a higher maximal activity, thereby causing behavioral contrast. Reprinted with the author's permission from Grossberg (2021).

off-surround of the horizontal motion direction cell. When the oblique velocity vectors are subtracted from the horizontal velocity vector through time, the resulting velocity vectors generate oscillatory motion vectors of the arm's motion directions relative to the horizontally moving body.

I will use a classical example of such a vector decomposition to clarify the above comments about how we perceive the relative motion of mommy's body and arm. This example is due to the Swedish psychophysicist Gunnar Johansson (Johannson, 1950). The display consists of two moving dots, one moving up and down, and the other moving left and right, in phase with each other (see box displaying Retinal motion in Figure 4.18) There is a common motion component (in green) which, when it is subtracted from the horizontal and vertical motions (in blue), leaves in-and-out relative motion vectors between the two dots (in red). In other words, the in-and-out (red) motion direction percepts of the dots are computed relative to the moving (green) reference frame that they together define.

This illusion illustrates how *peak shift* and *behavioral contrast* separate the relative motion direction (see the red directional Gaussian receptive field in Figure 4.18) from the common motion direction (see the blue directional Gaussian receptive field). The direction of the red Gaussian is shifted relative to the direction of the blue Gaussian (peak shift), and the red Gaussian has a higher peak activity than the blue Gaussian (behavioral contrast). This percept, as well as other famous motion illusions that involve moving reference frames, is explained and simulated in Grossberg, Leveille, and Versace (2011) and reviewed in Grossberg (2021).

As mommy walks, her leg that is further from the child is partly occluded to different degrees by her closer leg. A completed percept of the partially occluded leg is created and maintained using boundary and surface completion processes that realize *figure-ground separation* (e.g., Figures 2.2, 2.4, and 2.46; Grossberg, 1994, 1997, 2016, 2021; Kelly and Grossberg, 2000). The completed representations can then be recognized while the child's brain computes their motion directions and speeds.

The *chopsticks illusion* of Stuart Anstis (Anstis, 1990) illustrates these properties using simpler stimuli. Anstis studied a pair of rectangular black chopsticks moving in opposite directions on a white background, much as mommy's legs do. Figure-ground separation enables us to see one chopstick moving in front of the other, even though they are both really moving in a two-dimensional image plane, and intersect in a black X region that does not differentiate one chopstick from the other.

9 Learning "mommy walks left" while Observing Her Move: Nouns and Verbs

How does a child's brain learn both a perceptual category and a language category for "walk" and "walking"? In particular, how is the perceptual representation of the *verb* in "mommy walk(s)" represented in the brain?

To do this, a view, or succession of views, of mommy standing up with her legs on the ground in a characteristic walking pose is classified, despite variations in perceptual features that may differ across individuals who walk. This is accomplished using an adaptive filter much like the one that associates multiple view-specific recognition categories of an object with a view-invariant recognition category of that object (see Sections 2.17 and 2.18). Large receptive fields average across details about mommy's body, such as her clothing, to extract mommy's overall shape and its silhouette.

Multiple oriented scales, or filter sizes, from fine through coarse, initially process all incoming visual information (recall Figures 2.41–2.43). Higher-order processing stages select the scales that are most informative in different situations by associating all the scales with their predictive consequences. Only the informative scales will learn strong associations. Thus, as finer scales learn to categorize mommy's facial views, coarser scales learn to categorize her larger-scale actions like walking.

Suppose that the co-occurrence of two perceptual categories—of mommy's face and her walk pose—together trigger learning of a category that selectively fires when mommy walks. This *conjunctive* category can be associated via learning with the heard utterance "mommy walks" or "mommy is walking" via a bidirectional *associative map*.

A single pose of walking is often enough to recognize walking, just as a single pose of standing is enough to recognize that posture. Recognition of movements that cannot be effectively categorized using a single pose require the Where cortical stream.

10 Learning Movement Directions and Using Them to Track Moving Targets

As I noted in Section 4, interacting brain regions control eye movements that maintain foveation of an attended moving object like mommy. Suppose that the linear motion of mommy's body activates a long-range directional filter. Such a filter has an elongated shape that adds inputs from an object's motion signals over time that move in its preferred direction when they cross its receptive field (Figure 4.16). Arrays of such filters exist in the Where cortical stream, each tuned to a different preferred direction (Figure 4.18; Albright, 1984; Rodman and Albright, 1987). These filters compete across direction at each position to choose the filter, or nearby subset of filters, that are most activated by the moving object. They are a Where stream analog of the orientation columns in cortical area V1 of the What stream (Figure 3.18; Hubel and Wiesel, 1962, 1968, 1977).

Suppose that, when a directional filter is activated enough to fire for a sufficiently long time, its signals trigger learning of a directional motion category in the Where stream (Farivar, 2009). This category can then be associated, through learning, with a word or phrase in the What stream that best describes it, such as "left."

After the perceptual categories for recognizing "mommy," "walks," and "left" are learned, they can be associated with a phrase like "mommy walks left" uttered by an adult speaker, in the manner that I will now explain.

11 Learning to Say "mommy throws the ball" while Observing Her Movements

How does a baby or child learn to say "Mommy throws ball" while observing mommy doing that? The first part of the sentence, "Mommy throws," can be understood in much the same way as "Mommy walks." Suppose the child sees mommy in a profile view pull her arm back before thrusting it forward. Either extreme arm position may be sufficient to learn a category for "throw" in the What stream. The movement's motion can also be categorized in the Where stream.

If the arm that mommy uses to throw the ball is further away than her body, and thus partially occluded by her body during the movement, then a G-wave from the pulled back location of her arm to its thrust forward position can flow "behind" her body, much as G-waves can flow behind scenic clutter when tracking a predator or prey.

As mommy completes the throw, a ball emerges from her hand and moves in the same direction through space. Often, attention will focus on mommy's face before she throws the ball. When the ball is thrown, attention flows to her arm, and then to the ball, via a G-wave. This temporal sequence of recognition events is thus: mommy, throws, ball. Perceptual categories that classify these events are activated and stored in a *perceptual working memory* in their correct temporal order, and trigger learning of a sequence category, or *list chunk*, that classifies the entire sequence. A heard sentence category of "mommy throws ball" can simultaneously be stored in a *linguistic working memory*, and trigger learning of its own list chunk.

The active linguistic list chunk can then learn an association to the active perceptual list chunk, and conversely. These list chunks also send learned top-down signals to the working memory sequences that they categorize, which can then be read out from working memory in the correct order, under volitional control by a GO signal from the basal ganglia. Using these various working memory and learned item and sequence categories enables an observer, upon seeing these events, to elicit a descriptive sentence. Further details about these processes will now be described.

12 Learning to Say "Watch mommy throw the ball" before or as She Does So: Nouns and Verbs

A sentence such as "watch mommy throw the ball" or "look at mommy throw the ball" can prime a child to orient to mommy to experience her throwing the ball. Such priming is based on previous learning experiences:

> First, a baby or child learns the meaning of the command to "Watch" or "Look" from a teacher who points in the direction that the baby or child should look while one of these words is uttered. Learning this skill required the social cognitive ability to share *joint attention* between the pointing teacher and the student. How our brains accomplish this feat was discussed in Section 4. The learner's attention can hereby shift to foveate mommy in response to the word "Watch" or "Look" while a teacher looks at and points to mommy.

The teacher may also utter the word "mommy" to form the sentence "Watch mommy." How is the meaning of the sentence "watch mommy" learned? The words "Watch" and "mommy" get stored in working memory in the correct temporal order, and then trigger learning of a list chunk. Chapter 2 reviewed how working memories are designed and list chunks are learned. Section 13 will review these concepts again from the perspective of the current context.

As "Watch mommy" is getting stored in working memory and being learned as a list chunk, the baby or child is orienting toward mommy's face. The active list chunk can then learn to be associated with the active invariant face category of mommy. While it is active, the invariant face category of mommy reads out priming signals that subliminally activate all the view-selective and position-selective categories of mommy's face (Figure 2.25). The view category that codes the view and position in the visual field where mommy's face is observed will be activated to suprathreshold activities.

Because of its position selectivity, this view category in the What cortical stream can then read out a motor priming command to the corresponding positional representation of mommy's face in the Where cortical stream via a What-to-Where stream interaction (see View Category pathway to Target Position in Figure 2.26).

When the position of mommy's face is primed, activation of the list chunk that codes "watch mommy" can also activate a nonspecific GO signal from the basal ganglia to the Where cortical stream. Taken together, the positional prime and the GO signal fully activate the primed position and thereby elicit orienting movements to foveate mommy.

With this background, we can ask what parts of the brain are used to store and understand the sentence "Watch mommy throw the ball"? For starters, note that the meanings of the *verbs* "watch" and "throw" are understood by using cortical representations in the Where cortical stream. In contrast, the meanings of *nouns* like "mommy" and "ball" are understood by activating cortical representations in the What cortical stream. Both noun and verb word representations are stored in the temporal and prefrontal cortices of their respective streams.

Thus, understanding the meaning of the sentence "Watch mommy throw the ball" requires alternate switching between the noun and verb representations in the What and Where cortical streams, respectively. Words like the article "the" that helps to structure sentences are part of *syntactics* and discussed in Section 15.

In contrast, the branch of linguistics that is concerned with meaning is called *semantics*. Unlike the present chapter, traditional semantic studies do not link language with the real-word perceptual and affective experiences that they denote during language learning; for example, Jackendoff (2006). Without such associative links, language does not acquire meaning.

13 Storing Linguistic Items in Working Memory and Learning Their List Chunks

Before continuing with my analysis of how a baby or child can learn language meanings, I will review some basic concepts about what working memories are and how they work. As I explained in Chapter 3, working memories, and models thereof, temporarily store sequences of items or events as *item chunks*. An *item chunk* responds selectively when the distributed features that it represents are presented (e.g., a phoneme or musical note). Every time that we attentively listen to a sufficiently short series of words, they can be stored in a linguistic working memory, like the sentence: "Watch mommy throw the ball."

If certain sequences of words are repeated often enough, they can be learned by the process of *unitization* or *chunking*. If a series of words can be unitized into a single chunk, that chunk is called a *list chunk*. A *list chunk* selectively responds to prescribed sequences of item chunks that are stored in working memory. These processes occur in brain regions like ventrolateral prefrontal cortex (VLPFC) and dorsolateral prefrontal cortex (DLPFC); see Figure 4.1.

Interactions of list chunks with other brain regions, including perirhinal cortex (PRC), parahippocampal cortex (PHC), amygdala (AMYG), lateral hypothalamus (LH), hippocampus (HIPPO), and basal ganglia (BG), enable predictions and actions to be chosen that are most likely to succeed based on sequences of previously rewarded experiences. Figure 3.8 summarizes a model macrocircuit of the predictive Adaptive Resonance Theory, or pART, model of cognitive and cognitive-emotional dynamics that explains how these interactions work. The seven relevant prefrontal regions that are involved in storing, learning, and planning sequences of events are colored green at the top of Figure 3.8.

Grossberg (1978a, 1978b) introduced a neural model of working memory, which has been incrementally developed to the present time. In it, a sequence of inputs that occurs one at a time is stored as an evolving *spatial pattern* of activation of *item chunks* that code the cell populations of the working memory (Figure 3.6). This working memory model is called an Item-and-Order model because individual cell populations represent list *items*, and their temporal *order* of occurrence is stored by an *activity gradient* across the populations.

The Item-Order-Rank, or IOR, model generalizes the Item-and-Order model to enable storage in working memory of lists in which some items are repeated, as in the sequence of letters "ABACBD," or the sequence of repeated words "my true love is true" (Bradski, Carpenter, and Grossberg, 1994; Grossberg, 2018, 2022; Grossberg and Pearson, 2008; Silver et al., 2011). Other kinds of events can also be temporarily stored in working memory, such as the sequence of turns taken during navigation to a goal, of arm movements made during a dance, or of musical notes played in a melody. As noted in Chapter 3, a *single* canonical circuit design, suitably specialized, can store auditory, linguistic, spatial, motor, or musical sequences in multiple working memories that operate in parallel in the prefrontal cortex.

IOR working memories provide unified and principled explanations of many psychological and neurobiological data about working memory and list chunk dynamics. No less important is the fact that they explain why and how sequences of items and events that are stored in a working memory can be learned and stably remembered through time as list chunks (Grossberg, 2021, 2022).

Anyone interested in cognitive psychology and neuroscience needs to learn about Item-and-Order working memories because they are the *unique* kind of circuits that embody two simple postulates that are hard to deny and which enable their list chunks to be learned and stably remembered: the *LTM Invariance Principle* and the *Normalization Rule*. These postulates were used to derive mathematical equations for Item-and-Order working memories when I introduced them in Grossberg (1978a, 1978b).

Most importantly, the LTM Invariance Principle prevents storage of longer lists of events in working memory (such as MYSELF) from causing catastrophic forgetting of previously learned list chunks of its shorter embedded lists (such as MY, SELF, and ELF). The LTM Invariance Principle guarantees that if bottom-up inputs lead to storage of a familiar list chunk, say for the word MY, then continuing to present the word SELF to complete storage of the novel word MYSELF will not cause forgetting of the learned weights that activated the list chunk of MY.

The Normalization Rule just says that the *maximum total activity* that is stored across a working memory is independent of the number of activated cells. This Rule follows from the fact that the cells in an Item-and-Order working memory compete among themselves via a *recurrent shunting on-center off-surround network* (Figure 3.7). Such networks occur ubiquitously in our brains because they solve what I have called the *noise-saturation dilemma* (Grossberg, 1973, 2021). Solving this dilemma insists that cell activities in such a network can store the *relative* sizes, and thus importance, of the inputs that activated them, without flattening the pattern of activity with *saturation*—in response to large input sizes—or obscuring them in cellular *noise*—in response to small input sizes.

This is a "dilemma" because cells have only a finite number of excitable sites. If inputs are too large, they could in principle activate all the cellular sites, thereby causing saturation. But if inputs are too small, their activations could get contaminated by cellular noise. I derived recurrent shunting on-center off-surround networks from a thought experiment that shows how inhibition balances excitation to prevent this problem. See Appendix C in Grossberg (1980), or Chapter 2 in Grossberg (2021).

As I explained in Chapter 3, the LTM Invariance Principle and Normalization Rule also imply that only sufficiently short lists can be stored in working memory in a way that enables their performance in the correct temporal order. In particular, a stored *primacy gradient* of activity enables items to be recalled in the correct temporal order, whether in language or music. In a primacy gradient, the first item in the sequence is stored in working memory with the most activity, the second item in the sequence activates its item chunk with the next largest activity, and so on, until all items are stored (Figure 3.7). For example, the primacy gradient that stores a sequence "A-B-C" of items stores "A" with the highest activity, "B" with the second highest activity, and "C" with the least activity.

A stored *spatial* pattern in working memory is *recalled* as a *temporal* sequence of items when a *rehearsal wave*, or GO signal, uniformly, or nonspecifically, activates all the working memory cells (Figure 3.7). This rehearsal wave is emitted by the basal ganglia, or BG (Figures 3.15 and 3.19). The cell population with the highest activity is read out fastest because it exceeds its output threshold fastest. While it is read out, it self-inhibits its working memory representation via a recurrent inhibitory interneuron (Figure 3.7), a process that is called *inhibition-of-return* in the cognitive science literature (Posner et al., 1985). Inhibition-of-return prevents perseverative performance of the most recent item.

Because each rehearsed item representation self-inhibits (Figure 3.7), the cell population with the next largest activity can be read out, and the cycle repeats until the entire sequence is performed. I explained and quantitatively simulated many psychological and neurobiological data about working memory dynamics in laminar cortical circuits with my PhD student, Lance Pearson, using an Item-and-Order model in Grossberg and Pearson (2008), and reviewed our results in Grossberg (2021).

The flexibility of working memory makes it possible to assemble novel sentences on the fly, since all that matters, after multiple sources input to the working memory, is the final *activity gradient* that determines the temporal order in which the words in a phrase are performed. Phrases that are stored using a *primacy gradient* can then be performed in the order that they are loaded into working memory. Whether during linguistic, motor, or musical performance, such a grouping, or phrase, is stored before being unitized into list chunks that enable automatic future performance of whole sequences of words.

As noted in Figure 3.6, just three interacting processing levels are sufficient to store, learn, and perform long sequences of words or notes that include repeats, such as in the lyric "our true love was true." Our brains do not need, nor do they have, many processing levels to store, learn, and perform sequential behaviors. The pART neural architecture in Figure 3.8 illustrates that each brain region carries out different tasks with its own characteristic anatomy, in contrast to Deep Learning models that may include over one hundred networks in a hierarchy, each with similar connectivity (Srivastava, Greff, and Schmidhuber, 2015).

14 Basal Ganglia Gate Percepts, Concepts, Feelings, and Actions

As noted in Chapter 3, Section 4.3, all parts of our brain, including linguistic, cognitive, and motor working memories, are gated ON and OFF by the basal ganglia, which are part of multiple feedback loops running through all cortical representations of percepts, concepts, feelings, and actions (Alexander and Crutcher, 1990; Alexander DeLong, and Strick, 1986; Brown, Bullock, and Grossberg, 1999, 2004; Dubois et al., 2007; Grahn, Parkinson, and Owen, 2009; Grossberg, 2016, 2021; Hikosaka and Wurtz, 1983, 1989; Middleton and Strick, 2000; Silver et al., 2011). The lisTELOS model (Figure 3.15) explains, for example, how three different basal ganglia loops are used to store, choose, and perform sequences of saccadic eye movements. Similar circuits help to store, choose, and perform all kinds of linguistic, motor, and musical sequences.

15 Learning Definite and Indefinite Articles and How to Use Them in Sentences

The sentences that have been considered so far have not included definite or indefinite adjectives, also called *articles* (Villiers and Villiers, 1973). The full richness of English language meanings cannot, however, be understood without articles. To start, let us compare the meanings of the indefinite article "a" and the definite article "the" in the phrases "a ball" and "the ball."

The sentence "It is a ball" could refer to any ball. In contrast, the sentence "Watch the ball" refers to a particular ball. The following quote about **Definite and Indefinite articles**, from *Study.com*, explains this distinction in greater detail:

> "An article is a word used to modify a noun, which is a person, place, object, or idea. Technically, an article is an **adjective**, which is any word that modifies a noun. Usually adjectives modify nouns through description, but articles are used instead to point out or refer to nouns. There are two different types of articles that we use in writing and conversation to point out or refer to a noun or group of nouns: definite and indefinite articles.
>
> The definite article (the) is used before a noun to indicate that the identity of the noun is known to the reader. The indefinite article (a, an) is used before a noun that is general or when its identity is not known. There are certain situations in which a noun takes no article."

16 Using Definite and Indefinite Articles with Nouns and Verbs

Consider the variations: "It is the ball" or "That is the ball" versus "It is a ball" or "That is a ball," or the analogous sentences "Watch the ball" vs. "Watch a ball." The word "is" may be ambiguous in more than one sense: It can precede a *noun* (object word) or a *verb* (action word). For example, the phrase "is a" disambiguates "is" to precede a noun. In contrast, "is throwing" illustrates how "is" can precede a verb. You can say that "Mommy is throwing the ball" or "Mommy is throwing a ball" depending upon whether a particular ball is intended.

How does a baby or young child learn the difference between "Mommy throws a ball" and "Mommy is throwing a ball"? Or the difference between "Mommy throws the ball" and "Mommy is throwing the ball"?

Both kinds of sentences can refer to the same action. Replacing "throws" by "is throwing" can emphasize that the action as occurring now, and can be learned by imitation from a teacher while witnessing the event.

17 Articles and Their Nouns: Attentional Blocking, Unblocking, and Familiarity

How does the child's brain learn to translate the experience of *seeing* a/the ball being thrown into *saying* a verbal phrase like "a/the ball is being thrown"? Or saying a verbal phrase like "a big ball" or "the green ball" is being thrown?

One ongoing kind of scientific experiment that aims at clarifying such issues uses functional neuroimaging to study the brain activity patterns that are elicited by adjective–noun phrases (e.g., Chang et al., 2009; Fyshe et al., 2016, 2019; Kochari et al., 2021; Lange, Perret, and Laganaro, 2015). These studies, as well as other studies of how adjectives interact with nouns (e.g., Sandhofer and Smith, 2007) do not, however, describe how adjective–noun phrases interact with the perceptual, affective, and motoric processes that represent their meaning in the world.

For example, Fyshe et al. (2019, p. 4458) write: "The computational (rather than neuroscientific) study of language semantics has been greatly influenced by the idea that word meaning can be inferred by the context surrounding a given word, averaged over many examples of the word's usage … . For example, we might see the word ball with verbs like kick, throw, catch, with adjectives like bouncy or with nouns like save and goal. Context cues suggest meaning, so we can use large collections of text to compile statistics about word usage (e.g., the frequency of pairs of words), to build models of word meaning." Although "context cues suggest meaning," they only do so because humans who use these cues can link them to the previously experienced and remembered perceptual, affective, and motoric representations that they reference.

A related issue concerns the issue of whether the phrases "a ball" or "the ball" are learned as *list chunks*, or unitized representations, in their own right when a child listens to mommy speak about a perceptual event that involves a ball. If this is indeed the case, then how are articles ever dissociated from the particular nouns with which they co-occur, so that individual object words like "ball" can be flexibly associated with appropriate articles "a," "the," and so on?

This dissociation of object words from preceding articles can be explained by processes like *attentional blocking* (Grossberg, 1975, 2018; Grossberg and Levine, 1987; Grossberg and Merrill, 1992, 1996; Grossberg and Schmajuk, 1989; Kamin, 1968, 1969; Pavlov, 1927). Attentional blocking of a word or perceptual object can occur when it is predictively irrelevant. Otherwise expressed, an article can be "blocked" if it does not change the expected meaning of the noun that is associated with an object. The article is then suppressed and not attended, thereby enabling the noun, all by itself, to be associated with the object (Figure 4.19).

For completeness, let me also note that *unblocking* of a suppressed word or object can occur when it becomes predictively relevant again (Figure 4.20). In Figures 4.19 and 4.20, blocking and unblocking are illustrated by examples

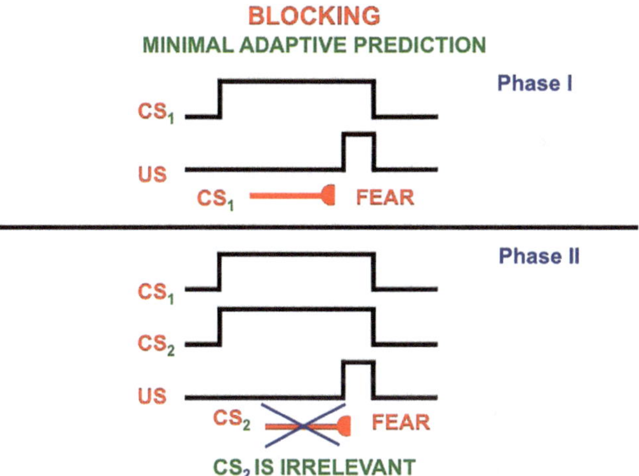

FIGURE 4.19 The blocking paradigm illustrates how cues that do not predict different consequences may fail to be attended. During the training Phase I, a conditioned stimulus (CS_1) occurs shortly before an unconditioned stimulus (US), such as a shock, on enough learning trials for CS_1 to become a conditioned reinforcer whose activation elicits conditioned fear. During Phase II, CS_1 is presented simultaneously with an equally salient, but unconditioned stimulus, CS_2, the same amount of time before the same unconditioned stimulus. This can be done on quite a few learning trials. Despite this training protocol, CS_2 does not learn to elicit fear. It has been *attentionally blocked* by CS_1. Reprinted with permission from Grossberg (2021).

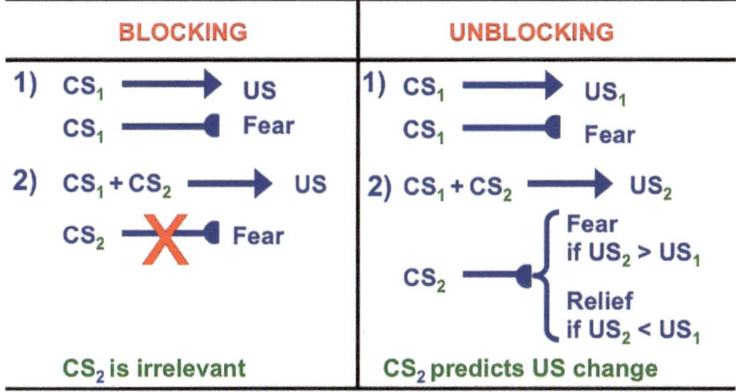

FIGURE 4.20 Unblocking of a conditioned stimulus CS_2 that is presented simultaneously with a previously conditioned stimulus CS_1 can occur if the stimulus pair is followed by a different unconditioned stimulus US_2 than the unconditioned stimulus US_1 that was used to condition CS_1. CS_2 can become either a source of conditioned fear or relief depending upon whether US_2 is more or less intense than US_1. Author created.

that occur during classical, or Pavlovian, conditioning, which was also discussed in Chapter 3. See Figures 1.30 and 3.22. Blocking and unblocking can be used to regulate how and whether attention will be focused upon a wide variety of cortical representations.

Since the word "ball" is always associated with the perceptual experience of a ball, it is predictively relevant in phrases like "a ball" and "the ball." But the articles "a" and "the" are not, at least at first. Rather, they may co-occur with many other words and are chosen via a one-to-many mapping from each article to the many words with which it co-occurs in sentences. When the articles are suppressed, the primacy gradient that stores the word "ball" in working memory can trigger learning of a list chunk, or category, of the word that can be associated with visual categories of the perceptual experiences of seeing a/the ball.

An article can remain predictively irrelevant and blocked until a predictive perceptual context is associated with a phrase like "a ball," when an unfamiliar ball is experienced, or "the ball" when the ball is a particular or familiar one. In these situations, the phrases "a ball" and "the ball" in working memory may trigger learning of their own list chunks.

Behavioral interactions like the following between a teacher and a young learner may help to understand how the meanings of these phrases are learned: Suppose that a child says "Mommy throws ball," and mommy says in return "This is *the* ball daddy bought." If experiences like this happen enough, the child can learn that "*the* ball" may refer to a particular or familiar ball and, as noted above, the phrase "the ball" may be learned as a list chunk in response to its recurring representation as a primacy gradient in working memory.

Keep in mind that the articles interact with both perceptual and cognitive processes: On the perceptual side, the choice of which article "a" or "the" to store in working memory depends upon first perceiving, or at least imagining, the object that the article modifies. The article "the" may refer to a particular, or familiar, ball, as in the sentence: "Mommy threw the ball." The article "a" may refer to any ball, especially an unfamiliar one, as in the sentence: "Pick a ball from the pile." On the cognitive side, the appropriate articles are inserted into phrases and sentences that are stored in a linguistic working memory, along with the nouns that they modify, before the stored items are performed in response to a volitional GO signal.

Adjectives and adverbs can influence what is perceived when constructing a sentence, or imagined when hearing the sentence; for example, "big ball," "quickly running," and so on. Adjective–noun and adverb–verb phrases can also correspond to experienced percepts and hearing them can trigger visual, or other perceptual, memories of such experiences. These words can be inserted in sentences in much the same way as articles are.

Consistent with this analysis, Chang et al. (2009) note: "Multiplicative composition models of the two-word phrase outperform additive models, consistent with the assumption that people use adjectives to modify the meaning of the noun, rather than conjoining the meaning of the adjective and noun." This type of result indicates that adjectives are bound together with their nouns to modify their meaning, but do not clarify how such phrases learn to represent that meaning in the first place.

18 Learning to Associate Visual Objects with Their Auditory Names

In order to better understand where in the brain, and how, an article like "a" or "the" is inserted into a phrase or sentence, more detail is needed about the processes whereby visual events like objects are unitized through learning into object categories. These visual categories are then associated with their learned auditory names in working memory. In this way, the meaning of a noun that describes an object can be learned when it is associated through visual-to-auditory learning with a learned visual category of the object (e.g., a ball). Biological neural networks that are capable of such learned mappings are described in greater detail below.

As this mapping is described in more detail below, it will become clear that the ability to fluently recognize an object as a ball does not, in isolation, determine whether the name "ball" should be modified by the article "a" ("That's a ball") or "the" ("That's the only ball that I have). It is often not the ball-like features that determine article choice about a previously learned object. Rather, it is whether or not a particular combination of ball-characterizing features is *familiar*, say due to a particular texture or design on it, or an unusual size–color combination, and the like; or whether an object that is an exemplar of a previously learned category is being used in a definite context; for example, "Watch mommy throw the ball."

In order to understand how these properties arise, I will now review, and ground in different brain regions, some of the steps that go into learning an object's visual recognition category, whether it is for a ball, mommy's face, or whatever. As noted in Section 6, the functional units of 3D vision are perceptual boundaries and surfaces. The visual boundaries and surfaces that enable our brains to perceive a ball are computed in the striate and prestriate visual cortices, including cortical areas V1, V2, and V4, in the What cortical stream (Figure 2.17; Gegenfurtner, 2003; Motter, 1993; Sereno et al., 1995).

After the processing of visual boundaries and surfaces is completed (e.g., in cortical area V4), they are then categorized at subsequent cortical processing stages. As noted in Sections 2 and 3, a particular view of an surface like mommy's face can be learned and recognized by a category within posterior inferotemporal cortex, or ITp. An invariant category that selectively responds to multiple views of mommy's face can be learned within anterior inferotemporal cortex, or ITa. Such an invariant category may reciprocally interact via bidirectional adaptive connections with all the view categories of mommy's face in ITp (Figure 2.25).

These visual object recognition categories, in turn, activate additional processes at higher cortical areas, such as those that code familiarity about objects, including anterior temporal cortex, anterior occipitotemporal sulcus, anterior fusiform gyrus, posterior superior temporal sulcus, and the precentral gyrus over the frontal cortex (Figure 4.11; Bar et al., 2001; Bonner and Price, 2013; Chao, Haxby, and Martin, 1999; Haxby et al., 2001; Huth et al., 2012; Kovács, 2020; Ramon and Gobbini, 2018; Rajimehr et al., 2009; Sugiura et al., 2011). Auditory object name categories, and facts about them, may also be computed in anterior temporal cortex, among other cortical areas (Bemis and Pylkkänen, 2013; Hamberger et al., 2005).

Visual recognition categories and auditory name categories can be linked through learning by an associative map. Neural network models of map learning between different brain regions or modalities include ARTMAP, for learning binary mappings; and fuzzy ARTMAP for learning binary or analog mappings, among other variants (Figure 2.24; Asfour et al., 1993; Bradski and Grossberg, 1995; Carpenter, 1997, 2003; Carpenter et al., 1992; Carpenter, Grossberg, and Reynolds, 1991, 1995; Carpenter, Martens, and Ogas, 2005; Carpenter, Milenova, and Noeske, 1998; Carpenter and Ravindran, 2008; Carpenter, Rubin, and Streilein, 1997; Granger et al., 2000; Grossberg and Williamson, 1999).

Figure 4.21 depicts two different kinds of associative maps: Many-to-one maps and one-to-many maps. The example in Figure 4.21 (left panel) of a many-to-one map shows how visual images of multiple different kinds of fruit are mapped into the same name "fruit." The example of a one-to-many map in Figure 4.21 (left panel) shows how a single image of a dog can be associated with many different words to describe the dog, ranging from the general word "animal" to the specific name of this particular dog "Rover." As more names are associated with the dog's image, our "knowledge" about the dog expands.

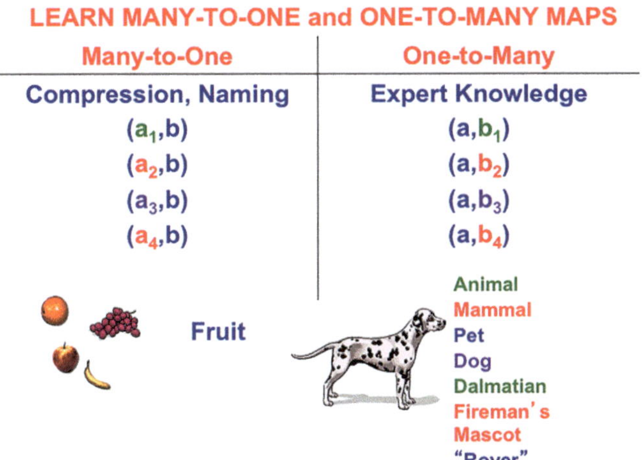

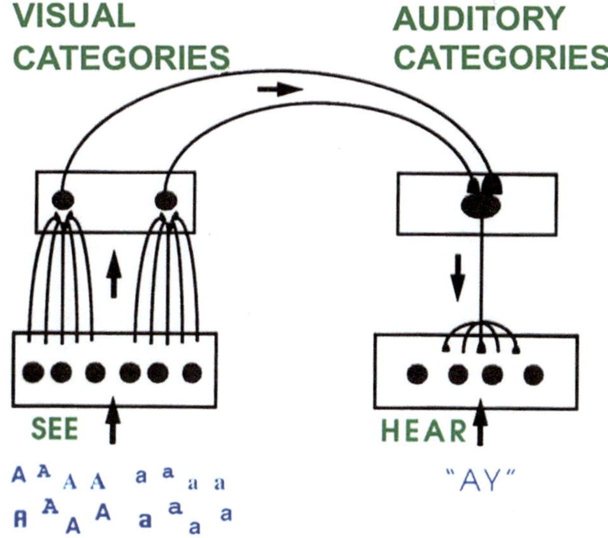

FIGURE 4.21 (left panel) Humans and other autonomous adaptive intelligent agents need to be able to learn both many-to-one maps and one-to-many maps. (right panel) Learning a many-to-one map from multiple visual fonts of a letter to the letter's name requires a stage of visual category learning that is linked to a stage of auditory name learning via a learned associative map. Figure 2.24 shows that a Map Field is needed to link learned visual categories with learning auditory categories via associative learning. Reprinted with the author's permission from Grossberg (2021).

Figure 4.21 (right panel) illustrates that learning of a many-to-one map requires two different stages of learning. In this example of visual-to-auditory learning, multiple fonts of a letter /A/ drive the learning of multiple visual categories that selectively respond to similar variations of each letter font. Because the letter fonts /A/ and /a/ are composed of qualitatively different visual features, they learn to activate different visual categories. Then these visual categories are all associated with the same auditory name of the letter via a Map Field (Figure 2.24).

Whereas several different kinds of neural networks can learn many-to-one maps, learning one-to-many maps incrementally and in real time requires the kind of match-based category learning of Adaptive Resonance Theory, or ART (Figures 2.20–2.22 and 4.4). Mismatch learning, as used by algorithms like back propagation and Deep Learning, will create mismatches with previously learned categories, and thus erasure of them, when learning a new category, and thereby prevent these algorithms from learning one-to-many maps.

19 Map Fields Are Working Memories

The discussions above clarify how observing a visually experienced sequence of events during which mommy throws a ball can enable that sequence of events to be stored, in the same order, in a linguistic working memory as a sequence of words describing that sequence. The sentence "mommy throws the ball" was mentioned as one example.

Putting together the discussions of working memories and Map Fields leads to the conclusion that a Map Field may also be working memory where linguistic sequences can be stored in response to sequential activation of their visual categories through time. It is also possible that a Map Field topographically inputs to a linguistic working memory, but that the Map Field itself just categorizes individual sequences of variable length into list categories, before inputting each list category into the linguistic working memory. Which alternative may hold in vivo in the prefrontal cortex still needs to be experimentally studied.

How to think of the choice of articles "a" or "the" in such sequences depends upon whether they are in response to heard speech that is uttered by someone else, as in a sentence like "Watch mommy throw the ball," or as self-generated speech in response to an externally viewed, or internally remembered, perceptual experience like "Mommy threw a ball." Since children learn at least their first languages by listening to teachers who know the language, the choice of article will depend upon the perceptual experiences to which the teachers' utterances correspond.

20 Conscious Speech Is Resonant Wave: Resonances Group Speech and Language

The sound spectrograms of the words that define speech and language often exhibit no silence intervals between the words, especially when the words are heard in a noisy room. How can sharp word boundaries be perceived even if the sound spectrum that represents the words exhibits no silence intervals between them?

ART proposes that "conscious speech is a resonant wave" and that "perceived silence is a temporal break in the rate at which the resonant wave evolves." The "resonant wave" operates on multiple levels, and between What and Where cortical processing streams, though time. It includes the sequential activation of temporally coordinated spectral-pitch-and-timbre resonances, stream-shroud resonances, feature-category resonances, and item-list resonances. An interval of "perceived silence" occurs when the resonances that support an individual linguistic event, such as feature-category and stream-shroud resonances, are reset to enable the next event to be perceived. These revolutionary claims shed light on how the cycle of bottom-up filtering and top-down matching using the ART Matching Rule coherently groups, and separates, familiar linguistic units through time, even if they are not separated by silence intervals in the input stream.

21 Phonemic Restoration: How Future Speech Influences Consciously Heard Past Speech and Its Meaning

The existence of a "resonant wave" illustrates how the conscious percepts of speech sounds can proceed from past to future, even though the speech sounds that are heard in the past may be influenced by future words. One striking phenomenon that illustrates how future speech can influence conscious percepts of past speech is called *phonemic restoration*. Phonemic restoration was described by Richard Warren and his colleagues in the 1970s (e.g., Warren and Sherman, 1974). I will explain how phonemic restoration can be explained by properties of the ART Matching Rule, and thus supports my explanation of how ART enables fast learning without catastrophic forgetting.

Figure 4.22 summarizes an example of phonemic restoration: Suppose that broadband noise (the # in Figure 4.22)

PHONEMIC RESTORATION
Warren and Sherman (1974)

DELIVERY DELIBERATION

Noise supports restoration

DELI#ERY DELI#ERATION

Silence does not support restoration

DELI ERY DELI ERATION

ART MATCHING RULE!

FIGURE 4.22 Phonemic restoration: The two words "delivery" and "deliberation" both begin with "deli." If the next letter is replaced by broad band noise, then the complete words are consciously heard. If silence is substituted instead, then two separate syllables are heard that are separated by silence. These percepts show the ART Matching Rule at work. See the text for details. Author created.

replaces the phonemes/v/ and/b/ in the words "delivery" and "deliberation," respectively. Despite the identical initial portion of these words ("deli- "), if the broadband noise is immediately followed by "ery" or "eration," listeners hear the/v/ or/b/ as being fully intact. However, if the noise is replaced by silence, then restoration does not occur.

These properties of phonemic restoration are also properties of top-down matching by the ART Matching Rule (Figures 2.21 and 2.22). For example, the ART Matching Rule explains why the noise in "deli- noise- [ery/ eration]" is not heard before the last portion of the word is presented. This is because if the resonance for the first part of the word has not developed fully before the last portion of the word is presented, then the last portion can influence the top-down expectations that determine the conscious percept. The ART Matching Rule converts the noise in "deli- noise- [ery/ eration]" into a conscious percept of/v/ or/b/ by selecting *expected* feature clusters for attentive processing while suppressing unexpected ones. In the "deli- noise- [ery/ eration]" example, spectral components of the noise are suppressed that are not part of the expected consonant sound.

If the/v/ or/b/ in "delivery/ deliberation" is replaced by silence, then the silence is consciously perceived as silence, unlike the cases where noise replaced silence. This property follows because the on-center of the ART Matching Rule is *modulatory*. The ART Matching Rule can select feature components that are consistent with its prototype, but it cannot create something out of nothing.

Thus, phonemic restoration properties, rather than being just an amusing curiosity, illustrate basic attentive matching properties of the ART Matching Rule that enable speech and language to be learned quickly, without risking catastrophic forgetting. During speech perception, the learned categories are list chunks of phonemes, syllables, and words, rather than the pitch-and-timbre categories that are learned during auditory streaming.

The conscious ARTWORD, or cARTWORD, model that I developed with my PhD student, Sohrob Kazerounian (Grossberg and Kazerounian, 2011), simulates phonemic restoration in real time. As Figure 4.23 illustrates, the model used laminar cortical circuits that specialize the canonical laminar cortical circuit that I and my collaborators used to explain and simulate psychological and neurobiological data about vision and object recognition; audition, speech, and language; and cognitive information processing.

A variation of the basic phonemic restoration experiment provides additional support for the ART Matching Rule by using a reduced set of spectral components in the noise stimulus. This leads to hearing a correspondingly degraded consonant sound. This result is predicted by the ART Matching Rule because the expected speech formants are not fully present in the available noise spectrum, and the ART Matching Rule cannot create something out of nothing. This percept was reported in 1981 by Arthur Samuel (Samuel, 1981a, 1981b). This experiment emphasizes that phonemic restoration is not merely a symbolic reconstruction of an expected sound. Rather, it creates a conscious phonetic experience of this sound.

Experiments that Makio Kashino published in 2006 (Kashino, 2006) provide additional evidence of phonemic restoration as a perceptual phenomenon, rather than a symbolic reconstruction, of an expected sound. Kashino's stimuli contained multiple deletions of the speech sounds "Do you understand what I am saying," where the speech would alternate with silence intervals every 50, 100, or 200 milliseconds. In these examples, stimuli with silence intervals that are not also filled with broadband noise sound disjoint and are either difficult or impossible to understand.

When the deleted speech signals are filled with broadband noise, however, the speech sounds more natural and continuous, and are easy to understand. Because the earlier studies of phonemic restoration relied on just a single excised portion of the speech signal, the utterance remained relatively intact and understandable, regardless of whether or not noise was presented instead of silence, making response strategies susceptible to bias. In the Kashino stimuli, however, it would be very difficult to alter response strategies, since a subject cannot force understanding of an otherwise unintelligible acoustic stream.

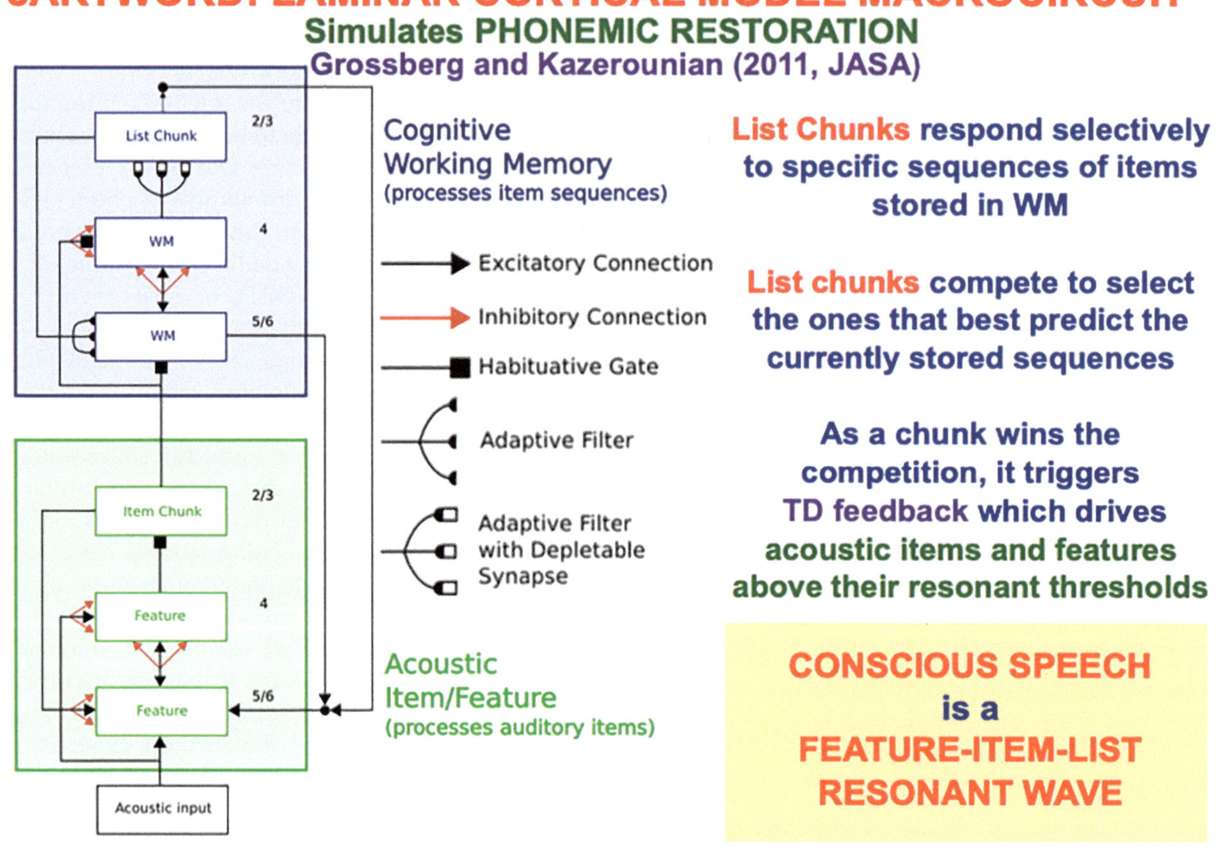

FIGURE 4.23 Macrocircuit of the cARTWORD laminar cortical model for conscious speech perception shows a hierarchy of levels responsible for the processes involved in speech and language perception. Each level is organized into laminar cortical circuits. Deep layers (6 and 4) process and store inputs, whereas superficial layers (2/3) group distributed patterns across the deeper layers into categories, or chunks. The lowest level processes acoustic features (cell activities F_i and E_i) and item chunks (cell activities $C_i^{(I)}$), whereas the higher level is responsible for storing of sequences of acoustic items in working memory (activities Y_i and X_i), and representing these stored sequences of these items as list chunks (activities $C_j^{(L)}$) in a Masking Field. Reprinted with the author's permission from Grossberg and Kazerounian (2011).

22 When "gray chip" Sounds Like "great ship": Backward Effect Changes Meaning

There are many examples of such "backward effects in time" in the speech and language perception literature. They are an important type of data because they challenge classical ideas about what the functional units of speech and language are, and how they are formed in the brain, while also clarifying how the formation of these functional units is influenced by the dynamically evolving speech context in which they are consciously heard. Many of these examples illustrate the process whereby our brains coherently group item sequences, using item-list resonances, into the emergent auditory objects that have meaning in speech and language.

I will end this chapter with a striking example of such dynamism in language. This example was published in 1978 by Bruno Repp, Alvin Liberman, Thomas Eccardt, and David Pesetsky (Repp et al., 1978). My colleagues and I began to outline an explanation of their data, and related data of their colleagues, using ART in 1986 (Cohen and Grossberg, 1986; Grossberg, 1986; Grossberg and Stone, 1986). We were able to do this at around the same time that Michael Cohen and I simulated how recognition categories can learn words of variable length using a Masking Field neural network (Cohen and Grossberg, 1986, 1987), a type of network that I first described in Grossberg (1978).

Many other types of modeling discoveries intervened after that (see sites.bu.edu/steveg) before I finally quantitatively simulated the Repp et al. (1978) speech data in a *Psychological Review* article with my PhD student Christopher Myers (Grossberg and Myers, 2000). At around the same time, other challenging speech psychophysical data were given a unified explanation, and quantitatively simulated, with other PhD students of mine, notably Ian Boardman, using the same ART concepts and mechanisms; for example, Boardman et al. (1999) and Grossberg, Boardman, and Cohen (1997).

Repp et al. (1978) used a recording of the sentence "Did anyone see the gray ship?" to show that variations in the durations of individual speech sounds and the silence intervals between them can produce strikingly different perceived consciously heard words, starting with the words "gray" and "ship" (Figure 4.24 (left column)). It is important to emphasize that both of these words are short, unambiguous, and familiar. Despite this fact, increasing the silence interval between the words "gray" and "ship" (see Figure 4.24 (right column, horizontal red arrow)) can cause listeners to perceive them as "gray chip" or—at even longer silence intervals—as "great chip."

Even this first result is remarkable: Why should *increasing* the silence interval between unambiguous words, and thereby separating them even more in time, cause the initial fricative noise in "chip" to leap backward in time over a silence interval of 100 milliseconds, thereby replacing the word "gray" with a word "great" that is not represented in the speech signal, with the result that "gray chip" is consciously heard as "great chip," along with a corresponding change in the perceived phrase's meaning?

In addition, increasing the duration of the initial fricative noise of the word "chip" (see Figure 4.24 (right column, vertical red arrow)) can induce a switch in the perception of "gray chip" to "great ship," thereby changing the conscious perception of both words, again with a change of meaning.

These results are unified, explained, and simulated, along with many of the other data that my many collaborators and I have studied, in Grossberg (2021).

PERCEPTUAL INTEGRATION OF ACOUSTIC CUES

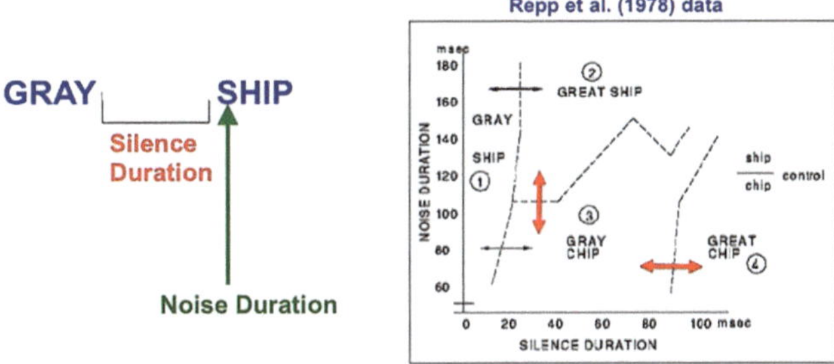

FIGURE 4.24 (left column) In experiments of Repp et al. (1978), the silence duration between the words GRAY and SHIP was varied, as was the duration of the fricative noise in S, with surprising results. (right column) The red arrows direct our attention to switches in which pairs of words are consciously heard as silence and noise durations increase, with no change in the input sounds. See the text for details. Reprinted with the author's permission from Grossberg (2021).

23 Concluding Remarks: "The limits of my language mean the limits of my world"

The famous quote in this section's header is due to Ludwig Wittgenstein in his classic *Tractatus Logigo-Philosophicus* (Wittgenstein, 1922). This simple sentence summarizes both the main contribution of the present chapter and its main limitation:

> The main contribution is to create a theoretical foundation for understanding how humans learn in real time to link language utterances to real-world perceptual, affective, and motoric experiences that give them meaning.
>
> The main limitation is that the open-ended nature of language enables it to bridge "the limits of [our] world." Providing an analysis of how we learn all the language meanings that we can express about our world will require many scientists working for years to complete. I hope that the foundation that I have provided in this chapter, and throughout the book, will be useful toward achieving this goal.

Explainable AI and Biological Intelligence

Contrasting deep learning with adaptive resonance

"It's becoming—whether you're an illustrator or a coder or a support person or somebody who is in the health care system—the ability to work well with AI and take advantage of it is now more important than understanding Excel or the internet."
Bill Gates

"[I]f machine learning programs like ChatGPT continue to dominate the field of AI, … It is at once comic and tragic … that so much money and attention should be concentrated on so little a thing—something so trivial when contrasted with the human mind."
Noam Chomsky

"Artificial intelligence is unprecedented in the history of civilization. Usually these things advance and we can say – yes, we understand it. This time it's different. That's why it's not satisfactory from my perspective."
Leon Chua

"Success in creating AI would be the biggest event in human history. Unfortunately, it might also be the last, unless we learn how to avoid the risks."
Stephen Hawking

Your Creative Brain and AI. Stephen Grossberg, Oxford University Press. © Oxford University Press (2026).
DOI: 10.1093/9780198965367.003.0005

1 Towards Explainable AI and Autonomous Adaptive Intelligence

1.1 From art, music, language, and meaning to XAI and AAI

Earlier chapters in this book have summarized biological neural network models whereby our brains make our minds, notably models that clarify how humans can consciously experience and know the beauty and meanings of art, music, and language. These capabilities are three of the greatest triumphs of human brains and civilizations. They exemplify the ability of our brains to realize the highest forms of *autonomous adaptive intelligence.*

The biological neural networks that model these capabilities are built from:

> a small number of *equations*, such as the equations for cell activations, or short-term memory (STM) traces; habituative chemical transmitters, or medium-term memory (MTM) traces; and adaptive weights for learning and memory, or long-term memory (LTM) traces;
>
> a somewhat larger number of *microcircuits*, such as non-recurrent and recurrent shunting on-center off-surround networks, gated dipole opponent processing networks, category learning networks, adaptive resonance networks, associative learning networks, and spectral timing networks; and
>
> *modal neural architectures* that combine specializations of these equations and microcircuits to support the different modalities of biological intelligence, such as 3D vision and figure-ground perception, cognitive information processing, cognitive–emotional interactions, speech and language learning, and adaptive sensory-motor control.

In a sense, that is all that there is to biological intelligence. The secret sauce that prevents this list from being disappointing, indeed enables these processes to support human creativity and even transcendence, is that their "intelligence" arises from *emergent properties* due to feedback, and often *resonant*, interactions between many neurons in brain networks. The emergent properties are the ultimate expression of the fact that "the whole is greater than its parts" in biological intelligence.

Because intelligent algorithms, machines, and robots are often needed to carry out large-scale applications in engineering, technology, and AI, many of the above kinds of neural networks have been applied in fielded applications over the past several decades. Such biologically inspired applications have the advantage that they facilitate harmonious man–machine interactions: Since the applications use networks that emulate how our brains work, they can be expected to be more easily understood, and to interact more seamlessly with, the humans who use and modify them.

Many of these applications may, in turn, transform how future art and music are made, and how we will store and remember language meanings. Familiar technologies have already transformed how art and music are made, whether due to the introduction of canvas on which to paint with oils or acrylics, or marble with which to create lifelike sculptures, or new instruments and computers on which to compose and perform new music.

Our use and understanding of language meanings has also been revolutionized in multiple ways, notably by the introduction of increasingly fast, capable, and inexpensive personal computers and word processing tools, and by information resources like Google and Amazon that have democratized the availability of all knowledge and goods, including art, music, and all the books that are currently available through any printed medium. These revolutions are arguably at least as significant as the invention of the printing press in 1440 AD by Johannes Gutenberg in Mainz, Germany (Childress, 2008).

This chapter will discuss *how* and *why* biologically inspired neural network models can contribute, and are contributing, to the development of the revolutionary computational paradigm of autonomous adaptive intelligence, whose ramifications will transform all aspects of society during this century and beyond, with applications ranging from smart appliances, self-organizing language tools, and humanoid companions to self-driving cars, automated medical diagnostics, and self-organizing mobile robots. Such developments may be expected if only because our human brains are the best examples that we know of systems capable of autonomous adaptive intelligence.

1.2 Deep Learning and ChatGPT: Deep problems and not intelligent

It is also important to explain why not all neural networks can achieve these goals. Indeed, the word AI is all too often used today to signify one particular neural network, called Deep Learning, whose computational properties prevent it from emulating any aspect of human intelligence, for reasons that I will explain below.

ChatGPT is another contribution to AI that is being heavily marketed by the company OpenAI, among others, and uses Deep Learning as its learning engine. As noted in The Street (https://www.thestreet.com/technology/new-chatgpt-update-dangers-of-ai-hype): "OpenAI dramatically changed the public discourse around artificial intelligence when it launched ChatGPT last year; for the first time, the public had access to an AI model that could seemingly do everything from penning essays to helping investors with stock picks. The resultant environment has been one of a feverish game of catch-up for other tech giants as excitement over the budding technology has consistently lifted tech stocks higher.

"The viral chatbot is now getting a major update. OpenAI said Sept. 25 that it is beginning to roll out new voice and image capabilities within ChatGPT, which allows the model to process visual and auditory, in addition to linguistic, data. Users can upload images and 'troubleshoot why your grill won't start, explore the contents of your fridge to plan a meal, or analyze a complex graph for work-related data,' OpenAI said in the blog post https://openai.com/blog/chatgpt-can-now-see-hear-and-speak."

If, however, these new capabilities are built on the original ChatGPT foundation, then they may fail to realize their ambitious goals. This is true because, as I will explain below, ChatGPT has neither goals nor values to guide its computations, and literally does not know what it is talking about, so it cannot understand, nor contribute to an understanding of, language meanings.

1.3 A useful contribution to AI needs to be trustworthy and reliable

Two necessary properties of societally acceptable applications of AI from the viewpoint of their usefulness to human users is that they be *trustworthy* and *reliable*. In order to be trustworthy, an AI algorithm needs to be *explainable*. For example, in order for us to trust medical databases that learn from their predictive successes and failures, we have to be able to check the factual basis on which each of their predictions is made, and the relevance of each prediction to the current informational and societal context. Only then can we intelligently decide whether to deploy them in particular cases. If they succeed consistently in predicting the correct medical outcomes for a sufficiently long time, we can then begin to trust them as valuable tools for medical care.

This chapter argues that because the dynamics and emergent properties of the models for perception, cognition, emotion, and action that the earlier chapters described are explainable, they provide an appropriate foundation for being specialized and implemented in a wide range of large-scale applications in engineering, technology, and AI. That is why many of them have already been applied to such applications for the past few decades.

Key to why these models are explainable is that, as I noted above, they include fast activations, or short-term memory (STM) traces, that interact with adaptive weights, or long-term memory (LTM) traces, to learn about and predict their environments. For example, visual and auditory perceptual models have explainable conscious STM representations of visual surfaces and auditory streams in surface-shroud resonances and stream-shroud resonances, respectively. As I explained in Chapter 2. we consciously see using surface-shroud resonances in order to be able to look at, and reach for, valued goal objects. STM traces define the cell activation patterns that occur during these resonances. Learning to accurately look and reach requires LTM traces. Likewise, Chapter 3 described the role of *stream-shroud resonances* in auditory processing, including music perception and performance.

Surface-shroud resonances and stream-shroud resonances are part of a more global organization of our brains that enables resonant states to learn quickly without forgetting just as quickly.

1.4 Adaptive Resonance Theory is trustworthy and reliable

Adaptive Resonance Theory, or ART, is currently the most advanced cognitive and neural theory that can attend, recognize, and predict objects and events in a changing world that is filled with unexpected events. ART is also a *self-organizing production system* that can process both familiar and novel environments using its attentional and orienting systems (Figures 2.21–2.23).

In particular, in response to an unfamiliar or novel input pattern, ART triggers hypothesis testing and memory search to discover if it has already learned a category that is similar enough to the input to resonate with it, and thereby learn from it. If, however, the novel input is too different—as determined by a vigilance parameter (see Sections 2.14 and 2.15 in Chapter 2)—from inputs that the best matching category has already learned about, then ART continues to search until it discovers and incrementally learns a new category. The net result is that ART can learn from arbitrary combinations of unsupervised and supervised learning trials, using only locally computable quantities, to rapidly classify large nonstationary databases, and does so without experiencing catastrophic forgetting. As my 1980 thought experiment that I published in Psychological Review showed (Grossberg, 1980), ART models are the *unique* class of solutions capable of *autonomously* learning to correct predictive errors in a changing world that is filled with unexpected events (see Section 2.8 of Chapter 2).

ART classifications and predictions are *trustworthy* because they are *explainable*. In particular, ART can learn to pay attention to *critical feature patterns* of cell activation using its STM traces (Figure 1.28). Critical feature patterns are the combinations of features that ART hypothesis testing has discovered to control predictive success in different environments. The critical feature patterns are also the ones that are learned via the LTM traces in ART bottom-up adaptive filters and top-down expectations (Figures 2.20 and 2.21) and that will control future classifications and predictions.

ART is *reliable* because it can learn quickly without experiencing catastrophic forgetting. Gail Carpenter and I mathematically proved that the ART Matching Rule dynamically stabilizes memories after fast learning occurs (Carpenter and Grossberg, 1987, 2003; Grossberg, 2003) It hereby offers a solution of the *stability-plasticity dilemma*: We can learn quickly (plasticity) without forcing catastrophic forgetting (stability) of what we already know.

A vivid example of how LTM adaptive weights contribute to explainability is achieved by the fuzzy ARTMAP algorithm (Figure 2.24; Carpenter and Grossberg, 1992; Carpenter, Grossberg, and Reynolds, 1995; Granger et al., 2000). The learned LTM traces of fuzzy ARTMAP define *fuzzy IF-THEN rules* that have a geometrical interpretation in terms of hyper-boxes. Those combinations of critical feature STM values that fit into these hyper-boxes depict successfully classified exemplars.

Because of all the above computational properties, ART has been successfully used in multiple large-scale real-world applications in engineering, technology, and AI, including applications to remote sensing, medical database prediction, social media data clustering, and engineering design retrieval systems—that include millions of parts defined by high-dimensional feature vectors, and that were used to design the Boeing 777 (Caudell et al., 1990, 1991, 1994; Escobedo, Smith, and Caudell, 1993); classification and prediction of sonar and radar signals; of medical, satellite, face imagery, social media data, and of musical scores; control of mobile robots and nuclear power plants; air quality monitoring; strength prediction for concrete mixes; signature verification; tool failure monitoring; chemical analysis from ultraviolent and infrared spectra; frequency-selective surface design for electromagnetic system devices; and power transmission line fault diagnosis, among others.

Also explainable are the MOTIVATOR model of reinforcement learning and cognitive-emotional interactions, and the VITE, DIRECT, DIVA, and SOVEREIGN models for reaching, speech production, and spatial navigation, all of which contribute to the design of a self-organizing algorithm or mobile robot capable of autonomous adaptive intelligence (Chapter 3). These biological models illustrate complementary computing (Chapter 1), and use local laws for match learning and mismatch learning.

1.5 Deep Learning does not equal the field of neural networks and AI

I will now contrast the explainable properties of ART with properties of other neural networks and AI systems that are not explainable, including Deep Learning and ChatGPT. These days, all too often, the mass media use the terms neural networks or AI interchangeably with the Deep Learning model. This attribution is made despite the fact that Deep Learning suffers from many foundational computational problems that ART never faced since I introduced it in 1976 (Grossberg, 1976a, 1976b, 1980), long before Deep Learning was popularized, and even before the back propagation algorithm used by Deep Learning began to be vigorously marketed by Rumelhart, Hinton, and Williams (1986).

In fact, much earlier, when I was a freshman at Dartmouth College in 1957 (https://en.wikipedia.org/wiki/Stephen_Grossberg), I introduced the paradigm of using neural networks to model how brains make minds; the foundational STM, MTM, and LTM laws that are still used around the world today for this purpose; and then rapidly discovered and developed neural networks to carry out many processes that make humans and other animals so adaptive and intelligent. These developments were accelerated with the collaboration of over hundred gifted PhD students and postdocs for forty years, once I was able to found and lead a graduate department and several interdisciplinary research institutes to accelerate the discovery process.

This history makes the glib identification of neural networks and AI with Deep Learning even more concerning to me, indeed deeply frustrating, especially in a world that prides itself for having technologies like Google that can search for any published scientific information in a matter of seconds.

1.6 Deep Learning is untrustworthy and unreliable

Two of the foundational computational problems of Deep Learning are that it is *untrustworthy* (because it is *not explainable*) and *unreliable* (because it can experience *catastrophic forgetting*).

I will explain why Deep Learning is not explainable below. Because Deep Learning is not explainable, even if it makes an accurate prediction, the user cannot explain what information was used to do so. The same is true if Deep Learning makes an incorrect prediction. In either case, the user never knows *why* the algorithm works or fails.

One cannot safely use an algorithm with this weakness in any life-or-death situation, notably to make a medical or financial prediction. If such a prediction turned out to be incorrect, with catastrophic consequences, and the victim asked the user on what basis the prediction was made, the answer would have to be: I do not know. That fact could trigger a lawsuit that could end the career of the user, and of all companies that promoted Deep Learning in a scientifically misleading manner.

In summary, Deep Learning does not solve the *Explainable AI Problem* (https://www.darpa.mil/research/programs/explainable-artificial-intelligence).

Its predictions consequently cannot be trusted.

Deep Learning uses a form of slow learning. Slow learning means that adaptive weights can change only a little on each learning trial, rather than converging fully on that trial to their new equilibrium. Thus, learning of a large database may require many thousands, or even millions, of learning trials.

Catastrophic forgetting means that, at any stage of training, an unpredictable part of the already learned memory can collapse. One then has to start all over again. Because Deep Learning can only use slow learning, and can experience catastrophic forgetting at any time, enormous computational resources may be exhausted before an entire database can be learned without a catastrophic forgetting mishap.

Deep Learning shares these problems with the back propagation algorithm, whose computational problems due to its use of *nonlocal weight transport during mismatch learning* were already well known in the 1980s (e.g., Grossberg, 1988). Back propagation lost popularity to alternative learning algorithms because of these problems, as well as because there were several more powerful neural network classifiers available that did not experience these problems, including Adaptive Resonance Theory, or ART. Back propagation was never "the best game in town."

Deep Learning became popular again after networks of very fast computer servers, and huge online databases on the World Wide Web, partially compensated for these weaknesses. I will now summarize the history of back propagation and Deep Learning in greater detail, as well as more of their computational problems, again while contrasting them with properties of ART.

1.7 How I decided to contrast ART with Deep Learning in 2021

I usually do not seek to criticize other models in print. I am optimistic enough to believe that, at least in the long run, the best and most powerful models will eventually be adopted. This will happen, I believe, if only out of enlightened self-interest, since the best and most powerful models will support the best basic and applied research, and earn the most professional rewards, including money. I also try to think positively about the scientific goals that I am trying to achieve. I avoid allowing negative feelings to diminish my inner positive sources of motivation, and use positive feelings and thoughts to sustain the hard work needed to make real scientific progress.

This strategy changed when I was told about a *Frontiers in Neurobotics* research topic about Explainable Artificial Intelligence by one of its editors, Jeffrey Krichmar, who I knew through earlier constructive scientific encounters. I thought that the Abstract in the announcement of the research topic expressed well some of the problems that I also saw with Deep Learning. The Abstract says:

> Though Deep Learning is the main pillar of current AI techniques and is ubiquitous in basic science and real-world applications, it is also flagged by AI researchers for its black-box problem: it is easy to fool, and it also cannot explain how it makes a prediction or decision ... In both ... biological brains and AI, intelligence involves decision-making using data that are noisy and often ambiguously labeled. Input data can also be incorrect due to faulty sensors. Moreover, during the skill acquisition process, failure is required to learn.

I was encouraged to write an article for Jeff's special issue that contrasted ART with Deep Learning. I thought that publishing such an article in a visible journal special issue about this topic might attract a large enough audience to make it worth the effort. The published article is Grossberg (2020).

I think that the article must have achieved some visibility because, after its publication, I was invited to give a number of keynote lectures around the world about this topic. Because this all happened during the COVID epidemic, these conferences were organized virtually over Zoom in cities as far flung as Ankara, Bangkok, Beijing, Boston, Campinas, Edinburgh, Glasgow, London, Marseille, Melbourne, Mumbai, Padua, New York, San Francisco, Santander, and Tehran.

Giving plenary lectures over Zoom had the additional benefit that some of them were videotaped and posted on the World Wide Web, where anyone could watch them for free. Videos of some of these lectures can be downloaded from my web page sites.bu.edu/steveg. If you search for "Deep Learning" on my web page, you will be led to videos of both lectures and interviews that I was invited to give on this topic.

1.8 Some history and properties of back propagation and Deep Learning

Deep Learning uses the back propagation algorithm to learn how to predict output vectors in response to input vectors. These models are based upon the Perceptron learning principles introduced by Frank Rosenblatt (Rosenblatt, 1958, 1962), who also introduced the term "back propagation." I had the pleasure of meeting and chatting with Frank when I lectured at Cornell University before his tragic death on his forty-third birthday in a boating accident on Chesapeake Bay.

Back propagation was developed between the 1970s and 1980s by people like Shun-ichi Amari (Amari, 1972), Paul Werbos (Werbos, 1974, 1994), and David Parker (Parker, 1985, 1986, 1987). It reached its modern form and was successfully simulated in applications by Paul Werbos (1974) as part of his PhD thesis from Harvard University. I had the pleasure of chatting with Paul before he was awarded his thesis because Dan Levine, a friend of Paul and one of my PhD students at MIT where I was then a professor, asked me to do so. Paul talked to me about problems he was having getting his thesis approved. As I recall, I recommended that he work out enough computational examples to illustrate the usefulness of the algorithm. He was duly awarded his PhD and went on to have a distinguished and multifaceted career.

The back propagation algorithm was later popularized in 1986 by an article of David Rumelhart, Geoffrey Hinton, and Ronald Williams (Rumelhart, Hinton, and Williams, 1986), but without adequate citations of its real discoverers. Schmidhuber (2020) has provided a detailed historical account of many additional scientists who contributed to this development.

Both back propagation and Deep Learning are defined by a feedforward network whose adaptive weights can be altered when, in response to an input vector, its adaptive filter generates an output vector that mismatches the correct, or desired, output vector. The heuristic statement that "failure is required to learn" in back propagation is computationally realized by computing a scalar error signal that calibrates the distance between the actual and desired output vectors. Thus, learning is *supervised* and requires that a desired output vector be supplied by a teacher on every learning trial, so that the error signal between the actual and desired output vectors can be computed.

Figure 5.1 from Carpenter (1989) provides a detailed macrocircuit of the back propagation algorithm. The Actual Output (from the processing stage at the top of the right column) is subtracted from the Target Output (from the processing stage at the top of the left column) to define an Error signal (at the processing stage below the Target Output stage). This Error signal, defined by δ_k, back-propagates along a pathway that ends at adaptive weights (black hemidisks in the processing stage labeled F_3). These adaptive weights occur at the ends of vertical pathways from the Hidden Units stage that is labeled F_2.

The location in the algorithm where the error signal is computed is not where the adaptive weights are computed at the ends of pathways within the adaptive filter (black hemidisks in the processing stage that is labeled F_2). *Weight transport* of the error signal across the network is thus needed to train the adaptive weights (Figure 5.1). This transport is "nonlocal" in the sense that there are no pathways in the model from where the error signal is computed along which it can locally flow to where the adaptive weights are computed.

As I noted above, *slow learning* occurs in back propagation. Technically, slow learning means that adaptive weights, or LTM traces, change only slightly on each learning trial to gradually reduce the error signals. *Fast learning*, in contrast, would zero the error signal after any single erroneous prediction. In back propagation, which uses mismatches to drive adaptive weight changes, fast learning could wash away already-learned memories because the new prediction that is being learned could mismatch and thus recode the information that was previously learned, notwithstanding its continued correctness in the original environment.

This kind of error-based supervised learning used in back propagation and Deep Learning has other foundational computational problems that were already known in the 1980s. One of the most consequential ones is a fact that I mentioned above: Even if learning is slow, Deep Learning can experience *catastrophic forgetting* (French, 1999; McCloskey and Cohen, 1989; Ratcliff, 1990). French (1999) traced this problem to the facts that all the inputs are processed through a shared set of learned weights, and the absence of any mechanism to selectively buffer previous learning that is still predictively useful. Catastrophic forgetting can, in fact, occur in any learning algorithm whose shared weight updates are based on the gradient of the error in response to the current batch of data points, while ignoring past batches.

In addition, the back propagation and Deep Learning algorithms explain no brain data, because they are based on biologically impossible computations like nonlocal weight transport. They cannot explain any psychological data due to their lack of perceptual, cognitive, emotional, or motor representations, all of which require information processing using rapidly changing activities, or short-term memory (STM) traces that are found throughout our brains.

Of particular importance is the fact, without STM, that there is no way for back propagation or Deep Learning to *pay attention* to information that may be predictively important in one environment, but irrelevant in another. The only residue of previous experiences lies in the changes that are learned by their adaptive weights (within the hemidisks in Figure 5.1). As I noted above, all future experiences are non-selectively filtered through this shared set of weights.

By contrast, ART can learn by either fast learning or slow learning. Gail Carpenter and I showed, for example, that an ART model can learn an entire database in one fast learning trial, without experiencing catastrophic forgetting (Carpenter and Grossberg, 1987, 1988). Many of us have had this kind of experience, in which we can learn the faces of family members and friends with one exposure, and remember them for a long time. ART and related biological neural network models that my colleagues and I have discovered and developed also provide unified and principled explanations of large psychological and neurobiological databases.

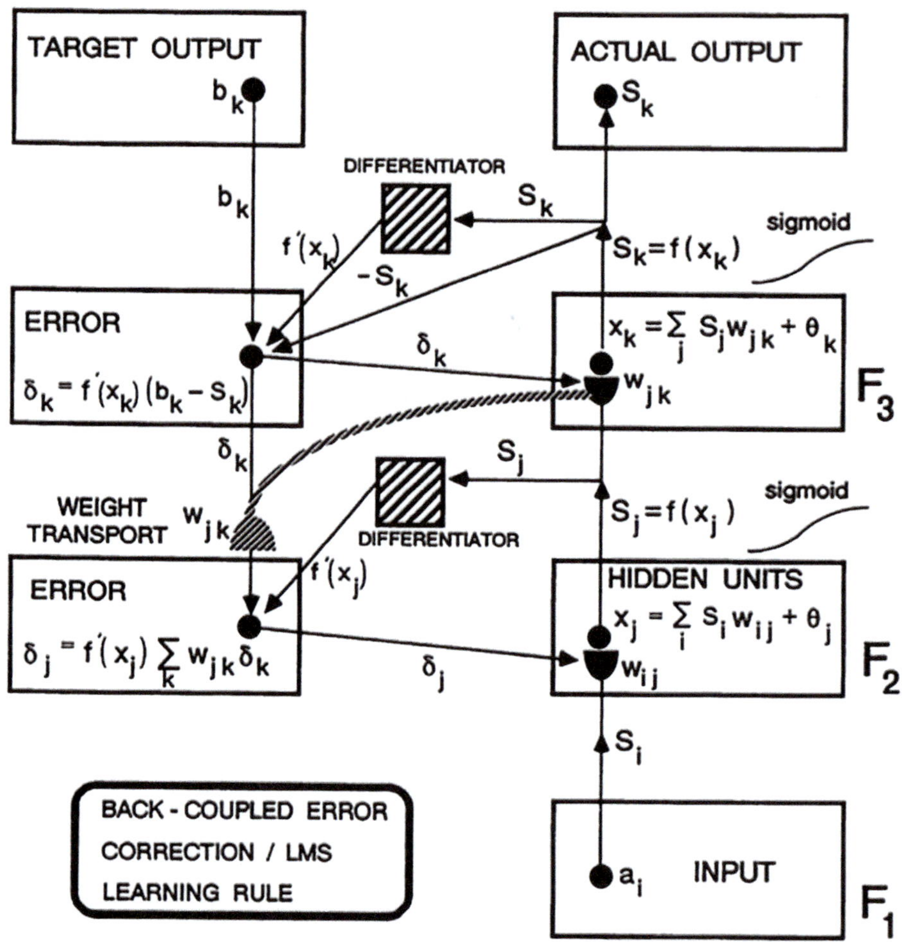

FIGURE 5.1 Circuit diagram of the back propagation model. Input vector a_i in level F_1 sends a sigmoid signal $S_i = f(a_i)$ that is multiplied by learned weights w_{ij} on their way to level F_2. These LTM-weighted signals are added together at F_2 with a bias term θ_j to define x_j. A sigmoid signal $S_j = f(x_j)$ then generates outputs from F_2 that activate two pathways. One pathway inputs to a Differentiator. The other pathway gets multiplied by adaptive weight w_{ji} on the way to level F_3. At level F_3, the weighted signals are added together with a bias term θ_k to define x_k. A sigmoid signal $S_k = f(x_k)$ from F_3 defines the Actual Output of the system. This Actual Output S_k is subtracted from a Target Output b_k via a back-coupled error correction step. The difference $b_k - S_k$ is also multiplied by the term $f'(x_k)$ that is computed at the Differentiator from level F_3. One function of the Differentiator step is to ensure that the activities and weights remain in a bounded range, because if x_k grows too large, then $f'(x_k)$ approaches zero. The net effect of these operations is to compute the Error $\delta_k = f'(x_k)(b_k - S_k)$ that sends a top-down output signal to the level just below it. On the way, each δ_k is multiplied by the bottom-up learned weights w_{jk} at F_3. These weights reach the pathways that carry δ_k via the process of *weight transport*. Weight transport is clearly a nonlocal operation relative to the network connections that carry locally computed signals. All the δ_k are multiplied by the transported weights w_{jk} and added. This sum is multiplied by another Differentiator step term $f'(x_i)$ from level F_2 to keep the resultant product δ_i bounded. δ_i is then back-coupled to adjust all the weights w_{ij} in pathways from level F_1 to F_2. Figure reprinted and text adapted with permission from Carpenter (1989).

Robert French wrote an article in 1999 (French, 1999) that reviewed multiple algorithmic refinements that were made to back propagation in an effort to at least partially overcome its property of catastrophic forgetting. Before reviewing various of these efforts, I will claim that they are reminiscent of the epicycles that were added to the Ptolemaic model of the solar system to help it to explain data about planetary motions more accurately. As we all now know, no amount of epicycle refinement could save the Ptolemaic model, because its core concepts were wrong. The need for epicycles is obviated by my ART "Copernican" model, which requires no epicycles to explain and predict large psychological and neurobiological databases.

Here are some of the efforts that have been made to overcome catastrophic forgetting problem: The article by Kirkpatrick et al. (2017) claims to "overcome this limitation and train networks that can maintain expertise on tasks that they have not experienced for a long time ... by selectively slowing down learning on the weights important for those tasks ... in supervised learning and reinforcement learning problems" (p. 3521).

The method that these authors used to carry out this process requires extensive external supervision, uses nonlocally computed mathematical quantities, and uses equations that implement a form of batch learning that operates offline. In particular, to determine "which weights are most important for a task," this method proceeds by "optimizing the parameters [by] finding their most probable values given some data *D*" by computing the "conditional probability $p(\theta/D)$ from the prior probability of the parameters $p(\theta)$ and the probability of the data $p(D/\theta)$ by using Bayes' rule" (p. 3522). Computing these probabilities requires nonlocal, batch access to multiple learning trials by an omniscient observer who can compute the aforementioned probabilities offline.

Additional external manipulation is needed because "[t]he true posterior probability is intractable so ... we approximate the posterior as a Gaussian distribution with mean given by the parameters θ_z^* and a diagonal precision given by the diagonal of the Fisher information matrix F" (p. 3522). This approximation leads to the problem of minimizing a functional that includes a parameter λ that "sets how important the old task is compared with the new one" (p. 3522).

Related approaches to evaluating the importance of a learned connection, or its "connection cost," include evolutionary algorithms that compute an evolutionary cost for each connection (e.g., Clune, Mouret, and Lipson, 2013). Evolutionary algorithms are inspired by Darwinian evolution (Darwin, 1859). They include a mechanism to search through various neural network configurations for the best weights whereby the model can solve a problem (Yao, 1999). Catastrophic forgetting in this setting is ameliorated by learning weights within one module without engaging other parts of the network. This approach experiences the same kinds of conceptual problems that Kirkpatrick et al. (2017) does.

Diffusion-based neuromodulation is another way to create different modules for different tasks by restricting task-specific learning to a local group of network nodes and connections. This method places "point sources at specific locations within an ANN [Artificial Neural Network] that emit diffusing learning signals that correspond to the positive and negative feedback for the tasks being learned" (Velez and Clune, 2017). In addition to conceptual problems about how these locations are chosen to try to overcome the catastrophic forgetting that obtains without diffusion, this algorithm seems thus far to have only been applied to a simple foraging task whose goal is "to learn which food items are nutritious and should be eaten, and which are poisonous and should not be eaten" across seasons where the nutritional value of the food items may change.

A related problem is solved by the ARTMAP neural network that is discussed below (Carpenter, Grossberg, and Reynolds, 1991) when it learns to distinguish highly similar edible and poisonous mushrooms (Lincoff, 1981) with high predictive accuracy, and does so without experiencing catastrophic forgetting or using neuromodulation. ARTMAP has also successfully classified much larger databases, such as the Boeing design retrieval system that is listed below.

The popularity of back propagation decreased as the above kinds of problems became increasingly evident during the 1980s. As I noted above, Deep Learning recently became popular again after the worst effects of slow and unstable learning were overcome by the advent of large networks of very fast computer servers and huge online databases on the World Wide Web (e.g., millions of pictures of cats), at least when these databases are presented without significant statistical biases. Although slow learning still requires many learning trials, these server farms enable large numbers of trials to learn from many input exemplars and thereby at least partially overcome the memory instabilities that can occur in response to biased small samples (Hinton et al., 2012; LeCun, Bangio, and Hinton, 2015). With these problems partially ameliorated, albeit not solved, many practitioners, including large companies like Apple and Google, have started to use Deep Learning in applications despite its foundational problems.

1.9 "Throw it all away and start over"

It is perhaps because the core problems of these algorithms have not been solved by them that Geoffrey Hinton, who played a key role in developing both back propagation and Deep Learning, said in an *Axios* interview on September 15, 2017 (LeVine, 2017) that he is "deeply suspicious of back propagation ... I don't think it's how the brain works. We clearly don't need all the *labeled data* ... My view is, *throw it all away and start over*" (italics mine).

The remainder of this chapter summarizes in greater detail how the problems of back propagation and Deep Learning that led Hinton to the conclusions in his Axios interview had been solved long ago by our biological neural networks. These solutions are embodied in explainable neural network models that were discovered by analyzing how human and other advanced brains realize autonomous adaptive intelligence. In addition to explaining and predicting many psychological and neurobiological data, these models have been used to solve outstanding problems in engineering, technology, and AI.

I will provide a summary below of some new facts, along with some facts that have earlier been described, in order to provide a summary in a single place.

Section 2 summarizes how explainable cognitive processes use ART circuits to learn to attend, recognize, and predict objects and events in environments whose statistical properties can rapidly and unexpectedly change.

Section 3 summarizes explainable models of biological vision and audition, such as the FACADE model of 3D vision and figure-ground perception, which propose how resonant dynamics support conscious perception and recognition of visual and auditory qualia.

Section 4 describes explainable models of cognitive-emotional interactions, such as the MOTIVATOR model, whose processes of reinforcement learning and incentive motivational learning enable attention to focus on valued goals and to release actions aimed at acquiring them.

Section 5 summarizes explainable motor models, such as the DIRECT and DIVA models, that can learn to control motor-equivalent reaching and speaking behaviors.

Section 6 combines these models with the GridPlaceMap model of spatial navigation to define the SOVEREIGN neural architecture that provides a unified foundation for autonomous adaptive intelligence of a mobile agent.

Section 7 provides a brief conclusion about the functional dynamics of multiple brain processes that are clarified by these models, as are unifying computational principles such as Complementary Computing and their use of Difference Vectors to control reaching, speaking, and navigation.

2 Adaptive Resonance Theory

2.1 Use ART: ART as a computational and biological theory

As I noted above, the problems of back propagation have been well known since the 1980s. An article that I published in 1988 (Grossberg, 1988) listed seventeen differences between back propagation and the biologically inspired Adaptive Resonance Theory, or ART, that I introduced in 1976 and that has been steadily developed by many researchers since then, notably Gail Carpenter. These differences can be summarized by the following bullets in which ART properties are summarized before being contrasted with (vs.) those of back propagation and Deep Learning.

- Real-time (online) learning vs. lab-time (offline) learning
- Learning in nonstationary unexpected world vs. in stationary controlled world
- Self-organized unsupervised or supervised learning vs. supervised learning
- Dynamically self-stabilized learning of arbitrary inputs vs. catastrophic forgetting
- Maintain plasticity vs. externally shut off learning when database gets too large
- Learning of arbitrary databases vs. statistical restrictions on learnable data
- Learn expectations vs. impose external cost functions
- Focus attention to selectively learn critical features vs. passive weight change
- Closed vs. open feedback loop between fast signaling and slower learning
- Top-down priming and selective processing vs. activating all memory resources
- Match learning vs. mismatch learning: Avoiding the noise catastrophe
- Fast and slow learning vs. slow learning: Avoiding the oscillation catastrophe
- Learning with hypothesis testing and memory search vs. passive weight change
- Direct access to globally best match vs. local minima
- Asynchronous learning vs. fixed duration, unstable slow learning
- Autonomous vigilance control vs. unchanging sensitivity during learning
- General-purpose self-organizing production system vs. passive adaptive filter.

Of particular relevance to the above quote from Hinton is the third of the seventeen differences between back propagation and ART; namely, that ART does not need *labeled data* to learn, and has not needed since I introduced it in 1976, thereby solving the problem that Hinton said still needed to be solved in 2017.

ART exists in two forms: First, as algorithms that are designed for use in large-scale applications to engineering, technology, and AI. Second, as an incrementally developed biological theory. In its latter form, ART is now the most advanced cognitive and neural theory about how our brains learn to attend, recognize, and predict objects and events in a changing world that is filled with unexpected events.

As of this writing, ART has explained and predicted far more psychological and neurobiological data than other available theories about these processes, and all of the foundational ART hypotheses from which ART was derived have been supported by subsequent psychological and neurobiological data. I have periodically written heuristic review articles (e.g., Grossberg, 2013, 2017a, 2017b, 2018, 2019b) that support this claim, and refer to articles downloadable from sites.bu.edu/steveg that have explained and predicted much more data since 1976 than these reviews.

2.2 Deriving ART from a universal thought experiment clarifies its range

ART was derived from a thought, or Gedanken, experiment in my 1980 *Psychological Review* article Grossberg (1980). This thought experiment does not require any scientific knowledge to understand or carry out. The thought experiment asks the question: How can a coding error be corrected if no individual cell knows that one has occurred? As my article noted: "The importance of this issue becomes clear when we realize that erroneous cues can accidentally be incorporated into a code when our interactions with the environment are simple and will only become evident when our environmental expectations become more demanding. Even if our code perfectly matched a given environment, we would certainly make errors as the environment itself fluctuates."

A good thought experiment is a story that logically follows from a few simple facts that we all know. The results of my thought experiment about autonomous learning and error correction are translated at every step of its story into processes operating autonomously in real time with only locally computed quantities.

The power of such a thought experiment is to show how when a few familiar facts that we all know from daily life act as environmental constraints on the evolution of our brains over millennia, then ART circuits naturally emerge. Because the facts that are used to derive ART as familiar environmental properties that do not mention mind or brain, ART is a *universal* solution of this autonomous learning and error correction problem. This universality property suggests that we may expect ART designs, in some form, to be embodied in all future devices that realize *autonomous adaptive intelligence*, whether biological or artificial.

Perhaps this is why ART has done well in benchmark studies where it has been compared with other algorithms, and has been used in many large-scale applications to engineering, technology, and AI that other neural algorithms can often not support at all, including an engineering design retrieval system that includes millions of parts defined by million-dimensional feature vectors, and that the Boeing Company used to design the Boeing 777 (Caudell et al., 1994; Escobedo, Smith, and Caudell, 1993).

This application at Boeing developed a CAD system for use on the Boeing factory floor. The simplest ART algorithm, ART1 (Carpenter and Grossberg, 1987), was used to rapidly learn and classify 18 million airplane parts, each of which was defined by a 1,000,000 dimensional vector. At the time it was being developed, I was told by a Boeing manager that the company anticipated savings of up to $80,000,000 in the design of new airplanes using this part retrieval system. This application illustrates that ART models can successfully classify arbitrarily large databases without degrading their foundational computational properties.

Other ART applications include classification and prediction of sonar and radar signals, of medical, satellite, face imagery, social media data, and of musical scores, control of mobile robots and nuclear power plants, cancer diagnosis, air quality monitoring, strength prediction for concrete mixes, solar hot water system monitoring, chemical process monitoring, signature verification, electric load forecasting, tool failure monitoring, fault diagnosis of pneumatic systems, chemical analysis from ultraviolent and infrared spectra, decision support for situation awareness, vision-based driver assistance, user profiles for personalized information dissemination, frequency-selective surface design for electromagnetic system devices, Chinese text categorization, semiconductor manufacturing, gene expression analysis, sleep apnea and narcolepsy detection, stock association discovery, viability of recommender systems, power transmission line fault diagnosis, million city traveling salesman problem, identification of long-range aerosol transport patterns, product redesign based on customer requirements, photometric clustering of regenerated plants of gladiolus, manufacturing cell formation with production data, and discovery of hierarchical thematic structure in text collections, among others. References and discussion of these and other applications and their biological foundations are found in Grossberg (2020).

As a result of these successes, ART became one of the standard neural network models to which practitioners would turn to design the software used in their applications. See the web site http://techlab.bu.edu/resources/articles/C5 of the CNS Tech Lab for a partial list of illustrative benchmark studies and technology transfers that Gail Carpenter assembled with her students up to 2010. Readers who would like a more recent summary of large-scale applications of ART in engineering may find one in the December 2019 issue of the journal *Neural Networks*. The following two articles by Donald Wunsch and his colleagues (da Silva, Elnabarawy, and Wunsch, 2019; Wunsch, 2019) from that special issue are of particular interest: https://arxiv.org/pdf/1910.13351.pdf and https://arxiv.org/pdf/1905.11437.pdf.

2.3 Explainable self-organizing production system in a nonstationary world

ART is more than a feedforward adaptive filter. Although "during the skill acquisition process, failure is required to learn" in any competent learning system, ART goes beyond the kind of learning that is due just to slow modifications of adaptive weights in a feedforward filter. Instead, ART is a *self-organizing production system* that can incrementally learn, using arbitrary combinations of unsupervised and supervised learning trials, to rapidly classify arbitrary nonstationary databases without experiencing catastrophic forgetting. In particular, as I noted above, ART can learn an entire database using fast learning on a single learning trial; for example, Carpenter and Grossberg (1987, 1988).

ART's predictions are explainable using both its activity patterns, or short-term memory (STM) traces, and its adaptive weights, or long-term memory (LTM) traces:

In every ART model, due to matching of bottom-up feature patterns with learned top-down expectations, an attentional focus emerges that selects the activity patterns of *critical features* that are deemed to be predictively important based on past learning. Already learned critical feature patterns may be refined, and new ones discovered, as incremental learning proceeds until, if the input stream is statistically stationary, the model converges to a final critical feature pattern.

These critical features are incorporated through learning into the adaptive weights, or LTM traces, of the bottom-up adaptive filters and top-down learned expectations of an ART network. Thus, by studying the critical feature patterns in STM of such a network, an observer automatically also knows what information has been incorporated into the learned bottom-up and top-down interactions that will determine future categorizations and predictions. Such explainable STM and LTM properties are among the reasons that ART algorithms can be used with confidence to help solve large-scale real-world problems.

2.4 Competition and learned expectations that obey ART Matching Rule

ART's good properties depend critically upon the fact that it supplements its feedforward, or *bottom-up*, adaptive filter circuits with two types of feedback interactions. The first type of feedback occurs in recurrent competitive, or lateral inhibitory, interactions at each processing stage. Such competitive interactions do not exist in back propagation or Deep Learning.

In our neural network models, these competitive interactions normalize activity patterns, a property that is often called *contrast normalization*. I first mathematically proved this property in recurrent shunting on-center off-surround networks in 1973 (Grossberg, 1973). See Grossberg (1980) for a review. Heeger (1992) applied the steady-state divisive normalization form factors of my networks, that arise from their shunting dynamics, to simulate neurophysiological data showing "normalization of cell responses in cat striate cortex," especially simple cells.

Our own work had earlier modeled stereo boundary fusion and long-range perceptual grouping by cortical interactions among simple cells, complex cells, hypercomplex cells, and the bipole grouping cells that I predicted in 1984 (Grossberg, 1984, 1987a, 1987b, Grossberg and Marshall, 1989; Grossberg and Mingolla, 1985a, 1985b). In 1984, bipole cells with the predicted properties were also reported in the neurophysiological experiments of von der Heydt, Peterhans, and Baumgartner (1984). Our work simulated this entire cortical hierarchy because it operates as a *unified functional system* for processing visual scenes. The simple-complex-hypercomplex-bipole neural system is a good example of the *hierarchical resolution of uncertainty* (also see Sections 3.3–3.5 of Chapter 2) which plays a key role in explaining how we consciously see properties of visual art, as I noted in Chapter 2.

At the level of feature processing, competitive interactions using shunting on-center off-surround networks help to choose the contextually most favored cortical representations and to normalize their total activity throughout the brain, including critical feature patterns.

At the level of category learning, competitive interactions help to choose the contextually most favored recognition categories.

ART also includes learned top-down expectations that are matched against bottom-up input patterns to focus attention using a circuit design that obeys what I have called the ART Matching Rule. The top-down pathways that realize the ART Matching Rule form a *modulatory on-center, off-surround network* (Figure 5.2). The off-surround network includes the competitive interactions that were mentioned in the last paragraph. This network realizes the following properties:

When a bottom-up input pattern is received at a processing stage, it can activate its target cells if no other inputs are received. When a top-down expectation is the only active input source, it can provide excitatory modulatory, or priming, signals to cells in its on-center, and driving inhibitory signals to cells in its off-surround. The on-center is modulatory because the off-surround also inhibits the on-center cells, and these two input sources are

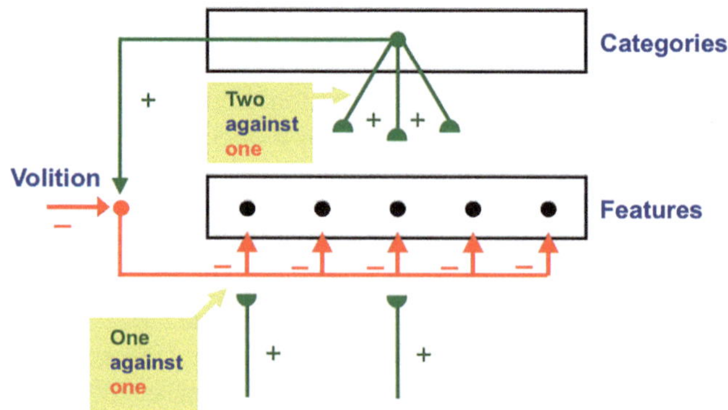

FIGURE 5.2 The ART Matching Rule circuit enables bottom-up inputs to fire their target cells, top-down expectations to provide excitatory modulation of cells in their on-center while inhibiting cells in their off-surround, and a convergence of bottom-up and top-down signals to generate an attentional focus at matched cells while continuing to inhibit unmatched cells in the off-surround. Reprinted with the author's permission from Grossberg (2017b).

approximately balanced (Figure 5.2). Such priming occurs in many experiences where we anticipate experiencing familiar examples of visual art, music, and language.

When a bottom-up input pattern and a top-down expectation are both active, cells that receive both bottom-up excitatory inputs and top-down excitatory priming signals can fire ("two-against-one"), while other cells in the off-surround are inhibited, even if they receive a bottom-up input ("one-against-one"). In this way, the only cells that fire are those whose features are "expected" by the top-down expectation. An attentional focus then starts to form at these cells, including when we begin to focus attention on examples of art, music, and language.

It should also be noted that the ART Matching Rule circuit occurs in multiple parts of the brain and in multiple species, in keeping with the general need to solve the stability-plasticity dilemma in all species that can continue to learn about changing environments. A happy memory for me is when Nobuo Suga came to lecture at our department about the neurophysiological results that his lab discovered about attention due to corticofugal signals in the bat auditory system (Gao and Suga, 1998; Suga, 2008; Suga and Ma, 2003; Suga, Xiao, Ma, and Ji, 2002). I still remember the joy I felt when Suga displayed an image of the bat circuit that does this. It clearly obeys the ART Matching Rule!

Experiments by other labs provided supportive data about how attention is focused on the cues to be learned during visual perceptual learning, as scientists like Merav Ahissar and Shaul Hochstein (Ahissar and Hochstein, 1998); Minami Ito, Gerald Westheimer, and Charles Gilbert (Ito, Westheimer, and Gilbert, 1998); and Zhong-Lin Lu and Barbara Dosher (Lu and Dosher, 2004) have shown. Similar results were obtained during somatosensory learning, as David Krupa, Asif Ghazanfar, and Miguel Nicolelis (Krupa, Ghazanfar, and Nicolelis, 1999) and Jayson Parker and Jonathan Dostrovsky (Parker and Dostrovsky, 1999) have shown.

ART learns how to focus attention upon the *critical feature patterns* that are learned in response to learning trials with multiple bottom-up input patterns, while irrelevant features and noise are suppressed. The critical features are the ones that contribute to accurate predictions as learning proceeds. This attentional selection process is one of the ways that ART successfully overcomes problems noted in the description of the *Frontiers* special issue on Explainable AI by enabling our brains to actively suppress or enhance "data that are noisy and often ambiguously labeled," as well as data that "can also be incorrect due to faulty sensors." Only reliably predictive critical feature combinations will eventually control ART decisions via its ability to pay attention to, and select, task-relevant information.

Back propagation and Deep Learning do not compute STM activation patterns, learned LTM top-down expectations, or attended STM patterns of critical features. Because back propagation and Deep Learning are just feedforward adaptive filters, they do not do any fast information processing using STM patterns, let alone attentive information processing that can selectively use critical features.

2.5 Self-organizing production system: Complementary computing

Because of this limitation, back propagation and Deep Learning can only correct an error using labeled data in which the output vector that embodies an incorrect prediction is mismatched with the correct prediction, thereby computing an error signal that uses weight transport to nonlocally modify the adaptive weights that led to the incorrect prediction (Figure 5.1).

This is not the case in either advanced brains or the biological neural networks like ART that model them. In ART, an unexpected outcome can be caused either by a mismatch of the predicted outcome with what actually occurs, or merely by the *unexpected nonoccurrence* of the predicted outcome. In either situation, a mismatch causes a burst of *nonspecific arousal* that calibrates how unexpected, or novel, the outcome is. This arousal burst can initiate *hypothesis testing* and *memory search*, which automatically leads to the discovery and choice of a better recognition category upon which to base predictions in the future. Figure 2.22 depicts how an ART system can carry out hypothesis testing and memory search to discover and learn a recognition category whereby to better represent a novel situation.

If a familiar input is presented when a category that codes different information is active, the ensuing mismatch triggers a memory search to activate the already-learned category that codes it. This is typically a very short cycle of just one reset event before the globally best matching category is chosen.

It is the combination of learned recognition categories and expectations, novelty responses, memory searches by hypothesis testing, and the discovery of rules that qualify ART as a self-organizing production system.

Why does ART need both an *attentional system* in which learning and recognition occur (levels F_1 and F_2 in Figure 2.22), and an orienting system (ρ in Figure 2.22) that drives memory search and hypothesis testing by the attentional system until a better matching, or entirely new, category is found there? This design enables ART to solve a problem that occurs in a system that learns only when a good enough excitatory *match* occurs (Figure 1.34).

How is anything new ever learned in a match learning system? In such a system, when there is a sufficiently bad match, why does not the learning system just become quiet, with nothing more happening? Instead, in ART, a sufficiently bad mismatch between an active top-down expectation and a bottom-up input pattern, say in response to a novel input, can drive a memory search that continues until the system discovers a new approximate match, which can either refine learning of an old category or begin learning of a new one. The later possibility occurs all the time when we shift our visual attention to inspect different parts of a visual work of art, or as we listen to music or language evolving through time. This memory search process engages both the attentional system and the orienting system of ART (Figure 2.22).

Thus, it is because an ART network has an attentional system and an orienting system, which have computationally *complementary* properties, that an ART network does not just get stuck doing nothing when a bad mismatch occurs. Given the importance of this result, I want to emphasize what it means to say that the properties of the attentional and orienting systems are complementary. In brief, the attentional system supports *top-down, conditionable, specific,* and *match* properties during a good enough attentive match, whereas the orienting system supports *bottom-up, unconditionable, nonspecific,* and *mismatch* properties during a bad enough mismatch (Figure 5.3).

In particular, a top-down expectation is a *top-down, conditionable,* and *specific* event that activates its target cells during a *match* (Figures 2.20 and 5.2). "Conditionable" means that the top-down pathways contain adaptive weights, or LTM traces, that can learn, or be conditioned, to encode a prototype of the recognition category that activates it. This prototype takes the form of a critical feature pattern. "Specific" means that each top-down expectation reads out its own learned critical feature pattern. Target cells are activated that "match" the prototype due to properties of the ART Matching Rule (Figure 5.2).

When a good enough match occurs between a bottom-up input pattern and a learned top-down expectation,

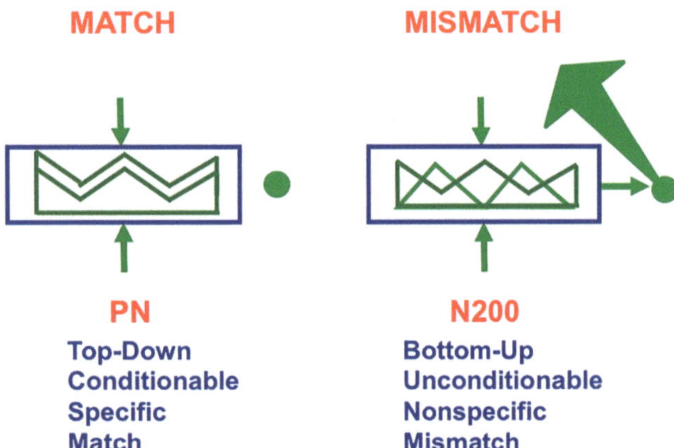

FIGURE 5.3 The PN and N200 event-related potentials are computationally complementary events that are computed within the attentional and orienting systems. See the text for details. Reprinted with the author's permission from Grossberg (2017b).

the initially attended features can reactivate the category via the bottom-up adaptive filter, and the activated category can reactivate the attended features via its top-down expectation. This self-reinforcing excitatory feedback cycle leads to a sustained *feature-category resonance* (Figure 2.20), which can support sustained attentional focusing and conscious recognition of visual objects and scenes.

One psychophysiological marker of such a resonant match during the read out of a learned top-down expectation is the processing negativity, or PN, event-related potential (Figure 5.3) that I predicted in 1978 (Grossberg, 1978b) and that Risto Näätänen reported experimentally in 1982 (Näätänen, 1982).

In contrast to the *top-down, conditionable, specific,* and *match* properties that occur during an attentive match, an orienting system mismatch is a *bottom-up, unconditionable, nonspecific,* and *mismatch* event (Figures 2.22c and 5.3): A mismatch occurs when *bottom-up* activation of the orienting system cannot be adequately inhibited by the bottom-up inhibition from the matched pattern (horizontal red arrow in Figure 2.22). The bottom-up signals to the orienting system are *unconditionable,* or not subject to learning. A mismatch-activated output from the orienting system equally, or *nonspecifically,* excites, or arouses, *all* the category cells because the orienting system does not know which categories read out the expectation that led to mismatch. Any category may be responsible, and may thus need to be reset by arousal. Finally, the orienting system is activated by a sufficiently big *mismatch.* These are four properties of the N200 event-related potential, or ERP (Figure 5.3; Näätänen, Simpson, and Loveless, 1982, Sams et al., 1985).

Nonspecific arousal inhibits the active category which read out the disconfirmed prediction that triggered the hypothesis testing cycle. Another category is then chosen that provides a better match, or is not yet conditioned to any category. This inhibition-then-choice sequence is due to how nonspecific arousal interacts with the habituated pathways that have been sustaining the active category. I have always considered how well this interaction works to be a Minor Mathematical Miracle. Chapter 13 of my Magnum Opus (Grossberg, 2021) explains in greater detail the model properties that cause this Minor Mathematical Miracle to occur.

In summary, during an ART memory search, sequences of mismatch (Figure 2.22b), arousal (Figure 2.22c), and reset (Figure 2.22d) events occur. These ART processes are reflected in the sequences of P120 (mismatch), N200 (arousal), and P300 (reset) Event-Related-Potentials, or ERPs, that I discovered experimentally with Jean-Paul Banquet in 1987 (Banquet and Grossberg, 1987). We measured these ERPs using electrodes that are attached to the scalp of human subjects. ERPs are thus a way to non-invasively measure cognitively interesting properties of brain dynamics in humans.

Measuring ART dynamics directly using neurophysiological methods would be considerably more challenging. In particular, the ART *attentional system* for visual category learning includes brain regions such as prestriate visual cortex, inferotemporal cortex, and prefrontal cortex, whereas the ART orienting system includes the nonspecific thalamus and the hippocampal system. See Carpenter and Grossberg (1993) and Grossberg and Versace (2008) for relevant data. Such a data summary is also provided in Grossberg (2021).

2.6 ART memory search and learning of attended critical feature patterns

This section explains in greater detail how the ART hypothesis testing and learning cycle works (see Figure 2.22), notably how ART searches use cycles of match-induced resonance and mismatch-induced reset to converge upon the next category to learn.

First, as in Figure 2.22a, an input pattern I activates feature detectors at level F_1 of the attentional system (via the vertical green arrow labeled + at the bottom of the figure), thereby activating the activity pattern X. Pattern X is drawn as a continuous yellow pattern that interpolates the activities between network cells. The height of the activity pattern over a given cell denotes the importance of its feature at that time. As this is happening, the input pattern I uses a parallel pathway (the horizontal green arrow labeled + at the bottom of the figure) to generate excitatory signals to the orienting system, which is denoted by an open triangle with the symbol ρ inside it. Along this pathway, I is multiplied by a gain ρ that is called the *vigilance* parameter, before the net signal ρ I inputs to the orienting system.

Activity pattern X outputs two types of signals: It transmits inhibitory signals to the orienting system (horizontal red arrow labeled—) while it also activates a bottom-up, excitatory, adaptive pathway from the feature pattern level F_1 to the category learning level F_2 (vertical green pathway labeled + ending in a hemidisk). The excitatory pathway transmits a signal pattern S to the category learning level F_2. Because the number of active pathways in I and the number of cells in X that are activated by I is approximately the same, the excitatory inputs from I are cancelled by the inhibitory inputs from X. The orienting system thus remains quiet while level F_2 is getting activated by S.

The bottom-up signals S in pathways from the feature level F_1 to the category level F_2 are multiplied by learned adaptive weights at the ends of these pathways to form the net input pattern T to F_2. The inputs T can initially activate many cells in F_2. However, once activated, these cells interact via recurrent lateral inhibitory signals that obey the membrane equations of neurophysiology, also called shunting interactions. The activity pattern in F_2 that survives this competition is contrast-enhanced and normalized, leading to storage in STM of an activity pattern across a small number of cells that received the largest inputs. The chosen cells represent the category Y that codes the feature pattern at F_1. In Figure 2.22a, a yellow triangle represents the chosen category.

The attended critical feature pattern at F_1 hereby learns to control what features are represented by the currently active category Y. As depicted in Figure 2.22b, the activated category Y then generates top-down signals U. These signals are also multiplied by adaptive weights to form a prototype, or critical feature pattern, V which encodes the expectation learned by the active F_2 category of what feature pattern to expect at F_1. This top-down expectation input V is added at F_1 cells using the ART Matching Rule (Figure 2.21), whereby a top-down, modulatory on-center, off-surround network focuses attention upon its critical feature pattern. This matching process, repeated over a series of learning trials, determines what critical features will be learned and chosen in response to each category in the network.

Inspecting an active critical feature pattern can "explain" what its category has learned, what features this category will prime in the future using top-down signals, and what features will control intermodal predictions that are formed via supervised learning and are read out by this category.

Also shown in Figure 2.22b is what happens if V mismatches I at some cells in F_1. Then a subset of the features in X—denoted by the STM activity pattern X* (the pattern of yellow features)—is selected at cells where the bottom-up and top-down input patterns match well enough. In this way, X* is active at features that are confirmed by V, at the same time that mismatched features (white region of the feature pattern) are inhibited. When X changes to X*, the total inhibitory signal from F_1 to the orienting system decreases, thereby setting the stage for initiation of a hypothesis testing cycle.

In particular, as illustrated by Figure 2.22c, if inhibition decreases sufficiently, the orienting system generates a nonspecific arousal burst (denoted by Reset) to F_2. This event mechanizes the intuition that "novel events are arousing." The vigilance parameter ρ, which is computed in the orienting system, determines how bad a match will be tolerated before nonspecific arousal is triggered. Each arousal burst initiates hypothesis testing and memory search for a better-matching category.

Whether the orienting system remains quiet or triggers nonspecific arousal is determined by a simple inequality that is controlled by the vigilance parameter ρ: If $-|X|$ is the total inhibitory signal emitted by activity pattern X to the orienting system, and $\rho|I|$ is the total excitatory input to the orienting system, then the orienting system triggers an arousal burst if $\rho|I| - |X| > 0$ and remains quiet if $0 \geq |I| - |X|$.

An arousal burst triggers a hypothesis testing and learning cycle as follows: First, arousal resets F_2 by inhibiting the active category cells Y (Y is crossed out in Figure 2.22c). After Y is inhibited, then, as denoted by the red x's over the top-down pathways in Figure 2.22c, the top-down expectation V shuts off too, thereby removing inhibition from all feature cells in F_1. As shown in Figure 2.22d, pattern X is thereby disinhibited and reinstated across F_1.

Category Y stays inhibited during the remainder of this hypothesis testing cycle, so that the normalized total activity of F_2 becomes available for X to activate a different category Y* at F_2. Category Y stays inhibited due to the habituation of the activity-dependent chemical transmitters that supported its previous activation (see Grossberg (2021) for details). If Y* does not lead to a sufficiently good match, then the hypothesis testing cycle continues until a better matching, or novel, category is chosen.

As learning dynamically stabilizes, no more mismatches that are big enough to activate the orienting system occur. Inputs I can then directly activate their globally best-matching categories via the bottom-up adaptive filter, without activating the orienting system, much as we can rapidly recognize familiar objects. I summarize psychological and neurobiological data that support each of these processing stages in Grossberg (2017b, 2021).

2.7 Catastrophic forgetting occurs when ART Matching Rule is removed

I will now provide examples of how learned top-down expectations prevent catastrophic forgetting, and how vigilance controls how specific or general the categories that are learned become (see Sections 2.14 and 2.15 in Chapter 2). The intuitive idea about how top-down expectations in ART avoid catastrophic forgetting is that ART learns critical feature patterns of LTM weights in both its bottom-up adaptive filters and its top-down expectations. ART can hereby focus attention upon predictively relevant data (Figure 2.20) while inhibiting irrelevant outlier features that could otherwise have caused catastrophic forgetting.

Gail Carpenter and I defined and simulated examples using the ART 1 model to show more clearly how top-down expectations prevent catastrophic forgetting (Carpenter and Grossberg, 1987). These examples illustrate how easy it is to cause catastrophic forgetting when top-down expectations that obey the ART Matching Rule are eliminated from ART. In fact, sequences of just four input patterns, suitably ordered, lead to catastrophic forgetting in the absence of top-down expectations.

Figure 5.4 summarizes the mathematical rules for generating sequences of just four input patterns A, B, C, and D, which lead to catastrophic forgetting when they are presented in the order ABCAD. Figure 5.5a summarizes a computer simulation that demonstrates catastrophic forgetting when top-down expectations are removed. Figure 5.5b illustrates how learning with stable memory occurs when top-down expectations are restored.

Note that pattern A in Figure 5.4 is a superset of each of the other patterns B, C, and D in the sequence. Pattern A is presented as the first and the fourth input in the sequence $ABCAD$. When it is presented as the first input, it is categorized by category node 1 in Figure 5.5a, but when it is presented as the fourth input, it is categorized by category node 2. An oscillatory recoding of A by nodes 1 and 2 occurs on each presentation of the sequence, so A is catastrophically recoded on every learning trial. The simulation in Figure 5.5b shows that restoring the ART Matching Rule prevents this kind of "superset recoding." Unstable coding can thus occur if a learned *subset* prototype gets recoded as a *superset* prototype when the superset input pattern is categorized by that category.

CODE INSTABILITY INPUT SEQUENCES

$$D \subset C \subset A$$

$$B \subset A$$

$$B \cap C = \varnothing$$

$$|D| < |B| < |C|$$

where $|E|$ is the number of features in the set E

Any set of input vectors that satisfy the above conditions will lead to unstable coding if they are periodically presented in the order

$$ABCAD$$

and the top-down ART Matching Rule is shut off

FIGURE 5.4 When the ART Matching Rule is eliminated by deleting an ART circuit's top-down expectations from the ART 1 model, the resulting competitive learning network experiences catastrophic forgetting even if it tries to learn any of arbitrarily many lists consisting of just four input vectors A, B, C, and D when they are presented repeatedly in the order ABCAD, assuming that the input vectors satisfy the constraints shown in the figure. Adapted with the author's permission from Carpenter and Grossberg (1987).

STABLE AND UNSTABLE LEARNING

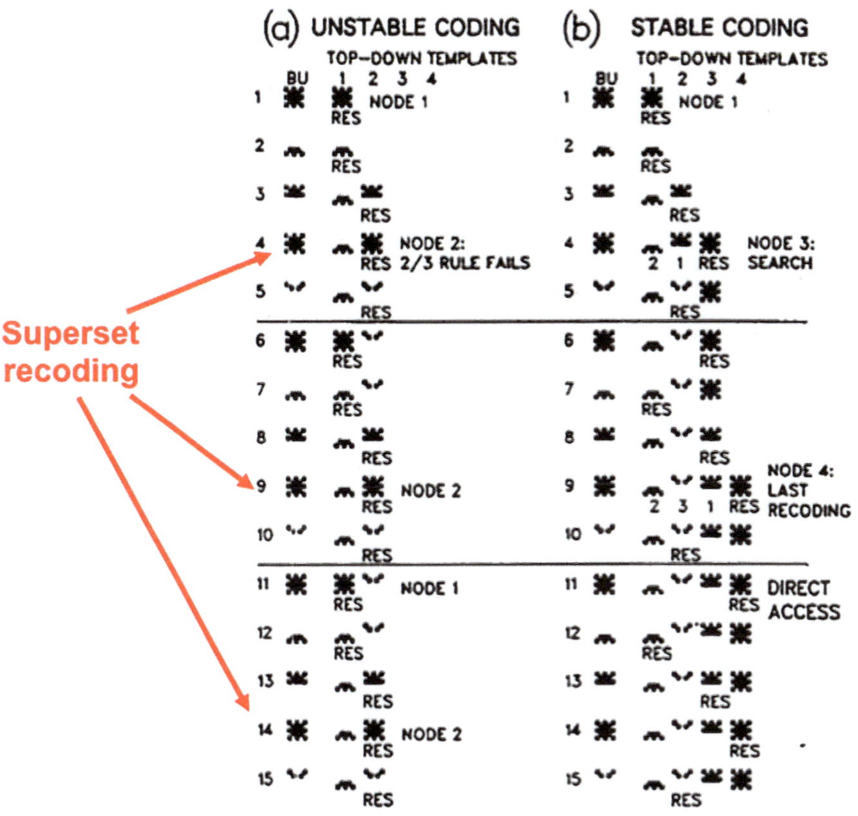

FIGURE 5.5 These computer simulations illustrate how (a) unstable learning and (b) stable learning occur in response to a particular sequence of input vectors A, B, C, D when they are presented repeatedly in the order ABCAD to an ART 1 model. Unstable learning with catastrophic forgetting of the category that codes vector A occurs when no top-down expectations exist, as illustrated by its periodic recoding by categories 1 and 2 on each learning trial. See the text for details. Adapted with the author's permission from Carpenter and Grossberg (1987).

2.8 Simulating vigilance control of concrete and general critical features

Figure 5.6 summarizes a computer simulation showing how different vigilance levels in a single ART model can lead to learning of both concrete and general category prototypes. Figure 5.6a shows how the letters are classified if vigilance is set at a smaller value 0.5. Figure 5.6b shows the same thing if vigilance is set at a larger value 0.8. In both cases, the network's learning rate is chosen to be high. Due to this difference of vigilance, a single ART system can learn categories that code a small number of general categories with only vaguely similar exemplars (Figure 5.6a) or a larger number of concrete categories with very similar prototypes (Figure 5.6b). If vigilance is set to its maximum value of 1, then no variability in a letter is tolerated, and every letter is classified into its own category. This is the limit of *exemplar* prototypes.

Figure 5.6 summarizes computer simulations published in Carpenter and Grossberg (1987) showing how the ART 1 model can learn to classify the letters of the alphabet. During alphabet learning in real life, the raw letters would not be directly input into the brain's recognition categories in the inferotemporal cortex. They would first be processed by multiple stages of the visual cortex (see Chapter 2 and Figure 2.17). How vigilance control works is, however, vividly shown by inputting letters directly to the ART classifier.

Going down the column in Figure 5.6a shows how the network learns in response to the first twenty letters of the alphabet when vigilance equals 0.5. Each row describes what categories and prototypes are learned through time. Black pixels represent prototype values equal to 1 at the corresponding positions. White pixels represent prototype values equal to 0 at their positions. Scanning down the learning trials 1, 2, ..., 20 shows that each prototype becomes more abstract as learning goes on. By the time letter T has been learned, only four categories have been learned with which to classify all twenty letters. The symbol RES, for resonance, under a prototype on each learning trial shows which category classifies the letter that was presented on that trial. In particular, category 1 classifies letters A, B, C, and D, among others, when they are presented, whereas category 2 classifies letters E, G, and H, among others, when they are presented.

Figure 5.6b shows that when the vigilance is increased to 0.8, nine categories are learned in response to the first twenty letters, instead of four. Letter C is no longer lumped into category 1 with A and B. Rather, it is classified by a new category 2 because it cannot satisfy vigilance when it is matched against the prototype of category 1. Together, Figures 5.6a and 5.6b show how, just by changing the sensitivity of the network to attentive matches and mismatches, it can either learn more

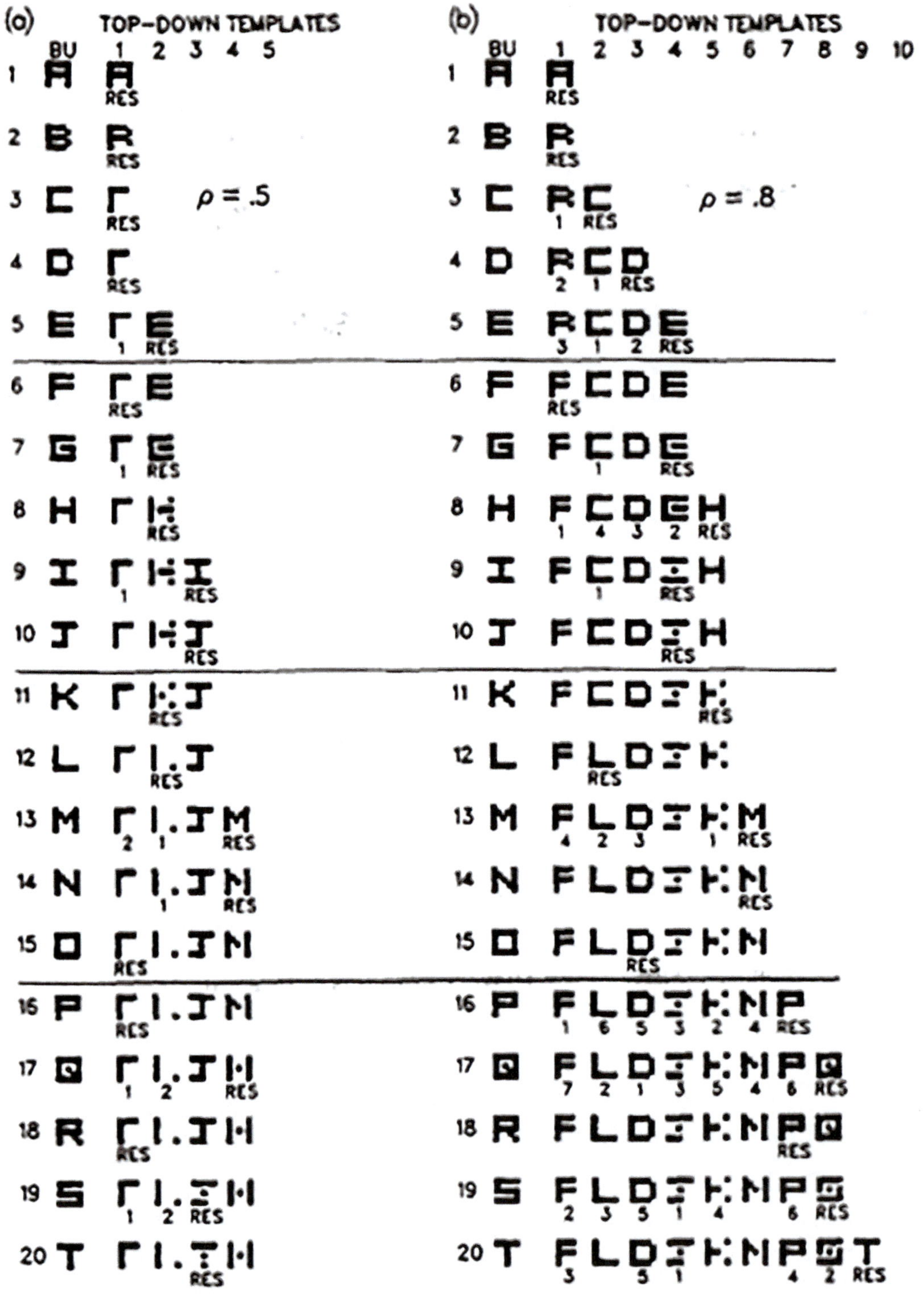

FIGURE 5.6 These computer simulations show how the alphabets A, B, C, … are learned by the ART 1 when vigilance is chosen to equal (a) 0.5, or (b) 0.8. Note that more categories are learned in (b) and that their learned prototypes more closely represent the letters that they categorize. Thus, higher vigilance leads to the learning of more concrete categories. See the text for details. Reprinted with the author's permission from Carpenter and Grossberg (1987).

abstract or more concrete prototypes with which to categorize the world.

Figure 5.6 also provides examples of how memory search works. During search, arousal bursts from the orienting system interact with the attentional system to rapidly reset mismatched categories, as in Figures 5.3 and 2.22c, and to thereby allow selection of better F_2 representations with which to categorize novel inputs at F_1, as in Figure 2.22d.

Memory search ends and resonance begins when the prototype of a familiar category matches well enough with a currently active input exemplar to satisfy the vigilance criterion. This prototype may then be refined by attentional focusing to incorporate the new information that is embodied in the exemplar. For example, in Figure 5.6a, the prototype of category 1 is refined when B and C are classified by it. Likewise, the prototype of category 2 is refined when G, H, and K are classified by it.

If, however, the input is too different from any previously learned prototype, then an uncommitted population of F_2 nodes is selected and learning of a new category is initiated. This is illustrated in Figure 5.6a when E is classified by category 2 and when I is classified by category 3. Search hereby uses vigilance control to determine how much a category prototype can change and, within these limits, protects previously learned categories from experiencing catastrophic forgetting.

The simulations in Figure 5.6 were carried out using *un*supervised learning. Variants of ARTMAP models can learn using arbitrary combinations of unsupervised and supervised learning trials. Fuzzy ARTMAP illustrates how this happens in Section 2.10.

2.9 Start with small initial bottom-up weights and large top-down weights

In a self-organizing system like ART that can learn in an unsupervised way, an important issue is: How does learning get started? This issue does not arise in systems, such as back propagation and Deep Learning, where the correct answer is provided on every supervised learning trial to back-propagate teaching signals that drive weights slowly toward target values (Figure 5.1). Here is how both bottom-up and top-down adaptive weights start out during ART unsupervised learning:

Bottom-up signals within ART adaptive filters from feature level F_1 (Figures 2.20 and 2.22a) are typically gated, before learning occurs, by small and randomly chosen adaptive weights before they activate category level F_2. When F_2 receives these gated signals from F_1, recurrent on-center off-surround signals within F_2 choose a small subset of cells that receive the largest inputs. This recurrent network also *contrast-enhances* the activities of the winning cells, suppresses the activities of other cells as part of the contrast-enhancement process, and *normalizes* the *total activity* across F_2 that is stored in STM. The small bottom-up inputs can hereby generate large enough activities in the winning F_2 cells for them to drive efficient learning in their abutting bottom-up synapses. This learning process *tunes* the adaptive weights until they match the prototype of all the bottom-up input patterns that activate each category, and *normalizes* their amplitudes.

Initial top-down learning faces a different problem: How does learning even get started? This is an issue because a top-down expectation of a newly chosen recognition category does not initially know the feature patterns that will learn to activate it. If, however, read out of the initial top-down expectation mismatches one of these feature patterns, then this mismatch will reset learning before it can even begin (cf. Figure 2.22c).

Given that the category has no idea what feature pattern is active, the top-down expectation must initially be able to match *any* feature pattern. This can only happen if *all* of its top-down adaptive weights initially have large values. As learning proceeds, these broadly distributed adaptive weights are *pruned* to match the critical feature pattern that are discovered and learned.

2.10 Fuzzy ARTMAP: Self-organizing production and rule discovery

The fuzzy ARTMAP model (Figure 2.24; Carpenter et al., 1992) is an ART architecture that can be trained by using arbitrary combinations of unsupervised or supervised learning. The model derives its name, in part, from the fact that it incorporates some of the operations of fuzzy logic, which enable it to learn vectors that code *analog* values. Another reason for its name is that it can learn an associative MAP from one ART category learning network, ARTa, to another ART category learning network, ARTb, where ARTa and ARTb can each be trained using unsupervised learning.

The choices of ARTa and ARTb include a huge variety of possibilities. For example, the categories learned by ARTa can represent visually processed objects, whereas the categories learned by ARTb can represent the auditorily processed names of these objects. An associative mapping from ARTa to ARTb, mediated by a map field F^{ab} (Figure 2.24), can then learn to predict the correct name of each object. This intermodal map learning illustrates how supervised learning can occur in ARTMAP.

In the present example, after map learning occurs, inputting a picture of an object into ARTa can predict its name via ARTb because each of ARTa and ARTb learns both bottom-up adaptive filter pathways *and* top-down expectation pathways. If a sufficiently similar picture has been learned in the past, its presentation to level x^a in ARTa can activate a visual recognition category in y^a. This category can then use the learned association from ARTa to ARTb to activate a category of the object's name in level y^b. Then a learned top-down expectation from the auditory name category can activate a representation in x^b of the auditory features that characterize the name.

The vectors that are categorized by ARTa and ARTb can be chosen very generally. For example, the categories learned by ARTa can include disease symptoms, treatments for them, and medical test results, whereas the categories learned by ARTb can represent the predicted probability of cure, or the length of stay in the hospital, in response to different combinations of these factors. Being able to predict this kind of outcome in advance can be invaluable in guiding hospital treatments and planning.

A hospital's willingness to trust such predictions is bolstered by the fact that the adaptive weights of fuzzy ARTMAP can, at any stage of learning, be translated into *fuzzy IF-THEN rules* that allow practitioners to understand the nature of the knowledge that the model has learned, as well as the amount of variability in the data that each of the learned rules can tolerate. As will be explained more completely in Section 2.12, these IF-THEN rules "explain" the knowledge upon which the predictions have been made. The learned categories themselves play the role of symbols that compress this rule-based knowledge and can be used to read out predictions based upon them. Fuzzy ARTMAP is thus a *self-organizing production and rule discovery system*, as well as a neural network that can learn symbols with which to predict changing environments.

Neither back propagation nor Deep Learning has any of these properties, in addition to being untrustworthy and unreliable. The next property, no less important, also has no computational counterpart in either of these algorithms.

2.11 Minimax learning via match tracking: Maximize generalization and minimize error

I will here use the notation in Figure 2.24 to elaborate on some other properties of fuzzy ARTMAP that I earlier discussed in Chapter 2. Vigilance is initially set to be as low as possible before learning begins. A low initial setting of vigilance enables the learning of large and general categories. General categories conserve memory resources, but may do so at the cost of allowing too many predictive errors. ARTMAP proposes how, in response to predictive errors, or *mismatches*, vigilance control provides a numerical criterion whereby ARTMAP can conjointly *maximize* category generality and *minimize* predictive errors by a process called *match tracking* that realizes a *minimax learning rule.*

Match tracking works as follows: Suppose that an incorrect prediction is made from ARTb in response to an input vector to ARTa on a supervised learning trial. In order for any prediction to have been made, the *analog match* in ARTa between, as described in Figure 2.24, the bottom-up input $A = (a, a^c)$ to x^a and the top-down expectation from y^a to x^a must exceed the vigilance parameter ρ_a at that time, as in Figure 2.23a. If the chosen category reads out a prediction that leads to a big enough mismatch in ARTb between the actual output vector that is read out by y^b, and desired output vector $B = (b, b^c)$ from the environment, then a *match tracking* signal travels from ARTb to ARTa via the map field F^{ab} (Figure 2.24). In order for these operations to work properly, the vectors $A = (a, a^c)$ and $B = (b, b^c)$ are normalized by complement coding, where $a^c = 1 - a$ and $b^c = 1 - b$.

The match tracking signal derives its name from the fact that it increases the vigilance parameter until it just exceeds the analog match value (Figure 2.23b). In other words, vigilance "tracks" the analog match value in ARTa. When vigilance exceeds the analog match value, it can activate the orienting system. A new bout of hypothesis testing and memory search in ARTa is then triggered to discover a better category in y^a with which to encode and predict the mismatched outcome.

Since each increase in vigilance causes a reduction in the generality of learned categories, increasing vigilance via match tracking causes the minimum reduction in category generality that can correct the predictive error by activating the orienting system. That is how the degree of category generality and predictive error realize *minimax learning* to carry out learning that maximizes category generality while it minimizes predictive error.

Since back propagation and Deep Learning do not have an orienting system, they cannot compute a parameter like vigilance. Indeed, these algorithms are not neural architectures at all, let alone architectures like ART and ARTMAP that possess computationally complementary attentional and orienting systems in order to autonomously process and learn about both expected and unexpected events in a changing world.

2.12 Learned fuzzy IF-THEN rules in fuzzy ARTMAP explain its categories

The mathematical equations that define fuzzy ARTMAP will not be reviewed here. For that, see at least one of the articles by Bradski and Grossberg (1995), Carpenter et al.

(1992), Carpenter, Grossberg, and Reynolds (1995), and Granger, Rubin, Grossberg, and Lavoie (2000). Here, I will just note that each adaptive weight vector has a geometric interpretation as a rectangle (or hyper-rectangle in higher dimensions) whose *corners* in each dimension represent the extreme values of the input feature which that dimension represents, and whose *size* represents the degree of fuzziness that input vectors which code that category can have and still remain within it.

If a new input vector falls outside the rectangle on a supervised learning trial, but does not trigger category reset and hypothesis testing for a new category due to the way vigilance has been chosen, then the rectangle is expanded to become the smallest rectangle that includes both the previous rectangle and the newly learned vector, unless this new rectangle becomes too large. The maximal size of such learned rectangles has an upper bound that increases as vigilance decreases, so that more general categories can be learned at lower vigilance.

Inspection of such hyper-rectangles provides immediate insight into both the feature vectors that control category learning and predictions, and how much feature variability is tolerated before category reset and hypothesis testing for another category will be triggered.

3 Explainable Visual and Auditory Percepts

Perhaps the most "explainable" representations in biological neural models are those that represent perceptual experiences, notably conscious visual and auditory percepts. I will briefly review various of the processes that have already been discussed in Chapter 2 and 3 from the perspective of the explainability of their perceptual representations.

3.1 Biological vision: Boundaries gate filling-in of depth-selective surfaces

As discussed in Chapter 2, the functional units of visual perception are 3D boundaries and surfaces. Visual boundaries are formed at the positions where perceptual groupings are completed in response to the spatial distribution of edges, textures, and shading in images and scenes. Visual surfaces are formed when brightnesses and colors fill in within these completed boundaries after the distorting effects of spatially non-uniform illumination levels are eliminated ("discounted the illuminant").

Multiple processing stages are needed to generate sufficiently complete, context-sensitive, and stable surfaces with which to look at and reach for desired objects in a scene, and to navigate within the scene. Such a *hierarchical resolution of uncertainty* is needed because the surface representations of external objects and scenes that are registered by the retinas of our eyes are noisy and incomplete.

They are noisy because of the scattering and absorption of light that occurs between the lens and the retina. They are incomplete because two types of occlusions prevent registration of light on our retinas: First, retinal photoreceptors do not occur at the positions where the optic nerve collects neural signals from photodetectors across the retinal surface and, second, light is not registered on the retina where nourishing retinal veins occlude light reception; see Figure 2.15.

A third kind of uncertainty at the retina are the often non-uniform intensities of light across space with which objects and scenes are illuminated. These gradients of illumination can cause a confusion between illumination levels and "real" object and scenic colors. These real colors are determined by the *relative* amount of light in each wavelength that are reflected from surfaces; that is, by their reflectances.

Because multiple processing stages are needed to complete perceptual boundaries and surfaces, our brains need a process capable of choosing those surfaces that are sufficiently complete, context-sensitive, and stable to regulate observable behaviors like looking and reaching. These surface representations are predicted to become consciously seen in prestriate visual cortical area V4. A *surface-shroud resonance* occurs between cortical area V4 and the posterior parietal cortex, or PPC, that "lights up" the final surface representation. This resonance makes the V4 surface representation consciously visible and enables it to be used to look at and reach for objects of interest.

This prediction is so important that I will mention again here some supportive experimental evidence, including clinical data about how parietal lesions—by disrupting the formation of surface-shroud resonances—lead to visual neglect (Bellmann, Meuli, and Clarke, 2001; Driver and Mattingley, 1998; Marshall, 2001; Mesulam, 1999), visual agnosia (Goodale and Milner, 1992; Goodale et al., 1991), impairments of sustained visual attention (Robertson et al., 1997; Rueckert and Grafman, 1998), and problems with motor planning, notably reaching deficits (Heilman et al., 1985; Mattingley et al., 1998), in support of the hypothesis that "we see in order to look and reach." Thus, the dual roles of parietal cortex—top-down focusing and maintenance of spatial attention in V4, and bottom-up control of motor intention and action—are both disrupted by parietal lesions.

The resonating surface representation in V4 during a surface-shroud resonance is *explainable* because of the conscious visual qualia that it supports. So too is the attentional shroud in PPC that covers, or shrouds, the surface representation with which it is resonating. Indeed, various psychophysical experiments have reported how attentional shrouds spread across surfaces (e.g., Cavanagh, Labianca, and Thornton, 2001; Pylyshyn, 1989; Tse, 2005; Tyler and Kontsevich, 1995).

3.2 Stream-shroud resonances for conscious hearing and auditory communication

Stream-shroud resonances occur in the auditory system, where they support conscious hearing and auditory communication, including speech and language. Stream-shroud resonances are predicted to be homologous to the surface-shroud resonances in the visual system, although visual surfaces represent physical space, whereas auditory streams represent frequency space. Parietal lesions that undermine stream-shroud resonances lead to clinical data that are similar to those in the visual system, including neglect, agnosia, attentional problems, and problems of auditory communication.

Figure-ground separation occurs in the auditory system as well as in the visual system. *Auditory streams* separate different acoustic sources in the environment so that they can be tracked and learned about, a process that is often called *auditory scene analysis* (Bregman, 1990). A classical example of streaming occurs in the *cocktail party problem* (Cherry, 1953), which notes how listeners can hear an attended speaker even if the frequencies in that speaker's voice are similar to the frequencies of nearby speakers, or are occluded by intermittent background noise. The ARTSTREAM neural model explains and simulates how this happens (Grossberg et al., 2004; Grossberg, 2021).

Auditory streams are explainable because they generate spatially distinct representations of frequency, pitch, and timbre that can be selectively attended, and consciously recognized and heard through time. Stream-shroud resonances over auditory streams support conscious hearing, just as surface-shroud resonances over visual surfaces support conscious seeing, and thus are also explainable.

4 Explainable Emotions during Cognitive-Emotional Interactions

4.1 Where cognition and emotion meet: Classical and operant conditioning

In addition to explainable representations of perception and cognition, explainable representations of emotion and cognitive-emotional interactions have also been modeled. These models explain how events in the world learn to activate emotional reactions, how emotions can influence the events to which attention is paid, and how emotionally salient events can learn to trigger responses aimed at acquiring valued goals. This kind of cognitive-emotional learning is often accomplished by either classical conditioning, also called Pavlovian conditioning (Pavlov, 1927), or operant conditioning, also called instrumental or Skinnerian conditioning (Skinner, 1938). See Figures 1.30 and 3.28 and the discussion in Chapter 3 of cognitive-emotional interactions.

The cognitive and emotional representations, summarized in Figure 3.28, that occur in anterior inferotemporal cortex (ITA), amygdala (AMYG), and orbitofrontal cortex (ORB) are explainable because presentation of a visual conditioned stimulus, or CS, will be highly correlated with selective activation of its ITA object category, just as presentation of a reinforcing unconditioned stimulus, or US, will be highly correlated with selective activation of its AMYG value category. Both CS and US, being observable events in the world, are explainable. Likewise, the ITA and AMYG representations that they activate can be measured by available neurophysiological methods.

The simultaneous activation of both ITA and AMYG will activate the ORB representation that responds selectively to this particular object-drive combination, and is thus also explainable. When the feedback loop between object (ITA), value (AMYG), and object-value (ORB) categories is closed by excitatory feedback signals from ORB to ITA, then this circuit goes into a *cognitive-emotional resonance* that supports conscious recognition of an emotion and of the object that has triggered it. Multiple electrode neurophysiological experiments that simultaneously record from ITa, AMYG, and ORB during these events can clarify their functions even more clearly. See Chapters 13 and 14 in Grossberg (2021) for a detailed discussion of neural models and supportive data about cognitive-emotional interactions.

In order for ITA, AMYG, and ORB to carry out their explainable functions, the basal ganglia (Reward Expectation Filter in Figure 3.28) is needed to carry out functions that are *computationally complementary* to those of the AMYG, thereby enabling the system as a whole to cope with both expected and unexpected events. In particular, in *expected* environmental contexts, the AMYG generates incentive motivational signals to the ORB object-value categories which enable ORG to emit signals that control previously learned actions. In contrast, in response to *unexpected* rewards, the basal ganglia generates Now Print signals that drive new learning. These Now Print signals release dopamine (DA) from regions like VTA to multiple brain regions, where they enable learning of new associations there.

4.2 Where cognition and emotion meet: Drive-value resonances

Further details about cognitive-emotional resonances are depicted in Figure 3.28b, which diagrams in greater detail the reciprocal excitatory adaptive connections between lateral hypothalamus (LH) and AMYG that are summarized in the box labeled Value Categories in Figure 3.28a. These reciprocal connections enable AMYG cells to become learned *value categories* that are associated with particular emotional qualia that are represented in LH. In particular, the AMYG interacts reciprocally with taste-drive cells in the LH at which taste and metabolite inputs converge; see the lower level in Figure 3.28b.

Bottom-up signals from activity patterns across these LH taste-drive cells activate competing value categories in the AMYG (upper level in Figure 3.28b). The AMYG value category that wins this competition has the largest input from taste-drive cells. This winning AMYG category can then learn to respond selectively to the combinations of taste-drive activity patterns that succeed in activating it. The winning AMYG category can also send adaptive top-down expectation signals back to the taste-drive cells that activated it and thereby learn a critical feature pattern of predictive taste-drive feature detectors.

When the reciprocal excitatory feedback pathways between the hypothalamic taste-drive cells and the AMYG value category are both active, they generate a *drive-value resonance* that supports the conscious emotion which corresponds to that drive. The activity pattern across LH taste-drive cells provides an explainable representation of the homeostatic factors that subserve the particular emotion corresponding to the winning AMYG category.

One final thought: Although a cognitive-emotional resonance will typically activate a drive-value resonance, a drive-value resonance can occur in the absence of a compatible external sensory cue. This is one way how we can "feel hungry" even if we are not in the presence of food.

5 Explainable Motor Representations

5.1 Computationally complementary What and Where stream processes

The above examples all describe processes that take place in the ventral, or What, cortical stream for perception and cognition (Figures 1.31 and 1.32; Mishkin, 1982; Mishkin, Ungerleider, and Macko, 1983). Only What stream representations are capable of resonating, and thus generating conscious representations, as examples of my general prediction that *all conscious states are resonant states* (Grossberg, 1980, 2017b, 2021). Criteria for a brain representation to be explainable are, however, weaker than those needed for it to become conscious. Accordingly, some properties of representations of the dorsal, or Where, cortical stream for spatial representation and action (Goodale and Milner, 1992) are explainable, even though they cannot support resonance or a conscious state.

This dichotomy between What and Where cortical representations is clarified by the properties summarized in Figure 1.31, which summarizes the fact that these cortical streams are computationally complementary (Grossberg 2000, 2013, 2021). As noted in Figure 1.31, the What stream uses *excitatory matching*, and *match learning* that occurs when a good enough excitatory match triggers a sustained adaptive resonance, notably a feature-category resonance (Figure 1.28). Such a feature-category resonance supports learning of recognition categories that solve the stability-plasticity dilemma. If the current input pattern does not match the currently active category well enough, then the orienting system is activated, leading to a bout of hypothesis testing (Figure 2.22) that carries out a memory search for a better matching category, whether already known or selected for new learning.

In contrast, the Where stream uses *inhibitory matching* to control reaching behaviors that are adaptively calibrated by *mismatch learning* (Figure 1.31). Mismatch learning can continually update the motor maps and gains that control our bodies throughout life.

The Vector Integration to Endpoint, or VITE, model of how arm movement trajectories are formed and executed (Bullock and Grossberg, 1988; Bullock, Grossberg, and Guenther, 1993) is a basic Where stream circuit that can be adaptively calibrated by mismatch learning in the more general Adaptive VITE, or AVITE, model (Gaudiano and Grossberg, 1991, 1992). Figure 3.9b summarizes the VITE model as part of the larger LIST PARSE model. Figure 5.7 summarizes the VITE model as a stand-alone circuit.

During performance, when an arm's *target position vector* (T)—or the position where the hand wants to move—does not equal its *present position vector* (P)—or the position where the hand is now—a *difference vector* (D) computes the direction and distance that the arm must move for the hand to reach its desired target. When the difference vector equals zero, then the hand is where it wants to be, so stops moving.

A difference vector provides explainable information about a movement since, by measuring it, one can predict the direction and distance of the next planned movement.

The inputs to D from T and from P come from different cell populations and are carried by different

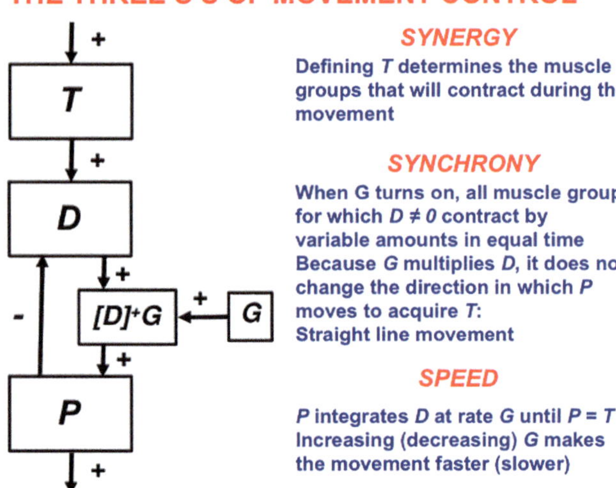

FIGURE 5.7 The Vector Integration to Endpoint, or VITE, model realizes the Three S's of arm movement control: Synergy, Synchrony, and Speed. See the text for details. Adapted with the author's permission from Bullock and Grossberg (1988).

pathways. Signals to T often come from visual representations, whereas P is often coded in motor coordinates. If a VITE model is representing a postural state where T and P represent the same position, then D should equal zero. How are signals from T and P to D calibrated to assure this? This requires learning that matches the gains of the pathways.

In particular, it can happen that a VITE circuit starts out with signals from T to D and from P to D that have different calibrations, so even when T = P, D does not equal zero. How can this calibration problem be solved via learning? This can be done by using the non-zero D vector as a *teaching signal* that changes the adaptive weights at the ends of the T to D axons until D does equal zero whenever T = P.

This kind of mismatch learning is different from the mismatch learning that takes place in back propagation and Deep Learning. One notable difference is that the mismatch learning that occurs in AVITE is defined by local signals, unlike the kind of nonlocal weight transport that occurs in back propagation and Deep Learning (Figure 5.1). Moreover, the archival articles that describe VITE used the model to quantitatively simulate lots of psychological and neurobiological data about reaching behaviors that include the existence of cortical difference vectors.

An inhibitory matching process cannot resonate. Hence, it cannot solve the stability-plasticity dilemma. That is why the parameters that control arm movements, or indeed other bodily movements that are controlled by difference vectors, can keep adapting throughout life to adjust to changes in bodily development, exercise, and aging.

5.3 Mismatch learning via circular reaction calibrates spatial and motor networks

The Vector Associative Map, or VAM, model of Gaudiano and Grossberg (1991, 1992) shows how the AVITE model can autonomously, in an unsupervised way, use mismatch learning changes the gains in the T-to-D pathways until they match those in the P-to-D pathways.

The VAM model does this using a *circular reaction* that generates reactive movements which create a situation in which both T and P represent the same position in space. I explained what a circular reaction is in Chapter 4 for the cases of auditory and visual circular reactions. Here I will show how it gets T and P to be equal when they represent the same position in space.

If the VAM model were properly calibrated, the excitatory T-to-D and inhibitory P-to-D signals that input to D in response to the same positions at T and P would cancel, causing D to equal zero, since then the model has already moved the arm to where it wants it to be. If D is not zero under these circumstances, then the signals are not properly calibrated. Figure 5.8 illustrates how VAM autonomously uses such non-zero D vectors as mismatch, or error, teaching signals that adaptively calibrate the T-to-D signals. As perfect calibration is achieved, D approaches zero, at which time mismatch learning self-terminates.

I noted in Chapter 4 that, during a circular reaction, babies endogenously babble, or spontaneously generate, hand/arm movements to multiple positions around their bodies; see the Endogenous Random Generator, or ERG+, in the upper-left panel of Figure 5.8. When the ERG+ turns on, it activates P, and causes an arm movement.

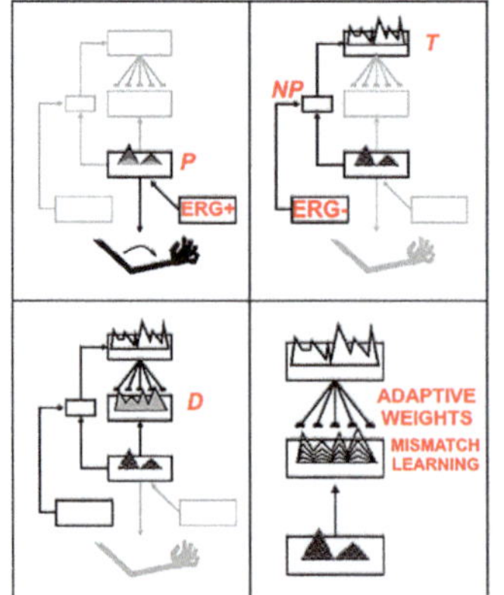

FIGURE 5.8 How a Vector Associative Map, or VAM, model uses mismatch learning during its development to calibrate inputs from a target position vector (T) and a present position vector (P) via mismatch learning of adaptive weights at the difference vector (D). See the text for details. Reprinted with the author's permission from Grossberg (2021).

Random activities in the ERG+ generate random movements that sample the workspace. When the ERG+ shuts off, the movement ends, and a postural state begins.

The offset of ERG+ also triggers the onset of an opponent ERG-. The ERG- opens a Now Print (NP) gate that allows P to be transmitted to T, as shown in the upper-right panel of Figure 5.8. Then both T and P send signals to form the difference vector D, as in the lower-left panel of Figure 5.8.

If D equals zero, the network is properly calibrated. If not, then D acts as an error signal that drives adaptive weights, or LTM traces (depicted by hemidisks), in the T-to-D pathway to change until D does equal zero, as in the lower-right panel of Figure 5.8. Computer simulations in Gaudiano and Grossberg (1991, 1992) demonstrate how the model samples the work space during the critical period and learns the correct LTM traces with which to enable later movements to be accurately made under volitional control, when ERG+ is replaced, after learning is completed, by a volitional GO signal G.

This kind of DV-mediated mismatch learning is just one of the kinds of mismatch learning that occurs in the Where cortical stream, as summarized in Figure 1.31. Mismatch learning allows our spatial and movement control systems to adjust continually to our changing bodies through time. A different kind of error-driven learning occurs in the hippocampus, cerebellum, and basal ganglia to control adaptively timed motivated attention, motor control, and reinforcement learning. All of these brain regions seem to have similar circuits and underlying biochemical mechanisms, notably mechanisms that include the metabotropic glutamate receptor system (mGluR), despite their very different behavioral functions (Grossberg, 2016; Grossberg and Pilly, 2014). I will now describe in greater detail various of these types of mismatch or error-based, learning.

5.4 Mismatch learning using local computations

Note that, unlike back propagation and Deep Learning, there is no nonlocal weight transport during the mismatch learning that occurs in Figure 5.8. Rather, the teaching signals that drive this learning are vectors T that are locally transported from vectors P along pathways that form part of the network's physical anatomy, regulated by opponent ERG+ and ERG- control signals.

5.5 ART and VAM: Synthesis of match and mismatch learning

There is also no collapse of recognition memory by catastrophic forgetting due to this kind of mismatch learning. As summarized in Figure 1.31, ART explains how perceptual and cognitive processes in the What ventral processing stream use excitatory matching and match-based learning to solve the stability-plasticity dilemma so that perceptual and cognitive learning can occur quickly without causing catastrophic forgetting. Excitatory matching also supports sustained resonance, and thus conscious awareness, when a good enough match with object information occurs.

Match-based recognition learning also supports learning of recognition categories at higher cortical regions that are increasingly invariant under changes in an object's views, positions, and sizes when it is registered on our retinas. The 3D ARTSCAN Search model (Figures 2.25 and 2.26) shows how such *invariant* category learning enables us to learn, recognize, and search for objects in the world without causing a combinatorial explosion of memories, much as we can recognize our mother's face from multiple perspectives, or find and recognize a favorite painting located deep within a museum. A great advantage of learning invariant object categories is that recognition is *positionally* invariant, whether we are viewing our mother's face in a scene or a painting in a museum. However, positionally invariant object category representations cannot, by themselves, be used to manipulate or approach objects at particular positions in space.

This huge dichotomy between the requirements for recognition and action is one basic reason why *complementary computing* is needed in our brain (Figure 1.31). Spatial and motor processes in the Where/How dorsal cortical processing stream can be used to manipulate objects in space, but they do so using neural circuits and systems whose designs are computationally complementary to the excitatory matching and match-based learning that are used to carry out perception and recognition in the What ventral processing stream. In particular, the Where stream uses inhibitory matching and mismatch learning, as in the VAM model, to continually update spatial maps and sensory-motor gains as bodily parameters change through time. Their difference vector computations cannot support an adaptive resonance and thus do not solve the stability-plasticity dilemma or support conscious awareness.

Each type of matching and learning in Figure 1.31 is thus insufficient to learn about the world *and* to effectively act upon it. But together they can. Perceptual and cognitive processes use excitatory matching and match-based learning to create self-stabilizing representations of objects and events that embody increasing expertise about the world. Complementary spatial and motor processes use inhibitory matching and mismatch learning to continually update spatial maps and sensory-motor gains to compensate for bodily changes throughout life. In whole-brain architectures (e.g., Figure 3.8), ART match learning provides a self-stabilizing perceptual and cognitive front end for conscious awareness and knowledge acquisition, and provides a stable platform upon which VAM mismatch learning enables our changing bodies to act effectively upon a changing world, without experiencing the catastrophic forgetting that afflicts back propagation and Deep Learning.

5.6 Motor equivalent reaching and speaking: DIRECT and DIVA models

The VITE model for moving a hand to a desired target position using a variable-speed reaching movement (Figure 5.7) has been substantially refined and developed over the years. These developments have shown, among other things, that models which explain how speaking may be controlled in response to auditory cues are homologs of VITE-like models for reaching in response to visual cues.

The variant of VITE for which this is true is the called the DIRECT model (Figures 4.2 and 5.9; Bullock, Grossberg, and Guenther, 1993), where the acronym DIRECT stands from Direction-to-Rotation Effector Control Transform. The DIRECT model is capable of the crucial property of *motor-equivalent* reaching, which includes the ability to reach a new target on the first try under visual guidance with either clamped joints (Figure 5.10c) or a tool (Figure 5.10b).

The conceptual importance of the ability of DIRECT to control tools in space cannot be overemphasized,

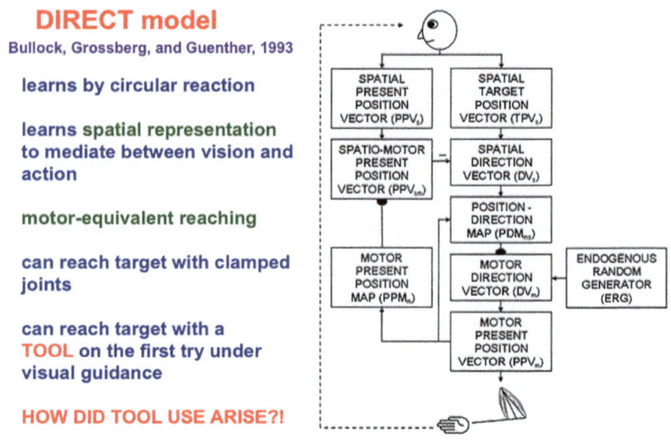

DIRECT model
Bullock, Grossberg, and Guenther, 1993
learns by circular reaction
learns spatial representation to mediate between vision and action
motor-equivalent reaching
can reach target with clamped joints
can reach target with a TOOL on the first try under visual guidance
HOW DID TOOL USE ARISE?!
SPATIAL PRESENT POSITION VECTOR (PPV)
SPATIAL TARGET POSITION VECTOR (TPV)
SPATIO-MOTOR PRESENT POSITION VECTOR (PPV)
SPATIAL DIRECTION VECTOR (DV)
POSITION-DIRECTION MAP (PDM)
MOTOR PRESENT POSITION MAP (PPM)
MOTOR DIRECTION VECTOR (DV)
ENDOGENOUS RANDOM GENERATOR (ERG)
MOTOR PRESENT POSITION VECTOR (PPV)

FIGURE 5.9 The DIRECT model learns, using a circular reaction that is energized by an Endogenous Random Generator, or ERG, to make motor-equivalent volitionally activated reaches. This circular reaction learns a spatial representation of a target in space. It can hereby make accurate reaches with clamped joints and on its first try using a tool under visual guidance; see Figure 5.10. Reprinted with the author's permission from Grossberg (2021).

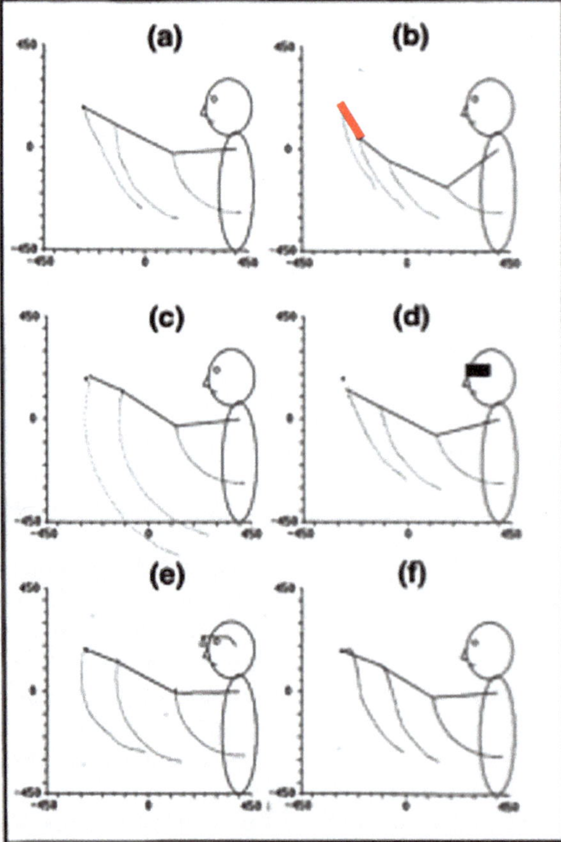

FIGURE 5.10 Computer simulations of DIRECT reaches with (b) a tool, (c) a clamped elbow, and (d) with a blindfold, among other constraints. Reprinted with the author's permission from Grossberg (2021).

since tool use is a critical competence that supported the rapid development of human societies. As shown in Figure 5.10b, without measuring tool length or angle with respect to the hand, the DIRECT model can move a tool's endpoint to touch a target's position correctly under visual guidance *on its first try*, in a single reach without later corrective movements, and without additional learning. In other words, the *spatial affordance* for tool use follows directly from the brain's ability to learn a circular reaction for motor-equivalent reaching. This is an essential competence in making visual art, where the tools in question may be paint brushes, palettes, canvases, mallets, and chisels.

The DIRECT ability to control tools emphasizes that motor control is not just a matter of combining visual and motor information. Instead, these visual and motor signals are first combined to learn a representation of the *space* around the actor. This fact is illustrated in Figure 5.9 by the multiple DIRECT processing stages that have a "spatial" representation. The spatial representations in DIRECT and related biological neural models of movement, such as LIST PARSE (Figure 3.9), enable these models to support learning, planning, and seamless performance of *sequences* of skilled actions in space, including the use of paint brushes by an artist to complete a painting, or coordinated body, arm, and leg movements by a dancer to carry out dance routines.

The DIRECT circuit is not restricted to reaching movements. During the development of the DIRECT model, an evolutionary rationale was noted for why the hand/arm and mouth/throat articulator systems may use homologous neural circuits. A key linking hypothesis notes the fact that eating preceded speech during human evolution (MacNeilage, 1998), and skillful eating requires movements that coordinate hand/arm movements to bring food into the mouth, with mouth/throat articulators to chew and swallow the food. This coordinated system includes motor-equivalent solutions for reaching, including using tools to hold food in space, and chewing to consume it. If the behaviors of reaching and eating are part of an integrated sensory-motor system, then the processes of reaching and speaking may be expected to use similar circuit mechanisms. By extension, so does dancing and singing.

This line of thinking turned out to be correct. An analog of the DIRECT model circuit, called the DIVA model (Figures 4.2 and 5.11; Guenther, 1995; Guenther, Ghosh, and Tourville, 2006), has successfully explained lots data about speech production. The acronym DIVA stands for Directions-Into-Velocities of Articulators. The motor-equivalent property of DIRECT translates in DIVA into the ability to compensate for coarticulation. The same speech articulators that control speech production also control the lyrics and melodies that are combined during singing.

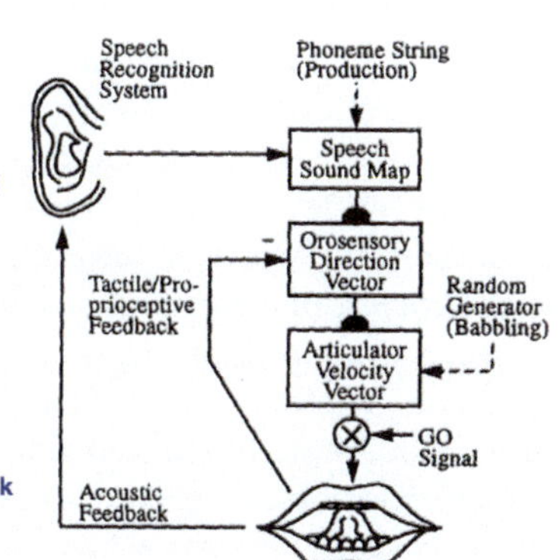

FIGURE 5.11 The DIVA model for the motor-equivalent control of speech articulators. DIRECT and DIVA models have homologous circuits to learn and control motor-equivalent reaching and speaking, with tool use and coarticulation resulting properties. See the text for details. Reprinted with permission from Grossberg (2021).

The property of motor equivalence in DIRECT enables a spatially defined target object to be reached using multiple arm movement trajectories. These multiple possibilities, which are important during both painting and dancing, derive from the fact that a human arm is defined by components with a higher dimensionality than the dimension of a target object in space; for example, a seven degree of freedom arm moving its hand along a desired path in three-dimensional space. Given that multiple trajectories are possible, how does a human or other primate choose among these possibilities to efficiently perform spatially defined arm movement tasks?

The DIRECT model fuses visual, spatial, and motor information to achieve this competence. An important part of this ability is to represent target information in space and to show how to transform it into motor commands; for example, the *spatial* target position vector, *spatial* present position vector, and *spatial* direction vector in Figure 5.9 get converted into a *motor* direction vector and a *motor* present position vector that commands the arm's present position. A position-direction map is interpolated between the spatial direction vector and the motor direction vector to make this spatial-to-motor coordinate transformation. In the reverse direction, a spatio-motor present position vector is interpolated between the motor present position map and the spatial present position vector to make this motor-to-spatial coordinate transformation.

DIRECT learns these transformations between spatial and motor representations by using a circular reaction, albeit an *intermodal* circular reaction that links vision to action. As I have noted above, during a visual circular reaction, babies endogenously babble, or spontaneously generate, hand/arm movements to multiple positions around their bodies.

These internally generated movements are energized by an Endogenous Random Generator, or ERG+. As the ERG+ energizes hand motions in the baby's workshop, the baby's eyes automatically, or reactively, look at her moving hands.

While her eyes are looking at her hands, the baby learns an associative map from her hand positions to the corresponding eye positions, *and* from her eye positions to the corresponding hand positions. Learning of the map between eye and hand in both directions constitutes the "circular" reaction. After map learning occurs, when a baby, child, or adult looks at a target position with its eyes, this eye position can activate a movement command via the learned associative map when a volitional GO signal turns on. This movement command will trigger a reach to the corresponding position in space. Because our bodies continue to grow as we develop from babies into children, teenagers, and adults, these maps continue updating themselves throughout life, which is made possible by the fact that they use difference vectors, like the spatial direction vector, to be learned.

As computer simulations in Bullock, Grossberg, and Guenther (1993) have demonstrated (Figure 5.10), after learning occurs, DIRECT can carry out accurate reaches of the hand to a target in space on the first try; of the endpoint of a tool to touch the target position accurately on the first try, without having to measure the length or the angle of the tool with respect to the hand, and of the hand to a target with the elbow joint held fixed. A nearly accurate reach can also be made under memory guidance when the actor is blindfolded.

A comparison of Figure 5.9 with Figure 5.11 shows that the DIRECT and DIVA models have homologous circuits to control arm movements and speech articulator movements, respectively. DIVA dynamics hereby closely emulate DIRECT dynamics, albeit in the auditory domain. In particular, DIVA models how learning and performance are initiated by an auditory circular reaction that occurs when babies endogenously babble simple sounds and hear the sounds that they create. When the motor commands that caused the sounds and the auditory representations of the heard sounds are simultaneously active in a baby's brain, it can learn an associative map between these auditory representations and the motor commands that produced them.

As Chapter 3 explained, in order for language imitation and subsequent language learning to be possible, speaker normalization must occur in response to heard sounds before they are mapped into motor commands. After a sufficient amount of map learning occurs, a child can then use the speaker-normalized map to imitate sounds from adult speakers, and thereby begin to learn how to speak using increasingly complicated speech and language utterances, again under volitional control. This circular reaction may also be used to learn how to sing a song.

Both DIRECT and DIVA contain explainable representations, just as VITE does. Their difference vectors explain the direction and distance of movements, whether in external space or the space of speech articulators. Chapter 12 of Grossberg (2021) provides a more complete exposition of DIRECT and DIVA, as well as of psychological and neurobiological data that VITE, DIRECT, and DIVA explain and simulate.

6 Explainable Autonomous Adaptive Intelligence: SOVEREIGN

6.1 Balancing reactive and planned movement while navigating to a goal

This chapter has thus far summarized examples of explainable representations that are computed by our brains during perception, cognition, emotion, and action. In addition to noting that the LTM traces of fuzzy ARTMAP can be represented as fuzzy IF-THEN rules which can be used to explain the basis of categorization and prediction dynamics in this model, the article has focused upon the critical role that activity, or STM, variables play in representing and controlling brain processes. All of the summarized STM representations are explainable.

By contrast, back propagation and Deep Learning compute only LTM traces, and no STM traces, indeed no neural architectures with specialized processes, as in the many figures that depict our biological neural networks throughout this book, and no dynamics for real-time information processing or learning.

Additional brain capabilities have been explained and simulated by the SOVEREIGN neural architecture (Gnadt and Grossberg, 2008), and its successor SOVEREIGN2 (Grossberg, 2019a). These architectures embody capabilities for autonomous adaptive perception, cognition, emotion, action, and *navigation* in changing environments. SOVEREIGN was designed to serve as an autonomous neural system for incrementally learning planned action sequences to navigate toward a rewarded goal. The acronym SOVEREIGN summarizes the model's main properties: Self-Organizing, Vision, Expectation, Recognition, Emotion, Intelligent, Goal-oriented Navigation.

SOVEREIGN illustrates how brains may, at several different organizational levels, regulate the all-important *balance between reactive and planned behaviors*. Balancing between reactive and planned movement during navigation occurs during the transition between exploratory behaviors in novel environments and planned behaviors that are learned as a result of previous exploration.

During initial exploration of a novel environment, many reactive movements may occur in response to unexpected or unfamiliar environmental cues (Leonard and McNaughton, 1990). These movements may initially appear to be locally random, as an animal or human orients toward and approaches many stimuli. As the surroundings become familiar, the animal or human learns to discriminate between objects likely to yield reward and those that yield punishment. More efficient routes to the goal are learned during this process. SOVEREIGN begins to model how *sequences* of such behaviors are released at appropriate times during autonomous navigation to realize valued goals.

The relevance of SOVEREIGN to art, music, and even language meanings derives from the fact that all of these activities occur in a spatial environment, including the ability to navigate to the locations in space where one will make a painting, or play a musical piece on an instrument. In the case of language meanings, many language meanings describe activities that take place in one spatial context or another.

As in DIRECT and DIVA, SOVEREIGN begins to learn approach-avoidance behavior via a circular reaction, in this case a perception–cognition–emotion–action cycle during which an action and its consequences elicit sensory cues that are associated with them. Rewards and punishments affect the likelihood that the same actions will be repeated in the future. If objects are not visible when navigation begins, multiple reactive exploratory movements may be needed to reach them. Eventually, these reactive exploratory behaviors are replaced by more efficient planned sequential trajectories within a familiar environment. Figure 5.12 illustrates this situation during learning to navigate from a start box to a rewarded goal box in a maze.

One of the main accomplishments of SOVEREIGN is to explain how erratic reactive exploratory behaviors, as in Figure 5.12a, lead to learning of the most efficient route whereby to acquire a valued goal, as in Figure 5.12b, without losing the ability to balance reactive and planned behaviors so that planned behaviors can be carried out where appropriate, while still retaining the ability to react quickly to novel challenges. These capabilities were demonstrated in SOVEREIGN by computer simulations showing how a simulated animal, or animat, could learn and execute efficient routes to a valued goal in a 3D virtual reality environment (Figure 5.13).

Learning was simulated in a cross maze (Figure 5.13a) that was seen by the animat as a virtual reality, 3D rendering of the maze as it navigated it through time. At the end of each corridor in the maze, a different visual cue was displayed (triangle, star, cross, and square). Sequences of virtual reality views on two navigational routes are shown in Figure 5.13b and Figure 5.13c, where the floor is green, the walls are blue, the ceiling is black, and the interior corners where pairs of maze corridors meet are in red. Figure 5.13b illustrates how the views change as the animat navigates straight down one corridor, whereas Figure 5.13c illustrates how the views change as the animal makes a turn from facing one corridor to facing a perpendicular one.

SOVEREIGN includes several interacting subsystems that model complementary properties of cortical What and Where processing streams that clarify similarities between mechanisms for arm movement control and navigation (Figure 5.14). This homology is important because it clarifies how arm, body, and leg movements are coordinated during tasks ranging from walking and running to painting, playing a musical instrument, and dancing.

Here is a more detailed overview of SOVEREIGN operations, without going into computational details:

> As the animat that SOVEREIGN controls explores an environment, visual inputs are processed by

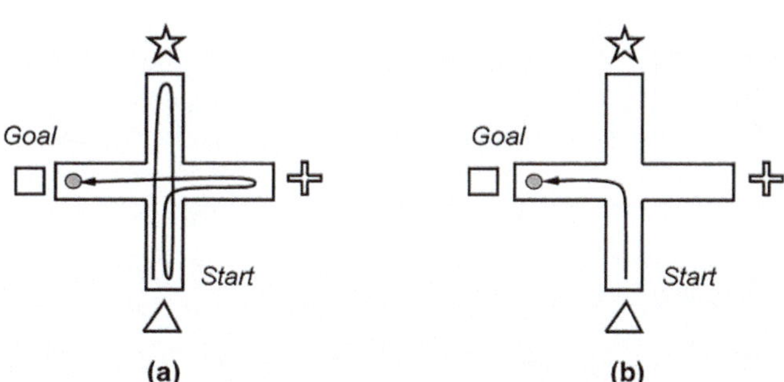

FIGURE 5.12 The animat that is simulated by the SOVEREIGN model learned to convert (a) inefficient exploration of the maze into (b) an efficient direct learned path to the goal. See the text for details. Reprinted with the author's permission from Grossberg (2021).

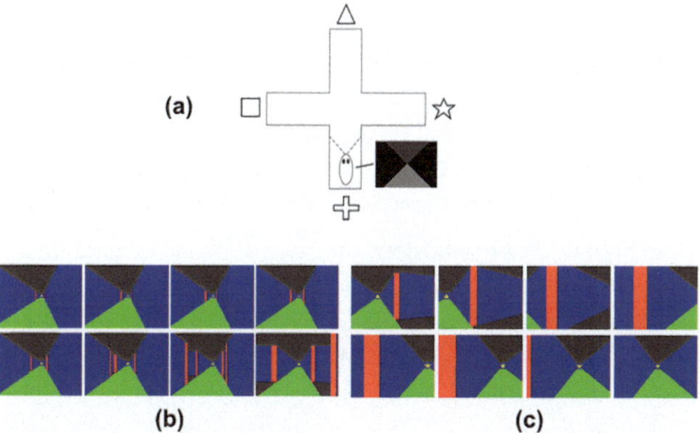

FIGURE 5.13 SOVEREIGN was tested using a virtual reality 3D rendering of a cross maze (a) with different visual cues at the end of each corridor. (b) and (c) illustrate views of the maze. See the text for details. Reprinted with the author's permission from Grossberg (2021).

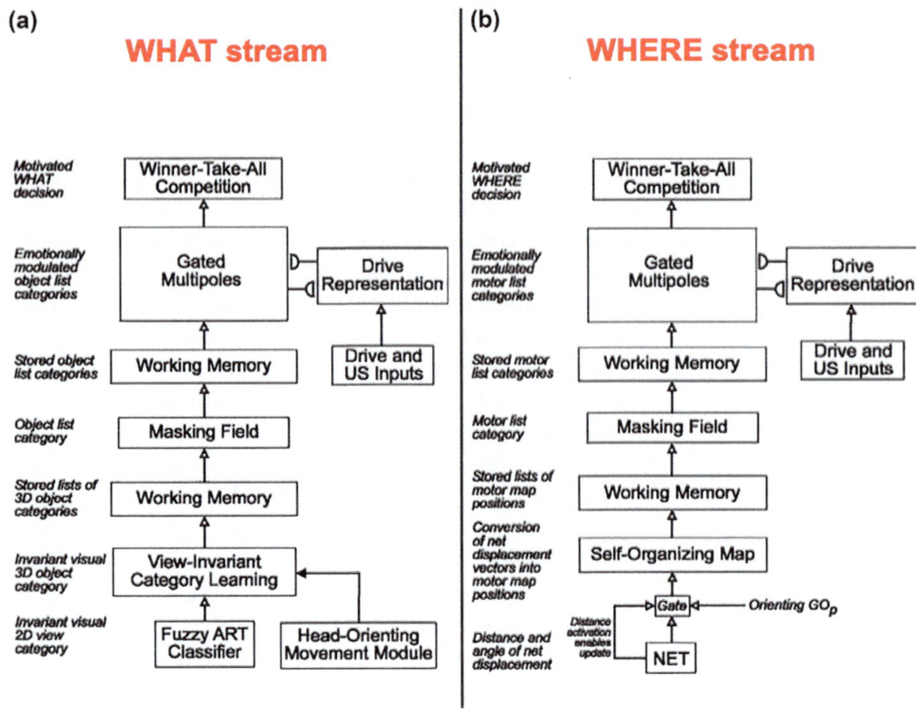

FIGURE 5.14 The SOVEREIGN neural architecture uses homologous processing stages to model the (a) What cortical stream and the (b) Where cortical stream, including their cognitive working memories and chunking networks, and their modulation by motivational mechanisms. See the text for details. Reprinted with permission from Grossberg (2021).

networks that are sensitive to visual form and motion in the What and Where cortical streams, respectively. Position-invariant and size-invariant recognition categories are learned by real-time incremental category learning in the What stream. Estimates of target position relative to the animat are computed in the Where stream, and can activate approach movements toward the target. Motion cues during locomotion, due to relative motion with respect to stationary environmental landmarks, can elicit head-orienting movements to bring a new target into view. A combination of *linear approach* movements within a maze corridor, and *rotational orienting* movements to look at different corridors, are alternately performed during animat navigation. An orienting movement can trigger rotation of the animat to enable exploration of a newly inspected corridor.

Cumulative estimates of each movement are derived from a combination of visual cues from the environment, and proprioceptive cues from cells that carry out *path integration of linear velocity* during a linear movement, and of *angular velocity* during a head turn. These cumulative measures estimate how far the animat has moved in a corridor, and the direction in which it is moving.

Sequences of movements are stored within a *motor* working memory, while sequences of visual categories are stored in a sensory working memory (Figure 5.14). These working memories trigger learning of sensory and motor sequence categories, or plans, which together control planned movements. Predictively effective chunk combinations are learned and selectively attended due to motivational amplification by reinforcement learning signals from drive representations.

These motivational signals were strengthened when the animat was rewarded for actions read out by those chunks during maze learning. Attended planning chunks control a gradual transition from variable reactive exploratory movements to efficient goal-oriented planned movements, and movement sequences. Volitional GO signals gate interactions between model subsystems and the release of overt behaviors. The model can control different motor sequences under different motivational states and learns more efficient sequences to rewarded goals as exploration proceeds.

6.2 Difference vectors during reaching and navigation

SOVEREIGN proposes how the circuit designs for spatial navigation are homologous to those that control reaching behaviors. In both cases, difference vectors (DVs) are computed to determine the direction and distance of a movement (Figure 5.15). During both navigation and reaching (Figure 5.7), in order to compute a difference vector D, both a target position vector T and a present position vector P first need to be computed. The target position T can be computed from visual information, for both arm movements and navigation. Because an arm is attached to the body, its present position P can be directly computed using outflow movement commands that explicitly code the commanded arm position. In contrast, when a body moves with respect to the world, no such immediately available present position command is available. During navigation, the ability to compute a difference vector D between a target position and the present position of the body requires more elaborate brain machinery that is partially illustrated in Figure 5.15. When D is computed in the navigational system using this more

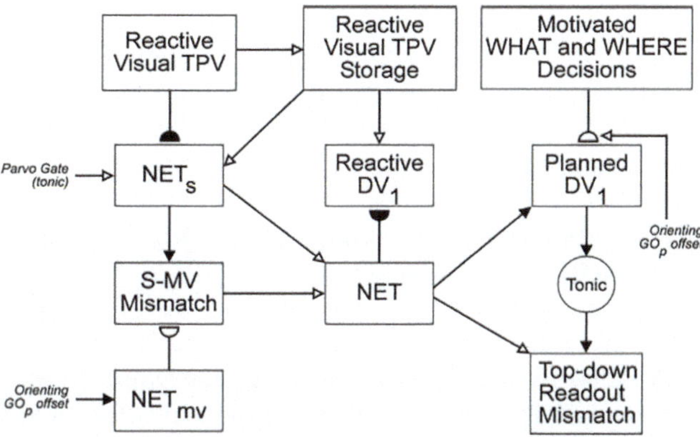

FIGURE 5.15 The main target position vector (TPV), difference vector (DV), and volitional GO computations in SOVEREIGN that bring together reactive and planned signals to control decision-making and action. See the text for details. Reprinted with the author's permission from Grossberg (2021).

elaborate machinery, it determines the direction and distance that the body needs to navigate to acquire the target.

During navigation, the vector NET (see Figure 5.15) replaces the outflow present position vector that is computed, and continually updated, during hand/arm movements. In Figure 5.15, target position information flows from the Reactive Visual TPV to the Reactive Visual TPV Storage module. The NET activity is subtracted from the Stored TPV via learned weights to compute the Reactive Difference Vector, or DV, which represents the angle and distance to move. The learned weights from the NET activity are necessary to calibrate the DV activity.

In familiar situations where planning of navigational movements can occur, learned top-down commands from the Motivated WHAT and WHERE Decisions activate the Planned DV, where NET movement signals are subtracted, yielding a planned angle and distance to move. The Reactive DV and Planned DV are the first motor control stages that can elicit head-orienting and body-approach movements. The NET estimates of head-orienting and body-approach displacement requires multiple stages of processing to be computed.

Finally, it is important to note that navigational difference vectors (DVs) are explainable, just as arm movement difference vectors are.

6.3 Hippocampal place cells: Computing present and target positions

Hippocampal *place cells* provide information about present position during spatial navigation by animals and humans. They selectively fire when an animal is in a particular position within an environment (Muller, 1996; O'Keefe and Dostrovsky, 1971; O'Keefe and Nadel, 1978).

I, in collaboration with several PhD students and postdocs, have worked over the years to develop biological neural network models of how spatial navigation occurs in vivo. One of the most advanced such models is called the GridPlaceMap model. The GridPlaceMap model explains and simulates how entorhinal grid cells and hippocampal place cells are learned during real-time navigation in realistic environments; for example, Grossberg and Pilly (2014).

Once our brains can compute a difference vector between an estimate of present position—as computed by a circuit like that in GridPlaceMap—and desired bodily position, a volitional GO signal can move the body toward the desired goal object, just as in the case of an arm movement. Once again, both navigational movement in the world and movement of limbs with respect to the body use a difference vector computational strategy.

It is also important during navigation to be able to control the kinds of movements that we, as bipeds, and various other animals, as quadrupeds, can make. In both cases, increasing the size of a GO signal can control movements with different gaits, such as walk or run in bipeds, and a walk, trot, pace, and gallop in quadrupeds (Figure 3.26; Pribe, Grossberg, and Cohen, 1997).

A considerable amount of additional machinery is needed to successfully navigate in the world. Successful navigation requires that humans and animals be able to solve basic problems of social cognition, such as how a student can learn a skill from a teacher whose behavior is observed in a different spatial location. The ability to share *joint attention* between actors in different locations is part of this competence. The CRIB (Circular Reactions for Imitative Behavior) model explains and simulates how this may be done (Grossberg and Vladusich, 2010).

Variants of joint attention also seem to occur in other animals. For example, "social" place cells in the hippocampus can fire in a bat as it observes another bat navigating a maze to reach a reward. In these experiments, the observer bat was motivated to do this so that it could subsequently navigate the same route to get the same reward. Under these circumstances, a social place cell can fire in the brain of the observing bat that corresponds to the position of the observed bat (Omer et al., 2018; Schafer and Schiller, 2018, 2020). The position of the observed bat can then function as a spatial target position vector to guide the navigation of the observer rat along the desired route.

These observations suggest a role for hippocampus, interacting with many other brain regions, in computing both present position vectors P and target position vectors T during spatial navigation, thereby enabling difference vectors D to be computed from them that can control navigational directions.

Various additional processes that have been modeled with a view toward achieving true adaptive autonomous intelligence in a mobile agent, whether biological or artificial, have been summarized in the SOVEREIGN and

SOVEREIGN2 architectures (Gnadt and Grossberg, 2008; Grossberg, 2019, 2020). These interactions among circuits to carry out aspects of perception, cognition, emotion, and action that have earlier been described in this book clarify how the "places" that are computed in brain regions like the hippocampus can become integrated into "social" and other higher forms of behavior.

All of the STM representations in these architectures are also explainable, in principle, using neurophysiological and functional neuroimaging methods.

7 Concluding Remarks

As summarized at the end of Section 1, this chapter outlines an explainable neural architecture for autonomous adaptive intelligence within stationary or mobile systems. Each of the chapters focuses on a different functional competence whereby biological brains, no less than artificial devices and robots, may be designed to achieve such autonomy. This book summarizes these competences in the service of explaining how we learn and consciously experience visual art, music, and language and its perceptual and affective meanings. My Magnum Opus (Grossberg, 2021) models a broader range of processes that explain how our brains, which are our best examples of autonomous adaptive intelligence, make our conscious minds in healthy individuals and clinical patients. In addition to its clarifying results on understanding brains and designing AI systems, achieving a computational understanding of autonomous adaptive intelligence may be expected during the coming century to have transformative effects on all aspects of society.

References

Chapter 1

Estes, W. K. (1950). Toward a statistical theory of learning. *Psychological Review*, 57, 94–107.

Felleman, D. J., and Van Essen, D. C. (1991). Distributed hierarchical processing in the primate cerebral cortex. *Cerebral Cortex*, 1, 1–47.

Gaudiano, P., and Grossberg, S. (1992). Adaptive vector integration to endpoint: Self-organizing neural circuits for control of planned movement trajectories. *Human Movement Science*, 11, 141–155.

Grossberg, S. (2021). *Conscious Mind, Resonant Brain: How Each Brain Makes a Mind*. New York: Oxford University Press.

Hebb, D. O. (1949). *The Organization of Behavior: A Neuropsychological Theory*. New York: John Wiley and Sons, Inc.

Hull, C. L. (1943). *Principles of Behavior: An Introduction to Behavior Theory*. Oxford, England: Appleton-Century.

Pavlov, I. P. (1927). *Conditional Reflexes: An Investigation of the Physiological Activity of the Cerebral Cortex*. Oxford, England: Oxford University Press.

Skinner, B. F. (1938). *The Behavior of Organisms: An Experimental Analysis*. New York: Appleton-Century-Crofts.

Suga, N., and Ma, X. (2003). Multiparametric corticofugal modulation and plasticity in the auditory system. *Nature Reviews Neuroscience*, 4, 783–794.

Suga, N., O'Neill, W. E., and Manabe, T. (1979). Harmonic-sensitive neurons in the auditory cortex of the mustache bat. *Science*, 203, 270–274.

Sullivan, E. V., and March, L. (2003). Hippocampal volume deficits in alcoholic Korsakoff's syndrome. *Neurology*, 61, 1716–1719.

Thorndike, E. L. (1927). The law of effect. *The American Journal of Psychology*, 39, 212–222.

Tolman, E. C. (1948). Cognitive maps in rats and men. *Psychological Review*, 55, 189–208.

Visser, P., Krabbendam, L., Verhey, F., Hofman, P., Verhoeven, W., Tuinier, S., Wester, A., Den Berg, Y. W. M. M. V., Goessens, L., Werf, Y., and Jolles, J. (1999). Brain correlates of memory dysfunction in alcoholic Korsakoff's syndrome. *Journal of Neurology, Neurosurgery, and Psychiatry*, 67, 774–778.

Warmflash, D. (2016). Santiago Ramon y Cajal and Camillo Golgi: The two fathers of neuroscience. *Visionlearning Vol. SCIRE*, 2(8): https://www.visionlearning.com/en/library/Inside-Science/58/Santiago-Ram%C3%B3n-y-Cajal-and-Camillo-Golgi/233.

Zhang, Y., Suga, N., and Yan, J. (1997). Corticofugal modulation of frequency processing in bat auditory system. *Nature*, 387, 900–903.

Chapter 2

Ahissar, M., and Hochstein, S. (1993). Attentional control of early perceptual learning. *Proceedings of the National Academy of Sciences USA*, 90, 5718–5722.

Ahissar, M., and Hochstein, S. (1997). Task difficulty and the specificity of perceptual learning. *Nature*, 387, 401–406.

Baer, J. Primary light group: Red, green, blue, 1964-65. (n.d.). http://www.moma.org/collection/works/79825.

Baer, J. (Winter, 1970–1971). Art & vision: Mach bands. *Aspen Magazine*, 8.

Banksy (2005). *Wall and Piece*. Century, London, UK.

Beck, J., and Prazdny, S. (1981). Highlights and the perception of glossiness. *Perception & Psychophysics*, 30, 407–410.

Bhatt, R., Carpenter, G., and Grossberg, S. (2007). Texture segregation by visual cortex: Perceptual grouping, attention, and learning. *Vision Research*, 47, 3173–3211.

Boersma, L. (1995, Fall). Jo Baer. *BOMB*, 53. http://bombmagazine.org/article/1888/jo-baer

Bressan, P. (2001). Explaining lightness illusions. *Perception*, 30, 1031–1046.

Brockmole, J. R., Castelhano, M. S., and Henderson, J. M. (2006). Contextual cueing in naturalistic scenes: Global and local contexts. *Journal of Experimental Psychology: Learning, Memory, and Cognition*, 32, 699–706.

Brooks, P. J., Tomasello, M., Dodson, K., and Lewis, L. B. (1999). Young children's overgeneralization with fixed transitivity verbs. *Child Development*, 70, 1325–1337.

Bullier, J., Hupé, J. M., James, A., and Girard, P. (1996). Functional interactions between areas V1 and V2 in the monkey. *Journal of Physiology (Paris)*, 90, 217–220.

Buschman, T. J., and Miller, E. K. (2007). Top-down versus bottom-up control of attention in the prefrontal and posterior parietal cortices. *Science*, 315, 1860–1862.

Cao, Y., and Grossberg, S. (2005). A laminar cortical model of stereopsis and 3D surface perception: Closure and da Vinci stereopsis. *Spatial Vision*, 18, 515–578.

Cao, Y., and Grossberg, S. (2012). Stereopsis and 3D surface perception by spiking neurons in laminar cortical circuits: A method of converting neural rate models into spiking models. *Neural Networks*, 26, 75–98.

Cao, Y., and Grossberg, S. (2014). How the Venetian blind percept emerges from the laminar cortical dynamics of 3D vision. *Frontiers in Psychology*, doi: 10.3389/fpsyg.2014.00694.

Cao, Y., Grossberg, S., and Markowitz, J. (2011). How does the brain rapidly learn and reorganize view- and positionally-invariant object representations in inferior temporal cortex? *Neural Networks*, 24, 1050–1061.

Caputo, G., and Guerra, S. (1998). Attentional selection by distractor suppression. *Vision Research*, 38, 669–689.

Carpenter, G. A., and Grossberg, S. (1987). A massively parallel architecture for a self-organizing neural pattern recognition machine. *Computer Vision, Graphics, and Image Processing*, 37, 54–115.

Carpenter, G. A., and Grossberg, S. (Eds.) (1991). *Pattern Recognition by Self-Organizing Neural Networks*. Cambridge, MA: MIT Press.

Carpenter, G. A., Grossberg, S., and Rosen, D. B. (1991a). ART 2-A: An adaptive resonance algorithm for rapid category learning and recognition. *Neural Networks*, 4, 493–504.

Carpenter, G. A., Grossberg, S., and Rosen, D. B. (1991b). Fuzzy ART: Fast stable learning and categorization of analog patterns by an adaptive resonance system. *Neural Networks*, 4, 759–771.

Casati, R. (2004). *Shadows: Unlocking their Secrets, from Plato to Our Time*. New York: Vintage Books.

Casati, R., and Cavanagh, P. (2019). *The Visual World of Shadows*. Cambridge, MA: MIT Press.

Chang, H.-C., Grossberg, S., and Cao, Y. (2014). Where's Waldo? How perceptual cognitive, and emotional brain processes cooperate during learning to categorize and find desired objects in a cluttered scene. *Frontiers in Integrative Neuroscience*, doi: 10.3389/fnint.2014.0043.

Chiu, Y. C., and Yantis, S. (2009). A domain-independent source of cognitive control for task sets: Shifting spatial attention and switching categorization rules. *The Journal of Neuroscience*, 29, 3930–3938.

Chun, M. M. (2000). Contextual cueing of visual attention. *Trends in Cognitive Sciences*, 4, 170–178.

Chun, M. M., and Jiang, Y. (1998). Contextual cueing: Implicit learning and memory of visual context guides spatial attention. *Cognitive Psychology*, 36, 28–71.

Collection: MOCA's first thirty years. (2010). http://www.moca-la.com/pc/viewArtWork.php?id=110

Conway, B. R., and Rehding, A. (2013). Neuroaesthetics and the trouble with beauty. *PLoS Biology*, 11(3), e1001504. doi:10.1371/journal.pbio.1001504.

Dee, H., and Santos, P. E. (2011). The perception and content of cast shadows: An interdisciplinary review. *Spatial Cognition and Computation*, 11, 226–253.

Desimone, R. (1998). Visual attention mediated by biased competition in extrastriate visual cortex. *Philosophical Transactions of the Royal Society of London*, 353, 1245–1255.

DeYoe, E. A., and Van Essen, D. C. (1988). Concurrent processing streams in monkey visual cortex. *Trends in Neurosciences*, 11, 219– 226.

Dosher, B. A., Sperling, G., and Wurst, S. A. (1986). Tradeoffs between stereopsis and proximity luminance covariance as determinants of perceived 3D structure. *Vision Research*, 26, 973–990.

Downing, C. J. (1988). Expectancy and visual-spatial attention: Effects on perceptual quality. *Journal of Experimental Psychology: Human Perception and Performance*, 14, 188–202.

Egusa, G. (1983). Effects of brightness, hue, and saturation on perceived depth between adjacent regions in the visual field. *Perception*, 12, 167–175.

Engel, A. K., Fries, P., and Singer, W. (2001). Dynamic predictions: Oscillations and synchrony in top-down processing. *Nature Reviews Neuroscience*, 2, 704–716.

Fang, L., and Grossberg, S. (2009). From stereogram to surface: How the brain sees the world in depth. *Spatial Vision*, 22, 45–82.

Fazl, A., Grossberg, S., and Mingolla, E. (2009). View-invariant object category learning, recognition, and search: How spatial and object attention are coordinated using surface-based attentional shrouds. *Cognitive Psychology*, 58, 1–48.

Foley, N. C., Grossberg, S., and Mingolla, E. (2012). Neural dynamics of object-based multifocal visual spatial attention and priming: Object cueing, useful-field-of-view, and crowding. *Cognitive Psychology*, 65, 77–117.

Frank Stella and the art of the protractor. (2000). [Video file]. https://www.sfmoma.org/watch/frank-stella-and-the-art-of-the-protractor/

Friedman, A. (1979). Framing pictures: The role of knowledge in automatized encoding and memory for gist. *Journal of Experimental Psychology: General*, 108, 316–355.

Gao, E., and Suga, N. (1998). Experience-dependent corticofugal adjustment of midbrain frequency map in bat auditory system. *Proceedings of the National Academy of Sciences*, 95, 12663–12670.

Gilchrist, A. L., Kossyfidis, C., Bonato, F., Agostini, T., Cataliotti, J., Li, X., Spehar, B., Annan, V., and Economou, E. (1999). An anchoring theory of lightness perception. *Psychological Review*, 106, 795–834.

Gove, A., Grossberg, S., and Mingolla, E. (1995). Brightness perception, illusory contours, and corticogeniculate feedback. *Visual Neuroscience*, 12, 1027–1052.

Gregoriou, G. G., Gotts, S. J., Zhou, H., and Desimone, R. (2009). High-frequency, long-range coupling between prefrontal and visual cortex during attention. *Science*, 324, 1207–1210.

Grossberg, S. (1976). Adaptive pattern classification and universal recoding, II: Feedback, expectation, olfaction, and illusions. *Biological Cybernetics*, 23, 187–202.

Grossberg, S. (1980). How does a brain build a cognitive code? *Psychological Review*, 87, 1–51.

Grossberg, S. (1984). Outline of a theory of brightness, color, and form perception. In E. Degreef and J. van Buggenhaut (Eds.), *Trends in Mathematical Psychology*. Amsterdam: North-Holland, pp. 59–85.

Grossberg, S. (1987). Cortical dynamics of three-dimensional form, color, and brightness perception, I: Monocular theory. *Perception and Psychophysics*, 41, 87–116.

Grossberg, S. (1994). 3-D vision and figure-ground separation by visual cortex. *Perception and Psychophysics*, 55, 48–120.

Grossberg, S. (1997). Cortical dynamics of three-dimensional figure-ground perception of two-dimensional figures. *Psychological Review*, 104, 618–658.

Grossberg, S. (1999). How does the cerebral cortex work? Learning, attention and grouping by the laminar circuits of visual cortex. *Spatial Vision*, 12, 163–186.

Grossberg, S. (2000). The complementary brain: Unifying brain dynamics and modularity. *Trends in Cognitive Sciences*, 4, 233–246.

Grossberg, S. (2001). Linking the laminar circuits of visual cortex to visual perception: Development, grouping, and attention. *Neuroscience and Biobehavioral Reviews*, 25, 513–526.

Grossberg, S. (2007). Towards a unified theory of neocortex: Laminar cortical circuits for vision and cognition. In P. Cisek, T. Drew, and J. Kalaska (Eds.), For *Computational Neuroscience: From Neurons to Theory and Back Again*. Amsterdam: Elsevier, pp. 79–104.

Grossberg, S. (2008). The art of seeing and painting. *Spatial Vision*, 21, 463–486.

Grossberg, S. (2009). Cortical and subcortical predictive dynamics and learning during perception, cognition, emotion and action. *Philosophical Transactions of the Royal Society of London B Biological Sciences*, 364, 1223–1234.

Grossberg, S. (2013). Adaptive Resonance Theory: How a brain learns to consciously attend, learn, and recognize a changing world. *Neural Networks*, 37, 1–47.

Grossberg, S. (2014a). How visual illusions illuminate complementary brain processes: Illusory depth from brightness and apparent motion of illusory contours. *Frontiers in Human Neuroscience*, doi: 10.3389/fnhum.2014.00854.

Grossberg, S. (2014b). The visual world as illusion: The ones we know and the ones we don't. In A. Shapiro and D. Todorovic (Eds.), *Oxford Compendium of Visual Illusions*. Oxford, United Kingdom: Oxford University press, in press.

Grossberg, S. (2016). Cortical dynamics of figure-ground separation in response to 2D pictures and 3D scenes: How V2 combines border ownership, stereoscopic cues, and Gestalt grouping rules. *Frontiers in Psychology: Perception Science*., 26 January 26, 2016.

Grossberg, S. (2017). Towards solving the Hard Problem of Consciousness: The varieties of brain resonances and the conscious experiences that they support. *Neural Networks*, 87, 38–95.

Grossberg, S. (2021a). A canonical laminar neocortical circuit whose bottom-up, horizontal, and top-down pathways control attention, learning, and prediction. *Frontiers in Systems Neuroscience*.

Grossberg, S. (2021b). *Conscious MIND, Resonant BRAIN: How Each Brain Makes a Mind*. New York: Oxford University Press.

Grossberg, S., and Hong, S. (2006). A neural model of surface perception: Lightness, anchoring, and filling-in. *Spatial Vision*, 19, 263–321.

Grossberg, S., and Huang, T.-R. (2009). ARTSCENE: A neural system for natural scene classification. *Journal of Vision*, 9(4), 6, 1–19. http://journalofvision.org/9/4/6/, doi:10.1167/9.4.6.

Grossberg, S., and Mingolla, E. (1985a). Neural dynamics of form perception: Boundary completion, illusory figures, and neon color spreading. *Psychological Review*, 92, 173–211.

Grossberg, S., and Mingolla, E. (1985b). Neural dynamics of perceptual grouping: Textures, boundaries, and emergent segmentations. *Perception and Psychophysics*, 38, 141–171.

Grossberg, S., and Gutowski, W. E. (1987). Neural dynamics of decision making under risk: Affective balance and cognitive-emotional interactions. *Psychological Review*, 94, 300–318.

Grossberg, S., and McLoughlin, N. (1997). Cortical dynamics of 3-D surface perception: Binocular and half-occluded scenic images. *Neural Networks*, 10, 1583–1605.

Grossberg, S., and Mingolla, E. (1987). Neural dynamics of surface perception: Boundary webs, illuminants, and shape-from-shading. *Computer Vision, Graphics, and Image Processing*, 37, 116–165.

Grossberg, S., and Raizada, R. (2000). Contrast-sensitive perceptual grouping and object-based attention in the laminar circuits of primary visual cortex. *Vision Research*, 40, 1413–1432.

Grossberg, S., and Schmajuk, N. A. (1987). Neural dynamics of attentionally-modulated Pavlovian conditioning: Conditioned reinforcement, inhibition, and opponent processing. *Psychobiology*, 15, 195–240.

Grossberg, S., and Seidman, D. (2006). Neural dynamics of autistic behaviors: Cognitive, emotional, and timing substrates. *Psychological Review*, 113, 483–525.

Grossberg, S., and Seitz, A. (2003). Laminar development of receptive fields, maps, and columns in visual cortex: The coordinating role of the subplate. *Cerebral Cortex*, 13, 852–863.

Grossberg, S., and Swaminathan, G. (2004). A laminar cortical model for 3D perception of slanted and curved surfaces and of 2D images: Development, attention and bistability. *Vision Research*, 44, 1147–1187.

Grossberg, S., and Todorovic, D. (1988). Neural dynamics of 1-D and 2-D brightness perception: A unified model of classical and recent phenomena. *Perception and Psychophysics*, 43, 241–277.

Grossberg, S., and Versace, M. (2008). Spikes, synchrony, and attentive learning by laminar thalamocortical circuits. *Brain Research*, 1218, 278–312.

Grossberg, S., and Williamson, J. R. (2001). A neural model of how horizontal and interlaminar connections of visual cortex develop into adult circuits that carry out perceptual groupings and learning. *Cerebral Cortex*, 11, 37–58.

Grossberg, S., and Yazdanbakhsh, A. (2005). Laminar cortical dynamics of 3D surface perception: Stratification, transparency, and neon color spreading. *Vision Research*, 45, 1725–1743.

Grossberg, S., Kuhlmann, L., and Mingolla, E. (2007). A neural model of 3D shape-from-texture: Multiple-scale filtering, boundary grouping, and surface filling-in. *Vision Research*, 47, 634–672.

Grossberg, S., Levine, D., and Schmajuk, N. A. (1988). Predictive regulation of associative learning in a neural network by reinforcement and attentive feedback. *International Journal of Neurology*, 21–22, 83–104.

Grossberg, S., Markowitz, J., and Cao, Y. (2011). On the road to invariant recognition: Explaining tradeoff and morph properties of cells in inferotemporal cortex using multiple-scale task-sensitive attentive learning. *Neural Networks*, 24, 1036–1049.

Grossberg, S., Mingolla, E., and Ross, W. D. (1997). Visual brain and visual perception: How does the cortex do perceptual grouping? *Trends in Neurosciences*, 20, 106–111.

Grossberg, S., Srinivasan, K., and Yazdanbakhsh, A. (2014). Binocular fusion and invariant category learning due to predictive remapping during scanning of a depthful scene with eye movements. *Frontiers in Psychology: Perception Science*, doi: 10.3389/fpsyg.2014.01457

Harnad, S. (1990). The symbol grounding problem. *Physica B*, 42, 335–346.

Hawthorne, C. W. (1938/1960). *Hawthorne on painting.* Mineola, New York: Dover.

Helmholtz, H. von. (1866). *Helmholtz's treatise on physiological optics.* New York: Optical Society of America.

Hensche, H. (1988). *The art of seeing and painting.* Thibodaux, Louisiana: Portier Gorman.

Hong, S., and Grossberg, S. (2004). A neuromorphic model for achromatic and chromatic surface representation of natural images. *Neural Networks*, 17, 787–808.

Huang, T.-R., and Grossberg, S. (2010). Cortical dynamics of contextually cued attentive visual learning and search: Spatial and object evidence accumulation. *Psychological Review*, 117, 1080–1112.

Hubel, D. H., and Wiesel, T. N. (1968). Receptive fields and functional architecture of monkey striate cortex. *Journal of Physiology*, 195, 215–243.

Hupé, J. M., James A. C., Girard, D. C., and Bullier, J. (1997). Feedback connections from V2 modulate intrinsic connectivity within V1. *Society for Neuroscience Abstracts*, 406.15, 1031.

Intraub, H. (1999). Understanding and remembering briefly glimpsed pictures: Implications for visual scanning and memory. In V. Coltheart (Ed.), *Fleeting Memories: Cognition of Brief Visual Stimuli.* Cambridge, MA: MIT Press, pp. 47–70.

Ito, M., Westheimer, G., and Gilbert, C. D. (1998). Attention and perceptual learning modulate contextual influences on visual perception. *Neuron*, 20, 1191–1197.

Jiang, Y., and Chun, M. M. (2001). Selective attention modulates implicit learning. *Quarterly Journal of Experimental Psychology*, 54A, 1105–1124.

Jiang, Y., and Wagner, L. C. (2004). What is learned in spatial contextual cueing: Configuration or individual locations? *Perception & Psychophysics*, 66, 454–463.

Julesz, B., and Schumer, R. A. (1981). Early visual perception. *Annual Review of Psychology*, 32, 575–627.

Kanizsa, G. (1955). Margini quasi-percettivi in campi con stimulazione omogenea. *Revista di Psychologia*, 49, 7–30.

Kanizsa, G. (1974). Contours without gradients or cognitive contours. *Italian Journal of Psychology*, 9, 93–113.

Kanizsa, G. (1976). Subjective contours. *Scientific American*, 234, 48–53.

Kanizsa, G. (1979). *Organization in Vision: Essays on Gestalt Perception.* New York: Praeger.

Kastner, S., and Ungerleider, L. G. (2001). The neural basis of biased competition in human visual cortex. *Neuropsychologia*, 39, 1263–1276.

Krupa, D. J., Ghazanfar, A. A., and Nicolelis, M. A. L. (1999). Immediate thalamic sensory plasticity depends on corticothalamic feedback. *Proceedings of the National Academy of Sciences*, 96, 8200–8205.

Kulikowski, J. J. (1978). Limit of single vision in stereopsis depends on contour sharpness. *Nature*, 275, 126–127.

Lamme, V. A. F. (2006). Towards a true neural stance on consciousness. *Trends in Cognitive Sciences*, 10, 494–501.

Leveille, J., Versace, M., and Grossberg, S. (2010). Running as fast as it can: How spiking dynamics form object groupings in the laminar circuits of visual cortex. *Journal of Computational Neuroscience*, 28, 323–346.

Livingstone, M. (2002). *Vision and Art: The Biology of Seeing.* New York: Henry N. Abrams, Inc.

Lleras, A., and von Mühlenen, A. (2004). Spatial context and top-down strategies in visual search. *Spatial Vision*, 17, 465–482.

Llinas, R., Ribary, U., Contreras, D., and Pedroarena, C. (1998). The neuronal basis for consciousness. *Philosophical Transactions of the Royal Society of London B*, 353, 1841–1849.

Lu, Z.-L., and Dosher, B. A. (2004). Perceptual learning retunes the perceptual template in foveal orientation identification. *Journal of Vision*, 4, 5. doi:10.1167/4.1.5.

Luck, S. J., Chelazzi, L., Hillyard, S. A., and Desimone, R. (1997). Neural mechanisms of spatial selective attention in areas V1, V2, and V4 of macaque visual cortex. *Journal of Neurophysiology*, 77, 24–42.

Lysander-1. (2015). http://www.guggenheim.org/newyork/collections/collection-online/artwork/3324

Mann, S. (2016). Perceptual systems, an inexhaustible reservoir of information and the importance of art. In P. Mann and R. Pepperell (Eds.), *Art & Perception.*

Matisse, H. (1947/1992). *Jazz.* Scranton, PA: George Braziller.

Mooney, C. M. (1957). Age in the development of closure ability in children. *Canadian Journal of Psychology*, 11, 219–226.

Mounts, J. R. W. (2000). Evidence for suppressive mechanisms in attentional selection: Feature singletons produce inhibitory surrounds. *Perception and Psychophysics*, 62, 969–983.

Mumford, D. (1992). On the computational architecture of the neocortex. II. The role of corticocortical loops. *Biological Cybernetics*, 66, 241–251.

Necker, L. A. (1832). Observations on some remarkable optical phaenomena seen in Switzerland; and on an optical phaenomenon which occurs on viewing a figure of a crystal or geometrical solid. *London and Edinburgh Philosophical Magazine and Journal of Science*, 1, 329–337.

Olitski, J. (1994). Clement greenberg in my studio. *American Art*, 8(3/4), 125–129.

Oliva, A. (2005). Gist of the scene. In L. Itti, G. Rees, and J. K. Tsotsos (Eds.), *Neurobiology of Attention.* Burlington, MA: Elsevier Academic Press, pp. 251–257.

Oliva, A., and Torralba, A. (2001). Modeling the shape of the scene: A holistic representation of the spatial envelope. *International Journal of Computer Vision*, 42, 145–175.

Olson, I. R., and Chun, M. M. (2002). Perceptual constraints on implicit learning of spatial context. *Visual Cognition*, 9, 273–302.

Palma, J., Versace, M., and Grossberg, S. (2012a). After-hyperpolarization currents and acetylcholine control sigmoid transfer functions in a spiking cortical model. *Journal of Computational Neuroscience*, 32, 253–280.

Palma, J., Grossberg, S., and Versace, M. (2012b). Persistence and storage of activity patterns in spiking recurrent cortical networks: Modulation of sigmoid signals by after-hyperpolarization currents and acetylcholine.

Frontiers in Computational Neuroscience, 6, 42. doi: 10.3389. fncom.2012.00042.

Parker, J. L., and Dostrovsky, J. O. (1999). Cortical involvement in the induction, but not expression, of thalamic plasticity. *The Journal of Neuroscience*, 19, 8623–8629.

Petter, G. (1956). Nuove ricerche sperimentalisulela totalizzazaione percettiva. *Rivista di psicologia*, 50, 213–227.

Perlovsky, L. I. (2010). Intersections of mathematical, cognitive, and aesthetic theories of mind. *Psychology of Aesthetics, Creativity, and the Arts*, 4, 11–17.

Perry, L. C. (1927). Reminiscences of Claude Monet from 1889 to 1909. *The American Magazine of Art*, 18, 119–126.

Pinna, B. (1987). Un effetto di colorazione. In V. Majer, M. Maeran, and M. Santinello, *Il laboratorio e la città. XXI Congresso degli Psicologi Italiani*, 158.

Pinna, B., and Grossberg, S. (2005). The watercolor illusion and neon color spreading: A unified analysis of new cases and neural mechanisms. *Journal of the Optical Society of America A*, 22, 2207–2221.

Poggio, G. F. (1991). Physiological basis of stereoscopic vision. In *Vision and Visual Dysfunction: Binocular Vision*. D. Regan (Ed.). CRC Press, Boca Raton, Florida.

Pollen, D. A. (1999). On the neural correlates of visual perception. *Cerebral Cortex*, 9, 4–19.

Posner, M. I. (1980). Orienting of attention. *Quarterly Journal of Experimental Psychology*, 32, 3–25.

Potter, M. C. (1976). Short-term conceptual memory for pictures. *Journal of Experimental Psychology: Human Learning and Memory*, 2, 509–522.

Potter, M. C., and Levy, E. I. (1969). Recognition memory for a rapid sequence of pictures. *Journal of Experimental Psychology*, 81, 10–15.

Raizada, R., and Grossberg, S. (2001). Context-sensitive bindings by the laminar circuits of V1 and V2: A unified model of perceptual grouping, attention, and orientation contrast. *Visual Cognition*, 8, 431–466.

Ramachandran, V. S., Ruskin, D., Cobb, S., Rogers-Ramachandran, D., and Tyler, C. W. (1994). On the perception of illusory contours. *Vision Research*, 34, 3145–3152.

Rankin, A. (Spring, 1987). Ross Bleckner. *BOMB Magazine*, 19.

Rao, R. P. N., and Ballard, D. H. (1999). Predictive coding in the visual cortex: A functional interpretation of some extra-classical receptive field effects. *Nature Neuroscience*, 2, 79–87.

Reynolds, J., Chelazzi, L., and Desimone, R. (1999). Competitive mechanisms subserve attention in macaque areas V2 and V4. *The Journal of Neuroscience*, 19, 1736–1753.

Richards, W., and Kaye, M. G. (1974). Local versus global stereopsis: Two mechanisms. *Vision Research*, 14, 1345–1347.

Roelfsema, P. R., Lamme, V. A. F., and Spekreijse, H. (1998). Object-based attention in the primary visual cortex of the macaque monkey. *Nature*, 395, 376–381.

Rubin, N. (2015). Banksy's graffiti art reveals insights about perceptual surface completion. *Art & Perception*, 3, 1–17.

Rust, G. (1988). *The Painted House*. New York City, NY: Alfred A. Knopf.

Schor, C. M., and Tyler, C. W. (1981). Spatio-temporal properties of Panum's fusional area. *Vision Research*, 21, 683–692.

Schor, C. M., and Wood, I. (1983). Disparity range for local stereopsis as a function of luminance spatial frequency. *Vision Research*, 23, 1649–1654.

Schor, C. M., Wood, I., and Ogawa, J. (1984). Binocular sensory fusion is limited by spatial resolution. *Vision Research*, 24, 661–665.

Sillito, A. M., Jones, H. E., Gerstein, G. L., and West, D. C. (1994). Feature-linked synchronization of thalamic relay cell firing induced by feedback from the visual cortex. *Nature*, 369, 479–482.

Singer, W. (1998). Consciousness and the structure of neuronal representations. *Philosophical Transactions of the Royal Society B*, 353, 1829–1840.

Somers, D. C., Dale, A. M., Seiffert, A. E., and Tootell, R. B. (1999). Functional MRI reveals spatially specific attentional modulation in human primary visual cortex. *Proceedings of the National Academy of Sciences USA*, 96, 1663–1668.

Steinman, B. A., Steinman, S. B., and Lehmkuhle, S. (1995). Visual attention mechanisms show a center-surround organization. *Vision Research*, 35, 1859–1869.

Thorell, L. G., De Valois, L. G., and Albrecht, D. G. (1984). Spatial mapping of monkey V1 cells with pure color and luminance stimuli, *Vision Research*, 24, 751–769.

Todd, J. T., and Akerstrom, R. A. (1987). Perception of three- dimensional form from patterns of optical texture. *Journal of Experimental Psychology: Human Perception and Performance*, 13, 242–255.

Todd, J. T., Oomes, A. H. J., Koenderink, J. J., and Kappers, A. M. L. (2004). The perception of doubly curved surfaces from anisotropic textures. *Psychological Science*, 15, 40–46.

Todd, J. T., and Thaler, L. (2010). The perception of 3D shape from texture based on directional width gradients. *Journal of Vision*, 10, 17, 1–13.

Tomasello, M., and Herron, C. (1988). Down the garden path: Inducing and correcting overgeneralization errors in the foreign language classroom. *Applied Psycholinguistics*, 9, 237–246.

Tse, P. U. (2005). Voluntary attention modulates the brightness of overlapping transparent surfaces. *Vision Research*, 45, 1095–1098.

Tyler, C. W. (1975). Spatial organization of binocular disparity sensitivity. *Vision Research*, 15, 583–590.

Tyler, C. W. (1983). Sensory processing of binocular disparity. In C. M. Schor and K. J. Cuiffreda (Eds.), *Vergence Eye Movements.* Boston: Butterworths, pp. 199–295.

Tyler, C. W., and Kontsevich, L. L. (1995). Mechanisms of stereoscopic processing: Stereoattention and surface perception in depth reconstruction. *Perception*, 24, 127–153.

Vanduffel, W., Tootell, R. B., and Orban, G. A. (2000). Attention-dependent suppression of meta-bolic activity in the early stages of the macaque visual system. *Cerebral Cortex*, 10, 109–126.

Van Tuijl, H. F. J. M. (1975). A new visual illusion: Neonlike color spreading and complementary color induction between subjective contours. *Acta Psychologica*, 39, 441–445.

Varin, D. (1971). Fenomini di contrasto e diffusione chromatica nell organizzazone spaziale del campo percettivo. *Revista di Psychologica*, 65, 101–128.

von der Heydt, R., Peterhans, E., and Baumgartner, G. (1984). Illusory contours and cortical neuron responses. *Science*, 224, 1260–1262.

Wagemans, J., Feldman, J., Gepshtein, S., Kimchi, R., Pomerantz, J. R., van der Helm, P. A., and van Leeuwen, C. (2012). A century of Gestalt psychology in visual perception II. Conceptual and theoretical foundations. *Psychological Bulletin*, 138, 1218–1252.

Wallach, H. (1948). Brightness constancy and the nature of achromatic colors. *Journal of Experimental Psychology*, 38, 310–324.

Wallach, H. (1976). *On perception*, New York: Quadrangle/The New Your Times Book Co.

Yazdanbakhsh, A., and Grossberg, S. (2004). Fast synchronization of perceptual grouping in laminar visual cortical circuits. *Neural Networks*, 17, 707–718.

Zavagno, D. (1999). Some new luminance-gradient effects. *Perception*, 28, 835–838.

Zavagno, D., Annan, V., and Caputo, G. (2004). The problem of being White: Texting the highest luminance rule. *Vision*, 16, 149–159.

Zeki, S. (1999). *Inner Vision: An Exploration of Art and the Brain*. Oxford, UK: Oxford University Press.

Chapter 3

Abdallah, S., and Plumbley, M. (2009). Information dynamics: Patterns of expectation and surprise in the perception of music. *Connection Science*, 21, 89–117.

Alexander, G. E., and Crutcher, M. D. (1990). Functional architecture of basal ganglia circuits: Neural substrates of parallel processing. *Trends in Neuroscience*, 1, 266–271.

Alexander, G. E., DeLong, M., and Strick, P. L. (1996). Parallel organization of functionally segregated circuits linking basal ganglia and cortex. *Annual Review of Neuroscience*, 9, 357–381.

Ames, H., and Grossberg, S. (2008). Speaker normalization using cortical strip maps: A neural model for steady state vowel categorization. *Journal of the Acoustical Society of America*, 124, 3918–3936.

Averbeck, B., Chafee, M., Crowe, D., and Georgopoulos, A. (2002). Parallel processing of serial movements in prefrontal cortex. *Proceedings of the National Academy of Sciences*, 99, 13172–13177.

Averbeck, B., Chafee, M., Crowe, D., and Georgopoulos, A. (2003). Neural activity in prefrontal cortex during copying geometrical shapes. I. Single cells encode shape, sequence, and metric parameters. *Experimental Brain Research*, 150, 127–141.

Barone, P., and Joseph, J. (1989). Prefrontal cortex and spatial sequencing in macaque monkey. *Experimental Brain Research*, 78, 447–464.

Beier, E. J., and Ferreira, F. (2018). The temporal prediction of stress in speech and its relation to musical beat perception. *Frontiers in Psychology*, April 3, 2018. https://www.frontiersin.org/articles/10.3389/fpsyg.2018.00431/full

Boardman, I., and Bullock, D. (1991). A neural network model of serial order recall from short-term memory. *Proceedings of the International Joint Conference on Neural Networks (IJCNN)*, II, 879–884.

Boardman, I., Grossberg, S., Myers, C., and Cohen, M. (1999). Neural dynamics of perceptual order and context effects for variable-rate speech syllables. *Perception & Psychophysics*, 6, 1477–1500.

Bohland, J., Bullock, D., and Guenther, F. (2010). Neural representations and mechanisms for the performance of simple speech sequences. *Journal of Cognitive Neuroscience*, 22, 1504–1529.

Bostan, A. C., Dum, R. P., and Strick, P. L. (2013). Cerebellar networks with the cerebral cortex and basal ganglia. *Trends in Cognitive Sciences*, 17, 241–254.

Bostan, A. C., and Strick, P. L. (2010). The cerebellum and basal ganglia are interconnected. *Neuropsychology Review*, 20, 261–270.

Bradski, G., Carpenter, G. A., and Grossberg, S. (1992). Working memory networks for learning multiple groupings of temporal order with application to 3-D visual object recognition. *Neural Computation*, 4, 270–286.

Bradski, G., Carpenter, G. A., and Grossberg, S. (1994). STORE working memory networks for storage and recall of arbitrary temporal sequences. *Biological Cybernetics*, 71, 469–480.

Bregman, A. S. (1990). *Auditory Scene Analysis: The Perceptual Organization of Sound*. Cambridge, MA: MIT Press.

Brown, G. (1911). The intrinsic factors in the act of progression in the mammal. *Proceedings of the Royal Society B Biological Sciences*, 84, 308–319.

Brown, J. W., Bullock, D., and Grossberg, S. (1999). How the basal ganglia use parallel excitatory and inhibitory learning pathways to selectively respond to unexpected rewarding cues. *The Journal of Neuroscience*, 19, 10502–10511.

Brown, J. W., Bullock, D., and Grossberg, S. (2004). How laminar frontal cortex and basal ganglia circuits interact to control planned and reactive saccades. *Neural Networks*, 17, 471–510.

Buhmann, J., Moens, B., Lorenzoni, V., and Leman, M. (2017). Shifting the musical beat to influence running cadence. In E. Van Dyck (Ed.), *Proceedings of the 25th Anniversary Conference of the European Society for the Cognitive Sciences of Music*, July 31–August 4, 2017. Ghent, Belgium. https://biblio.ugent.be/publication/8530848.

Bullock, D., and Grossberg, S. (1988). Neural dynamics of planned arm movements: Emergent invariants and speed-accuracy properties during trajectory formation. *Psychological Review*, 95, 49–90.

Bullock, D., and Grossberg, S. (1991). Adaptive neural networks for control of movement trajectories invariant under speed and force rescaling. *Human Movement Science*, 10, 3–53.

Bullock, D., Grossberg, S., and Guenther, F. H. (1993). A self-organizing neural model of motor equivalent reaching and tool use by a multijoint arm. *Journal of Cognitive Neuroscience*, 5, 408–435.

Bullock, D., and Rhodes, B. (2003). Competitive queuing for serial planning and performance. In M. Arbib (Ed.), *Handbook of Brain Theory and Neural Networks*, Second Edition. Cambridge, MA: MIT Press, pp. 241–244.

Buonomano, D. V., Baxter, D. A., and Byrne, J. H. (1990). Small networks of empirically derived adaptive elements simulate some higher-order features of classical conditioning. *Neural Networks*, 3, 507–523.

Buonomano, D. V., and Mauk, M. D. (1994). Neural network model of the cerebellum: Temporal discrimination and the timing of motor responses. *Neural Computation*, 6, 38–55.

Burger, B., Thompson, M. R., Luck, G., Saarikallio, S. H., and Toivianinen, P. (2014). Hunting for the beat in the body: On period and phase locking in music-induced movement. *Frontiers in Human Neuroscience*, 8, 1-16.

Cannell, M. B., Cheng, H., and Lederer, W. J. (1995). The control of calcium release in heart muscle. *Science*, 268, 1045–1049.

Carew, T. J., Walters, E. T., and Kandel, E. R. (1981). Classical conditioning in a simple withdrawal reflex in *Aplysia Californica*. *The Journal of Neuroscience*, 1, 1426–1437.

Carpenter, G. A. (1977a). A geometric approach to singular perturbation problems with applications to nerve impulse equations. *Journal of Differential Equations*, 23, 335–367.

Carpenter, G. A. (1977b). Periodic solutions of nerve impulse equations. *Journal of Mathematical Analysis and Applications*, 58, 152–173.

Carpenter, G. A. (1979). Bursting phenomena in excitable membranes. *SIAM Journal on Applied Mathematics*, 36, 334–372.

Carpenter, G. A., and Grossberg, S. (1987a). A massively parallel architecture for a self-organizing neural pattern recognition machine. *Computer Vision, Graphics, and Image Processing*, 37, 54–115.

Carpenter, G. A., and Grossberg, S. (1991). *Pattern Recognition by Self-organizing Neural Networks*. Cambridge, MA: MIT Press.

Chapin, H., Zanto, T., Jantzen, K., Kelso, J., Steinberg, F., and Large, E. W. (2010). Neural responses to complex auditory rhythms: The role of attending. *Frontiers in Psychology: Auditory Cognitive Neuroscience*, 1, 224. https://www.frontiersin.org/articles/10.3389/fpsyg.2010.00224/full

Cohen, M. A., and Grossberg, S. (1986). Neural dynamics of speech and language coding: Developmental programs, perceptual grouping, and competition for short-term memory. *Human Neurobiology*, 5, 1–22.

Cohen, M. A., and Grossberg, S. (1987). Masking fields: A massively parallel neural architecture for learning, recognizing, and predicting multiple groupings of patterned data. *Applied Optics*, 26, 1866–1891.

Cohen, M. A., Grossberg, S., and Wyse, L. L. (1995). A spectral network model of pitch perception. *Journal of the Acoustical Society of America*, 98, 862–879.

Contreras-Vidal, J.L., Grossberg, S., and Bullock, D. (1997). A neural model of cerebellar learning for arm movement control: Cortico-spino-cerebellar dynamics. *Learning and Memory*, 3, 475–502.

Cowan, N. (2001). The magical number 4 in short-term memory: A reconsideration of mental storage capacity. *Behavioral and Brain Sciences*, 24, 87–114.

Damm, L., Varoqui, D., de Cock, V. C., Dalla Bella, S., and Bardy, B. (2020). Why do we move to the beat? A multi-scale approach, from physical principles to brain dynamics. *Neuroscience and Biobehavioral Reviews*, 112, 553–584.

Deutsch, D. (1992). Paradoxes of musical pitch. *Scientific American*, August 1992, 88–95.

Deutsch, D. (Ed.) (2013). *The Psychology of Music*, Third Edition. New York: Elsevier/Academic Press.

Dranias, M., Grossberg, S., and Bullock, D. (2008). Dopaminergic and non-dopaminergic value systems in conditioning and outcome-specific revaluation. *Brain Research*, 1238, 239–287.

Dowling, W. J., and Tighe, T. J. (2014). *Psychology and Music: The Understanding of Melody and Rhythm*. Taylor and Francis, ebook; https://www.taylorfrancis.com/books/9781315808116.

Doyon, J., Penhune, V. B., and Ungerleider, L. G. (2003). Distinct contribution of the cortico-striatal and cortico-cerebellar systems to motor skill learning. *Neuropsychologia*, 41, 252–262.

Eccles, J. C., Ito, M., and Szentagothai, J. (1967). *The Cerebellum as a Neuronal Machine*. New York: Springer-Verlag.

Egermann, H., Pearce, M. T., Wiggins, G. A., and McAdams, S. (2013). Probabilistic models of expectation violation predict psychophysiological emotional response to live concert music. *Cognitive, Affective, and Behavioral Neuroscience*, 13, 533–553.

Ellias, S. A., and Grossberg, S. (1975). Pattern formation, contrast control, and oscillations in the short-term memory of shunting on-center off-surround networks. *Biological Cybernetics*, 20, 69–98.

Farrell, S., and Lewandowsky, S. (2004). Modeling transposition latencies: Constraints for theories of serial order memory. *Journal of Memory and Language*, 51, 115–135.

Fiala, J. C., Grossberg, S., and Bullock, D. (1996). Metabotropic glutamate receptor activation in cerebellar Purkinje cells as substrate for adaptive timing of the classically conditioned eye blink response. *The Journal of Neuroscience*, 16, 3760–3774.

Franklin, D. J., and Grossberg, S. (2017). A neural model of normal and abnormal learning and memory consolidation: Adaptively timed conditioning, hippocampus, amnesia, neurotrophins, and consciousness. *Cognitive, Affective, and Behavioral Neuroscience*, 17, 24–76. https://link.springer.com/article/10.3758/s13415-016-0463-y

Fujioka, T., Zendel, B., and Ross, B. (2010). Endogenous neuromagnetic activity for mental hierarchy of timing. *The Journal of Neuroscience*, 30, 3458–3466.

Funahashi, S., Inoue, M., and Kubota, K. (1997). Delay-period activity in the primate prefrontal cortex encoding multiple spatial positions and their order of presentation. *Behavioural Brain Research*, 84, 203–223.

Gao, Z., Davis, C., Thomas, A. M., Economo, M. N., Abrego, A. M., Svoboda, K., DeZeeuw, C. I., and Li, N. (2018). A cortico-cerebellar loop for motor planning. *Nature*, 563, 113–116.

Gjerdingen, R. O. (1994). Apparent motion in music? *Music Perception*, 11, 335–370.

Gjerdingen, R. O. (1989). Using connectionist models to explore complex musical patterns. *Computer Music Journal*, 13, 67–75.

Grahn, J., and Brett, M. (2007). Rhythm and beat perception in motor areas of the brain. *Journal of Cognitive Neuroscience*, 19, 893–906.

Grahn, J., and Brett, M. (2009). Impairment of beat-based rhythm discrimination in Parkinson's disease. *Cortex*, 45, 54–61.

Grahn, J., and Rowe, J. (2009). Feeling the beat: Premotor and striatal interactions in musicians and non-musicians during beat perception. *The Journal of Neuroscience*, 29, 7540–7548.

Grahn, J., and Rowe, J. (2013). Finding and feeling the musical beat: Striatal dissociations between detection and prediction of regularity. *Cerebral Cortex*, 23, 913–921.

Grossberg, S. (1970). Some networks that can learn, remember, and reproduce any number of complicated space-time patterns, II. *Studies in Applied Mathematics*, 49, 135–166.

Grossberg, S. (1971). On the dynamics of operant conditioning. *Journal of Theoretical Biology*, 33, 225–255.

Grossberg, S. (1972a). A neural theory of punishment and avoidance, I: Qualitative theory. *Mathematical Biosciences*, 15, 39–67.

Grossberg, S. (1972b). A neural theory of punishment and avoidance, II: Quantitative theory. *Mathematical Biosciences*, 15, 253–285.

Grossberg, S. (1974). Classical and instrumental learning by neural networks. In R. Rosen and F. Snell (Eds.), *Progress in Theoretical Biology*. New York: Academic Press, pp. 51–141.

Grossberg, S. (1975). A neural model of attention, reinforcement, and discrimination learning. *International Review of Neurobiology*, 18, 263–327.

Grossberg, S. (1976a). Adaptive pattern classification and universal recoding, I: Parallel development and coding of neural feature detectors. *Biological Cybernetics*, 23, 121–134.

Grossberg, S. (1976b). Adaptive pattern classification and universal recoding, II: Feedback, expectation, olfaction, and illusions. *Biological Cybernetics*, 23, 187–202.

Grossberg, S. (1978a). A theory of human memory: Self-organization and performance of sensory-motor codes, maps, and plans. In R. Rosen and F. Snell (Eds.), *Progress in Theoretical Biology*. New York: Academic Press, Vol. 5, pp. 233–374.

Grossberg, S. (1978b). Behavioral contrast in short-term memory: Serial binary memory models or parallel continuous memory models? *Journal of Mathematical Psychology*, 3, 199–219.

Grossberg, S. (1980a). Biological competition: Decision rules, pattern formation, and oscillations. *Proceedings of the National Academy of Sciences*, 77, 2338–2342.

Grossberg, S. (1980b). How does a brain build a cognitive code? *Psychological Review*, 87, 1–51.

Grossberg, S. (1982). Processing of expected and unexpected events during conditioning and attention: A psychophysiological theory. *Psychological Review*, 89, 529–572.

Grossberg, S. (1984). Some psychophysiological and pharmacological correlates of a developmental, cognitive, and motivational theory. In R. Karrer, J. Cohen, and P. Tueting (Eds.), *Brain and Information: Event Related Potentials*. New York: New York Academy of Sciences, pp. 58–142.

Grossberg, S. (1986). The adaptive self-organization of serial order in behavior: Speech, language, and motor control. In E. C. Schwab and H. C. Nusbaum (Eds.), *Pattern Recognition by Humans and Machines*, Volume. 1: *Speech perception*. New York: Academic Press, pp. 187–294.

Grossberg, S. (1999). Pitch-based streaming in auditory perception. In N. Griffith and P. Todd (Eds.), *Musical Networks: Parallel Distributed Perception and Performance*. Cambridge, MA: MIT Press, pp. 117–140.

Grossberg, S. (2003). Resonant neural dynamics of speech perception. *Journal of Phonetics*, 31, 423–445.

Grossberg, S. (2012). Neural Dynamics of Autistic Behaviors: Learning, Recognition, Attention, Emotion, Timing, and Social Cognition. In V. B. Patel, V. R. Preedy, and C. R. Martin (Eds.), *The Comprehensive Guide to Autism*. London: Springer.

Grossberg, S. (2013a). Adaptive Resonance Theory: How a brain learns to consciously attend, learn, and recognize a changing world. *Neural Networks*, 37, 1–47.

Grossberg, S. (2013b). Recurrent neural networks. *Scholarpedia*. http://www.scholarpedia.org/article/Recurrent_neural_networks

Grossberg, S. (2016). Neural dynamics of the basal ganglia during perceptual, cognitive, and motor learning and gating. In J.-J. Soghomonian (Ed.), *The Basal Ganglia: Novel Perspectives on Motor and Cognitive Functions*. Berlin: Springer, pp. 457–512.

Grossberg, S. (2017a). Acetylcholine neuromodulation in normal and abnormal learning and memory: Vigilance control in waking, sleep, autism, amnesia, and Alzheimer's disease. *Frontiers in Neural Circuits*, November 2, 2017. https://www.frontiersin.org/articles/10.3389/fncir.2017.00082/full

Grossberg, S. (2017c). Towards solving the Hard Problem of Consciousness: The varieties of brain resonances and the conscious experiences that they support. *Neural Networks*, 87, 38–95.

Grossberg, S. (2018). Desirability, availability, credit assignment, category learning, and attention: Cognitive-emotional and working memory dynamics of orbitofrontal, ventrolateral, and dorsolateral prefrontal cortices. *Brain and Neuroscience Advances*, May 8, 2018. https://journals.sagepub.com/doi/full/10.1177/2398212818772179

Grossberg, S. (2019). The embodied brain of SOVEREIGN2: From space-variant conscious percepts during visual search and navigation to learning invariant object categories and cognitive-emotional plans for acquiring valued goals. *Frontiers in Computational Neuroscience*. Published online: June 25, 2019. https://www.frontiersin.org/articles/10.3389/fncom.2019.00036/full

Grossberg, S. (2020b). Developmental designs and adult functions of cortical maps in multiple modalities: Perception, attention, navigation, numbers, streaming, speech, and cognition. *Frontiers in Neuroinformatics*, February 6, 2020. https://www.frontiersin.org/articles/10.3389/fninf.2020.00004/full

Grossberg, S. (2021). *Conscious MIND, Resonant BRAIN: How Each Brain Makes a Mind*. New York: Oxford University Press.

Grossberg, S. (2023). How children learn to understand language meanings: A neural model of adult–child multimodal interactions in real-time. *Frontiers in Psychology*, August 2, 2023. Section on Cognitive Science, Volume 14. https://www.frontiersin.org/journals/psychology/articles/10.3389/fpsyg.2023.1216479/full

Grossberg, S., and Kazerounian, S. (2011). Laminar cortical dynamics of conscious speech perception: A neural model of phonemic restoration using subsequent context in noise. *Journal of the Acoustical Society of America*, 130, 440–460.

Grossberg, S., and Kazerounian, S. (2016). Phoneme restoration and empirical coverage of Interactive Activation and Adaptive Resonance models of human speech processing, *Journal of the Acoustical Society of America*, 140, 1130; https://doi:org/10.1121/1.4946760.

Grossberg, S., and Levine, D. S. (1987). Neural dynamics of attentionally modulated Pavlovian conditioning: Blocking, inter-stimulus interval, and secondary reinforcement. *Applied Optics*, 26, 5015–5030.

Grossberg, S., and Merrill, J. W. L. (1992). A neural network model of adaptively timed reinforcement learning and hippocampal dynamics. *Cognitive Brain Research*, 1, 3–38.

Grossberg, S., and Merrill, J. W. L. (1996). The hippocampus and cerebellum in adaptively timed learning, recognition, and movement. *Journal of Cognitive Neuroscience*, 8, 257–277.

Grossberg, S., and Paine, R. W. (2000). A neural model of corticocerebellar interactions during attentive imitation and predictive learning of sequential handwriting movements. *Neural Networks*, 13, 999–1046.

Grossberg, S., and Pearson, L. (2008). Laminar cortical dynamics of cognitive and motor working memory, sequence learning and performance: Toward a unified theory of how the cerebral cortex works. *Psychological Review*, 115, 677–732.

Grossberg, S. and Pepe, J. (1970). Schizophrenia: Possible dependence of associational span, bowing, and primacy vs. recency on spiking threshold. *Behavioral Science*, 15, 359–362.

Grossberg, S., and Repin, D. (2003). A neural model of how the brain represents and compares multi-digit numbers: Spatial and categorical processes. *Neural Networks*, 16, 1107–1140.

Grossberg, S., and Rudd, M. (1989). A neural architecture for visual motion perception: Group and element apparent motion. *Neural Networks*, 2, 421–450.

Grossberg, S., and Rudd, M. E. (1992). Cortical dynamics of visual motion perception: Short-range and long-range apparent motion. *Psychological Review*, 99, 78–121.

Grossberg, S. and Schmajuk, N. A. (1987). Neural dynamics of attentionally-modulated Pavlovian conditioning: Conditioned reinforcement, inhibition, and opponent processing. *Psychobiology*, 15, 195–240.

Grossberg, S., and Schmajuk, N. A. (1989). Neural dynamics of adaptive timing and temporal discrimination during associative learning. *Neural Networks*, 2, 79–102.

Grossberg, S., Boardman, I., and Cohen, C. (1997). Neural dynamics of variable-rate speech categorization. *Journal of Experimental Psychology: Human Perception and Performance*, 23, 418–503.

Grossberg, S., Bullock, D., and Dranias, M. (2008). Neural Dynamics Underlying Impaired Autonomic and Conditioned Responses Following Amygdala and Orbitofrontal Lesions. *Behavioral Neuroscience*, 122, 1100–1125.

Grossberg, S., Govindarajan, K. K., Wyse, L. L., and Cohen, M. A. (2004). ARTSTREAM: A neural network model of auditory scene analysis and source segregation. *Neural Networks*, 17, 511–536.

Grossberg, S., Pribe, C., and Cohen, M. A. (1997). Neural control of interlimb oscillations, I: Human bimanual coordination. *Biological Cybernetics*, 77, 131–140.

Guenther, F. H. (1995). Speech sound acquisition, coarticulation, and rate effects in a neural network model of speech production. *Psychological Review*, 102, 594–621.

Hanslick, E. (1854). *On the Musically Beautiful: A Contribution Towards the Revision of the Aesthetics of Music*. Indianapolis, IN: Hackett.

von Helmholtz, H. L. F. (1954). *On the Sensations of Tone as a Physiological Basis for the Theory of Melody*. A. J. Ellis (Ed. and Trans.) New York: Dover. (Original work published 1885).

Hennig, H., Fleishmann, R., Fredebohm, A., Hagmayer, Y., Nagler, J., Witt, A., Theis, F. J., and Geisel, T. (2011). The nature and perception of fluctuations in human musical rhythms. *PLOS ONE*. https://doi.org/10.1371/journal.pone.0026457.

Henson, R. N. A. (1998). Short-term memory for serial order: The start-end model. *Cognitive Psychology*, 36, 73–137.

Hesmondhalph, D. (2013). *Why Music Matters*. Hoboken, NJ: Blackwell.

Hikosaka, O., and Wurtz, R. H. (1983). Visual and oculomotor functions of monkey substantia nigra pars reticulata. IV. Relation of substantia nigra to superior colliculus. *Journal of Neurophysiology*, 49, 1285–1301.

Hikosaka, O., and Wurtz, R. H. (1989). The basal ganglia. In R. Wurtz and M. Goldberg (Eds.), *The Neurobiology of Saccadic Eye Movements*. Amsterdam: Elsevier, pp. 257–281.

Hintzen, A., Pelzer, E. A., and Tittgemeyer, M. (2018). Thalamic interactions of cerebellum and basal ganglia. *Brain Structure and Function*, 223, 569–587.

Hodgkin, A. L. (1964). *The Conduction of the Nervous Impulse*. Springfield: C.C. Thomas.

Hodgkin, A. L., and Huxley, A. F. (1952). A quantitative description of membrane current and its application to conduction and excitation in nerve. *Journal of Physiology*, 117, 500–544.

Houghton, G. (1990). The problem of serial order: A neural network model of sequence learning and recall. In R. Dale, C. Mellish, and M. Zock (Eds.), *Current Research in Natural Language Generation*. London: Academic Press, pp. 287–319.

Howell, P., West, R., and Cross, I. (Eds.) (1991). *Representing Musical Structure*. New York: Academic Press.

Hubel, D. H., and Wiesel, T. N. (1962). Receptive fields, binocular interaction, and functional architecture in the cat's visual cortex. *The Journal of Physiology*, 160, 106–154.

Hubel, D. H., and Wiesel, T. N. (1963). Shape and arrangement of columns in cat's striate cortex. *The Journal of Physiology*, 165, 559–568.

Inoue, M., and Mikami, A. (2006). Prefrontal activity during serial probe reproduction task: Encoding, mnemonic, and retrieval processes. *Journal of Neurophysiology*, 95, 1008–1041.

Ito, M., and Kano, M. (1982). Long-lasting depression of parallel fiber-Purkinje cell transmission induced by conjunctive stimulation of parallel fibers and climbing fibers in the cerebellar cortex. *Neuroscience Letters*, 33, 253–258.

Ivanov, P. C., Hu, K., Hilton, M. F., Shea, S. A., and Stanley, H. E. (2007). Endogenous circadian rhythm in human motor activity uncoupled from circadian influences on cardiac dynamics. *Proceedings of the National Academy of Sciences*, 104, 20702–20707.

Ivry, R. B., Spencer, R. M., Zelaznik, H. N., and Diedrichsen, J. (2003). The cerebellum and event timing. *Annals of the New York Academy and Sciences*, 978, 302–317.

Jacobs, J. P., and Bullock, D. (1998). A two-process model for control of legato articulation across a wide range of tempos during piano performance. *Music Perception*, 16, 169–199.

Jones, M. R. (1976). Time, our lost dimension: Toward a new theory of perception, attention, and memory. *Psychological Review*, 83, 323–335.

Kamin, L. J. (1968). Attention-like processes in classical conditioning. In M. R. Jones (Ed.), *Miami Symposium on the Prediction of Behavior: Aversive Stimulation*. Miami: University of Miami Press.

Kazerounian, S., and Grossberg, S. (2014). Real-time learning of predictive recognition categories that chunk sequences of items stored in working memory. *Frontiers in Psychology: Language Sciences*, doi: 10.3389/fpsyg.2014.01053 https://www.ncbi.nlm.nih.gov/pmc/articles/PMC4186345/

Kelso, J. A. S. (1981). On the oscillatory basis of movement. *Bulletin of the Psychonomic Society*, 18, 63.

Kelso, J. A. S. (1984). Phase transitions and critical behavior in human bimanual coordination. *American Journal of Physiology: Regulatory, Integrative, and Comparative Physiology*, 246, Rl000–RI004.

Kennerley, S. W., Sakai, K., and Rushworth, M. F. (2004). Organization of action sequences and the role of the pre-SMA. *Journal of Neurophysiology*, 91, 978–993.

Kermadi, I., and Joseph, J. P. (1995). Activity in the caudate nucleus of monkey during spatial sequencing. *Journal of Neurophysiology*, 74, 911–933.

Klein, R. M. (2000). Inhibition of return. *Trends in Cognitive Sciences*, 4, 138–146.

Krumhansl, C. L. (2000). Rhythm and pitch in music cognition. *Psychological Bulletin*, 126, 159–179.

Kung, S-J., Chen, J. L., Zatorre, R. J., and Penhune, V. B. (2013). Interacting cortical and basal ganglia networks underlying finding and tapping to the musical beat. *Journal of Cognitive Neuroscience*, 25, 401–420.

Lagrois, M.-É., Palmer, C., and Peretz, I. (2019). Poor synchronization to musical beat generalizes to speech. *Brain Sciences*, 9, 157. https://www.mdpi.com/2076-3425/9/7/157

Large, E. W. (2010). Neurodynamics of music. In M. R. Jones et al. (Eds.), *Music Perception*. Springer Handbook of Auditory Research. New York: Springer Science+Business Media, LLC, p. 36.

Large, E. W., Herrera, J. A., and Velasco, M. J. (2015). Neural networks for beat perception in musical rhythm. *Frontiers in Systems Neuroscience*, November 25, 2015. 10.3389/fnsys.2015.00159.

Lenc, T., Keller, P. E., Varlet, M., and Nozaradan, S. (2018). Neural tracking of the musical beat is enhanced by low-frequency sounds. *Proceedings of the National Academy of Sciences*, 115, 8221–8226.

Lerdahl, F., and Jackendoff, R. S. (1983). *A Generative Theory of Tonal Music*. Cambridge, MA: MIT Press.

Levitin, D. J. (2006). *This Is Your Brain on Music: The Science of a Human Obsession*. New York, NY: Dutton.

London, J. M. (2004). *Hearing in Time: Psychological Aspects of Musical Meter*. New York, NY: Oxford University Press.

López- López, J. R., Shacklock, P. S., Balke, C. W., and Wier, W. G. (1995). Local calcium transients triggered by single L-type calcium channel currents in cardiac cells. *Science*, 268, 1042–1045.

Masters, J. R. (2002). HeLa cells 50 years on: The good, the bad, and the ugly. *Nature Reviews Cancer*, 2, 315–319.

Millenson, J. R., Kehoe, E. J., and Gormenzano, 1. (1977). Classical conditioning of the rabbit's nictitating membrane response under fixed and mixed CS-US intervals. *Learning and Motivation*, 8, 351–366.

Miller, G. A. (1956). The magical number seven, plus or minus two: Some limits on our capacity for processing information. *Psychological Review*, 63, 81–97.

Mink, J. (1996). The basal ganglia: Focused selection and inhibition of competing motor programs. *Progress in Neurobiology*, 50, 381–425.

Mink, J. W., and Thach, W. T. (1993). Basal ganglia intrinsic circuits and their role in behavior. *Current Opinion in Neurobiology*, 3, 950–957.

Murdock, B. B. (1962). The serial position effect of free recall. *Journal of Experimental Psychology*, 64, 482–488.

Naccache, L., and Dehaene, S. (2001). The priming method: Imaging unconscious repetition priming reveals an abstract representation of number in the parietal lobes. *Cerebral Cortex*, 11, 966–974.

Nachev, P., Kennard, C., and Husain, M. (2008). Functional role of the supplementary and pre-supplementary motor areas. *Nature Reviews Neuroscience*, 9, 856–869.

Nguyen, T., Gibbings, A., and Grahn, J. (2018). Rhythm and beat perception. In R. Bader (Ed.), *Springer Handbook of Systematic Musicology*. New York: Springer, pp. 507–521.

Nieder, A., and Miller, E. (2004a). Analog numerical representations in rhesus monkeys: Evidence for parallel processing. *Journal of Cognitive Neuroscience*, 16, 889–901.

Nieder, A., and Miller, E. (2004a). Analog numerical representations in rhesus monkeys: Evidence for parallel processing. *Journal of Cognitive Neuroscience*, 16, 889–901.

Nieder, A., and Miller, E. (2004b). A parieto-frontal network for visual numerical information in the monkey. *Proceedings of the National Academy of Sciences*, 101, 7457–7462.

Ninokura, Y., Mushiaske, H., and Tanji, J. (2004). Integration of temporal order and object information in the monkey lateral prefrontal cortex. *Journal of Neurophysiology*, 91, 555–560.

Page, M. P. A., and Norris, D. (1998). The primacy model: A new model of immediate serial recall. *Psychological Review*, 105, 761–781.

Patel, A. D., and Iversen, J. R. (2014). The evolutionary neuroscience of musical beat perception: The Action Simulation for Auditory Prediction (ASAP) hypothesis. *Frontiers in Systems Neuroscience*, 8, 1–14.

Penhune, V. B., and Doyon, J. (2005). Cerebellum and M1 interaction during early learning of timed motor sequences. *Neuroimage*, 26, 801–812.

Peterson, G. E., and Barney, H. L. (1952). Control methods used in a study of the vowels. *Journal of the Acoustical Society of America*, 24, 175–184.

Phillips-Silver, J., and Trainor, L. J. (2007). Hearing what the body feels: Auditory encoding of rhythmic movement. *Cognition*, 105, 533–546.

Piaget, J. (1963). *The Origins of Intelligence in Children*. New York: Norton.

Posner, M. I. (1980). Orienting of attention. *Quarterly Journal of Experimental Psychology*, 32, 3–25.

Posner, M. I., and Cohen, Y. (1984). Components of visual orienting. In H. Bourma and D. Bouwhuis (Eds.), *Attention and Performance*. Erlbaum, Vol. X, pp. 531–556.

Posner, M. I., Rafal, R. D., Choate, L. S., and Hamilton, J. V. (1985). Inhibition of return: Neural basis and function. *Cognitive Neuropsychology*, 2, 211–228.

Pribe, C., Grossberg, S., and Cohen, M. A. (1997). Neural control of interlimb oscillations, II: Biped and quadruped gaits and bifurcations. *Biological Cybernetics*, 77, 141–152.

Rajendran, V. G., Harper, N. S., Garcia-Lazaro, J. A., Lesica, N. A., and Schnupp, J. W. H. (2017). Midbrain adaptation may set the stage for the perception of musical beat. *Proceedings of the Royal Society B*, 284, 20171455. https://royalsocietypublishing.org/doi/full/10.1098/rspb.2017.1455

Rajendran, V. G., Teki, S., and Schnupp, J. W. H. (2018). Temporal processing in audition: Insights from music. *Neuroscience*, 389, 4–18.

Ramnani, N., and Passingham, R. E. (2001). Changes in the human brain during rhythm learning. *Journal of Cognitive Neuroscience*, 13, 952–966.

Repp, B. H. (2005). Sensorimotor synchronization: A review of the tapping literature. *Psychonomic Bulletin & Review*, 12, 969–992.

Repp, B. H. (2006a). Musical synchronization. In E. Altenmüller, M. Wiesendanger, and J. Kesselring (Eds.), *Music, Motor Control, and the Brain*. Oxford, UK: Oxford University Press, pp. 55–76.

Repp, B. H. (2006b). Rate limits of sensorimotor synchronization. *Advances in Cognitive Psychology*, 2, 163–181.

Repp, B. H., and Su, Y.-H. (2013). Sensorimotor synchronization: A review of recent research (2006-2012). *Psychonomic Bulletin & Review*, 20, 403–452.

Rickard, T. C., Romero, S. G., Basso, G., Wharton, C., Flitman, S., and Grafman, J. (2000). The calculating brain: An fMRI study. *Neuropsychologia*, 38, 325–335.

Rumelhart, D. E., and Zipser, D. (1985). Feature discovery by competitive learning. *Cognitive Science*, 9, 75–112.

Sakai, K., Ramnani, N., and Passingham, R. E. (2002). Learning of sequences of finger movements and timing: Frontal lobe and action-oriented representation. *Journal of Neurophysiology*, 88, 2035–2046.

Schaefer, R. S., and Overy, K. (2015). Motor responses to a steady beat. *Annals of the New York Academy of Sciences, The Neurosciences and Music V*, 40–44.

Schulkin, J., and Raglan, G. B. (2014). The evolution of music and human social capability. *Frontiers in Neuroscience*, published September 17, 2014, https://www.frontiersin.org/articles/10.3389/fnins.2014.00292/full

Shima, K., and Tanji, J. (2000). Neuronal activity in the supplementary and presupplementary motor areas for temporal organization of multiple movements. *Journal of Neurophysiology*, 84, 2148–2160.

Silver, M. R., Grossberg, S., Bullock, D., Histed, M. H., and Miller, E. K. (2011). A neural model of sequential movement planning and control of eye movements: Item-order-rank working memory and saccade selection by the supplementary eye fields. *Neural Networks*, 26, 29–58.

Slater, J., Ashley, R., Tierney, A., and Kraus, N. (2018). Got rhythm? Better inhibitory control is linked with more consistent drumming and enhanced neural tracking of the musical beat in adult percussionists and nonpercussionists. *Journal of Cognitive Neuroscience*, 30, 14–24.

Smith, M. C. (1968). CS-US interval and US intensity in classical conditioning of the rabbit's nictitating membrane response. *Journal of Comparative and Physiological Psychology*, 3, 679–687.

Styns, F., van Noorden, L., Moelants, D., and Leman, M. (2007). Walking on music. *Human Movement Science*, 26, 769–785.

Sowinski, J., and Bella, S. D. (2013). Poor synchronization to the beat may result from deficient auditory-motor mapping. *Neuropsychologia*, 51, 1952–1963.

Tal, I., Large, E. W., Rabinovitch, E., Wei, Y., Schroeder, C. E., Poeppel, D., and Golumbic, E. Z. (2017). Neural entrainment to the beat: The "missing'pulse" phenomenon. *The Journal of Neuroscience*, 37, 6331–6341.

Tan, S.-L., Pfordresher, P., and Harré, R. (2018). *Psychology of Music: From Sound to Significance*, Second Edition. New York: Rutledge.

Thach, W. T. A. (1998). A role for the cerebellum in learning movement coordination. *Neurobiology of Learning and Memory*, 70, 177–188.

Thompson, W. F. (2009). *Music, Thought, and Feeling: Understanding the Psychology of Music*. New York: Oxford University Press.

Toiviainen, P., Burunat, I., Brattico, E., Vuust, P., and Alluri, V. (2019). The chronnectome of musical beat. *Neuroimage*, September 2019. https://www.sciencedirect.com/science/article/pii/S1053811919307827?via%3Dihub

Trainor, L. J. (2007). Do preferred beat rate and entrainment to the beat have a common origin in movement? *Empirical Musicology Review*, 2, 17–20.

Vladusich, T., Lafe, F., Kim, D.-S., Tager-Flusberg, H., and Grossberg, S. (2010). Prototypical category learning in high-functioning autism. *Autism Research*, 3, 226–236.

Wallin, N. L., Merker, B., and Brown, S. (2000). *The Origins of Music*. Cambridge, MA: MIT Press.

Willshaw, D. J., and von der Malsburg, C. (1976). How patterned neural connections can be set up by self-organization. *Proceedings of the Royal Society of London*, 194, 431–445.

Woodruff-Pak, D. S., Papka, M., and Ivry, R. B. (1996). Cerebellar involvement in eyeblink classical conditioning in humans. *Neuropsychology*, 10, 443–458.

Yamanishi, J., Kawato, M., and Suzuki, R. (1980). Two coupled oscillators as a model for the coordinated

finger tapping by both hands. *Biological Cybernetics*, 37, 219–225.
Yao, Y., Choi, J., and Parker, I. (1995). Quantal putts of intracellular Ca2+ evoked by inositol trisphosphate in Xenopus oocytes. *Journal of Physiology*, 482, 533–553.
Zatorre, R. J., Chen, J. L., and Penhune, V. B. (2007). When the brain plays music: Auditory-motor interactions in music perception and production. *Nature Reviews Neuroscience*, 8, 547–558.

Chapter 4

Albright, T. D. (1984). Direction and orientation selectivity of neurons in visual area MT of the macaque. *Journal of Neurophysiology*, 52, 1106–1130.
Alexander, G. E., and Crutcher, M. D. (1990). Functional architecture of basal ganglia circuits: Neural substrates of parallel processing. *Trends in Neurosciences*, 1, 266–271.
Alexander, G. E., DeLong, M., and Strick, P. L. (1986). Parallel organization of functionally segregated circuits linking basal ganglia and cortex. *Annual Review of Neuroscience*, 9, 357–381.
Anstis, S. M. (1990). Imperceptible intersections: The chopstick illusion. In A. Blake and T. Troscianko (Eds.), *AI and the Eye*. New York: John Wiley and Sons, pp. 105–117.
Asfour, Y. R., Carpenter, G. A., Grossberg, S., and Lesher, G. W. (1993). Fusion ARTMAP: A neural network architecture for multi-channel data fusion and classification. In *Proceedings of the World Congress on Neural Networks*, Portland, II, 210–215. Hillsdale, NJ: Erlbaum Associates.
Aziz-Aadeh, L. (2013). Embodied semantics for language related to actions: A review of fMRI and neuropsychological research. In Y. Coello and A. Bartolo (Eds.), *Language and Action in Cognitive Neuroscience*. New York: Psychology Press, pp. 278–283.
Bar, M., Tootell, B. H., Schacter, D. L., Greve, D. N., Fischl, B., Mendola, J. D., Rosen, B. R., and Dale, A. M. (2001). Cortical mechanisms specific to explicit visual object recognition. *Neuron*, 29, 529–535.
Bemis, D. K., and Pylkkänen, L. (2013). Basic linguistic composition recruits the left anterior temporal lobe and left angular gyrus during both listening and reading. *Cerebral Cortex*, 23, 1859–1873.
Bergen, B. K. (2012). *Louder than Words: The New Science of How the Mind Makes Meaning*. New York: Basic Books.
Berzhanskaya, J., Grossberg, S., and Mingolla, E. (2007). Laminar cortical dynamics of visual form and motion interactions during coherent object motion perception. *Spatial Vision*, 20, 337–395.
Bonner, M. F., and Price, A. R. (2013). Where is the anterior temporal lobe and what does it do? *Journal of Neuroscience*, 33, 4213–4215.
Boroditsky, L. (2000). Metaphoric structuring: Understanding time through spatial metaphors. *Cognition*, 675, 1–27.
Boroditsky, L., and Ramscar, M. (2002). The roles of body and mind in abstract thoughts. *Psychological Science*, 13, 185–189.
Boulenger, V., Hauk, O., and Pulvemuller, F. (2009). Grasping ideas with the motor system: Semantic somatotopy in idiom comprehension. *Cerebral Cortex*, 19, 1905–1914.
Bradski, G., Carpenter, G. A., and Grossberg, S. (1994). STORE working memory networks for storage and recall of arbitrary temporal sequences. *Biological Cybernetics*, 71, 469–480.
Bradski, G., and Grossberg, S. (1995). Fast learning VIEWNET architectures for recognizing 3-D objects from multiple 2-D views. *Neural Networks*, 8, 1053–1080.
Brown, J. W., Bullock, D., and Grossberg, S. (1999). How the basal ganglia use parallel excitatory and inhibitory learning pathways to selectively respond to unexpected rewarding cues. *Journal of Neuroscience*, 19, 10502–10511.
Brown, J. W., Bullock, D., and Grossberg, S. (2004). How laminar frontal cortex and basal ganglia circuits interact to control planned and reactive saccades. *Neural Networks*, 17, 471–510.
Browning, A., Grossberg, S., and Mingolla, M. (2009a). A neural model of how the brain computes heading from optic flow in realistic scenes. *Cognitive Psychology*, 59, 320–356.
Browning, A., Grossberg, S., and Mingolla, M. (2009b). Cortical dynamics of navigation and steering in natural scenes: Motion-based object segmentation, heading, and obstacle avoidance. *Neural Networks*, 22, 1383–1398.
Bremner, A.J., Bryant, P.E. and Mareschal, D. (2005). Object-centred spatial reference in 4-month-old infants. *Infant Behaviour and Development*, 29, 1–10.
Bruner, J. (1985). The role of interaction formats in language acquisition. In J. P. Forgas (Ed.), *Language and Social Situations*, Chapter 2, pp. 31–46.
Bullock, D., Cisek, P., and Grossberg, S. (1998). Cortical networks for control of voluntary arm movements under variable force conditions. *Cerebral Cortex*, 8, 48–62.
Bullock, D., and Grossberg, S. (1988). Neural dynamics of planned arm movements: Emergent invariants and speed-accuracy properties during trajectory formation. *Psychological Review*, 95, 49–90.
Bullock, D., and Grossberg, S. (1991). Adaptive neural networks for control of movement trajectories invariant under speed and force rescaling. *Human Movement Science*, 10, 3–53.
Bullock, D., Grossberg, S., and Guenther, F. H. (1993). A self-organizing neural model of motor equivalent reaching and tool use by a multijoint arm. *Journal of Cognitive Neuroscience*, 5, 408–435.
Burman, J. T. (2021). The genetic epistemology of Jean Piaget. In W. Pickren, P. Hegarty, C. Logan, W. Long, P. Pettikainen, and A. Rutherford (Eds.), *Oxford Research Encyclopedia of the History of Modern Psychology*. New York: Oxford University Press. https://doi.org/10.1093/acrefore/9780190236557.013.521
Cao, Y., Grossberg, S., and Markowitz, J. (2011). How does the brain rapidly learn and reorganize view- and positionally-invariant object representations in inferior temporal cortex? *Neural Networks*, 24, 1050–1061.
Carey, S. E. (2022). Becoming a cognitive scientist. *Annual Review of Developmental Psychology*, 4, 1–19.

Carey, S., Zaitchik, D., and Bascandziev, I. (2015). Theories of development: In dialog with Jean Piaget. *Developmental Review*, 38, 36–54.

Carpenter G. A. (1997). Distributed learning, recognition, and prediction by ART and ARTMAP neural networks. *Neural Networks*, 10, 1473–1494.

Carpenter, G. A. (2003). Default ARTMAP. *Proceedings of the International Joint Conference on Neural Networks*, 1396–1401.

Carpenter, G. A., Grossberg, S., Markuzon, N., Reynolds, J. H., and Rosen, D. B. (1992). Fuzzy ARTMAP: A neural network architecture for incremental supervised learning of analog multidimensional maps. *IEEE Transactions on Neural Networks*, 3, 698–713.

Carpenter, G. A., Grossberg, S., and Mehanian, C. (1989). Invariant recognition of cluttered scenes by a self-organizing ART architecture: CORT-X boundary segmentation. *Neural Networks*, 2, 169–181.

Carpenter, G. A., Grossberg, S., and Reynolds, J. H. (1991). ARTMAP: Supervised real-time learning and classification of nonstationary data by a self-organizing neural network. *Neural Networks*, 4, 565–588.

Carpenter, G. A., Grossberg, S., and Reynolds, J. H. (1995). A fuzzy ARTMAP nonparametric probability estimator for nonstationary pattern recognition problems. *IEEE Transactions on Neural Networks*, 6, 1330–1336.

Carpenter, G. A., Martens, S., and Ogas, O. J. (2005). Self-organizing information fusion and hierarchical knowledge discovery: A new framework using ARTMAP neural networks. *Neural Networks*, 18, 287–295.

Carpenter, G. A., Milenova, B. L., and Noeske, B. W. (1998). Distributed ARTMAP: A neural network for fast distributed supervised learning. *Neural Networks*, 11, 793–813.

Carpenter, G. A., and Ravindran, A. (2008). Unifying multiple knowledge domains using the ARTMAP information fusion system. *Proceedings of the 11th International Conference on Information Fusion*, Cologne, Germany, June 30–July 3, 2008.

Carpenter, G. A., Rubin, M. A., and Streilein, W. W. (1997). ARTMAP-FD: Familiarity discrimination applied to radar target recognition. *Proceedings of the International Conference on Neural Networks*, 3, pp. 1459–1464, IEEE Press, Piscataway, NJ.

Casasanto, D., and Dijkstra, K. (2010). Motor action and emotional memory, *Cognition*, 115, 179–185.

Chang, K. K., Cherkassky, V. L., Mitchell, T. M., and Just, M. A. (2009). Quantitative modeling of the neural representation of adjective-noun phrases to account for fMRI activation. *Proceedings of the 47th Annual Meeting of the ACL and the 4th IJCNLP of the AFNLP*, pp. 638–646, Suntec, Singapore, August 2–7, 2009.

Chao, L. L., Haxby, J. V., and Martin, A. (1999). Attribute-based neural substrates in temporal cortex for perceiving and knowing about objects. *Nature Neuroscience*, 2, 913–919.

Chatterjee, A. (2008). The neural organization of spatial thought and language. *Seminars in Speech and Language*, 29, 226–238.

Chatterjee, A. (2010). Disembodying cognition. *Language and Cognition*, 2, 79–116.

Cohen, M. A., Grossberg, S., and Stork, D. G. (1988). Speech perception and production by a self-organizing neural network. In Y. C. Lee (Ed.), *Evolution, Learning, Cognition, and Advanced Architectures*. Hong Kong: World Scientific Publishers, pp. 217–231.

Colston, H. L. (2019). *How Language Makes Meaning: Embodiment and Conjoined Autonymy*. United Kingdom: Cambridge University Press.

Deák, G. O., Flom, R. A., and Pick, A. D. (2000). Effects of gesture and target on 12 and 18-month-olds' joint visual attention to objects in front of or behind them. *Developmental Psychology*, 36, 511–523.

Dove, G. O. (2009). Beyond perceptual symbols: A call for representational pluralism. *Cognition*, 110, 412–431.

Dranias, M., Grossberg, S., and Bullock, D. (2008). Dopaminergic and non-dopaminergic value systems in conditioning and outcome-specific revaluation. *Brain Research*, 1238, 239–287.

Dubois, B., Burn, D., Goetz, C., Aarsland, D., Brown, R. G., Broe, G. A., Dickson, D., Duyckaerts, C., et al. (2007). Diagnostic procedures for Parkinson's disease dementia: Recommendations from the Movement Disorder Society Task Force. *Movement Disorders*, 22, 2314–2324.

Duncker, K. (1929/1938). Induced motion. In W. D. Ellis (Ed.), *A Sourcebook of Gestalt Psychology*. London: Routledge & Kegan Paul, 1938. (Originally published in German, 1929).

Elder, D., Grossberg, S., and Mingolla, E. (2009). A neural model of visually guided steering, obstacle avoidance, and route selection. *Journal of Experimental Psychology: Human Perception and Performance*, 35, 1501–1531.

Emery, N. J., Lorincz, E. N., Perrett, D. I., Oram, M. W., and Baker, C. I. (1997). Gaze following and joint attention in rhesus monkeys (macaca mulatta). *Journal of Comparative Psychology*, 111, 286–293.

Exner, S. (1875). Ueber das Sehen von Bewegungen und die Theorie des zusammengesetzen Auges. *Sitzungsberichte Akademie Wissenschaft Wien*, 72, 156–190.

Farivar, R. (2009). Dorsal-ventral integration in object recognition. *Brain Research Reviews*, 61, 144–153.

Fazl, A., Grossberg, S., and Mingolla, E. (2009). View-invariant object category learning, recognition, and search: How spatial and object attention are coordinated using surface-based attentional shrouds. *Cognitive Psychology*, 58, 1–48.

Fernardino, L. Constant, L. L., Binder, J. R., Blindauer, K., Hiner, B., Spangler, K., and Desai, R. H. (2013). Where is the action: Action sentence processing in Parkinson's disease. *Neuropsychologia*, 51, 1510–1517.

Fiala, J. C., Grossberg, S., and Bullock, D. (1996). Metabotropic glutamate receptor activation in cerebellar Purkinje cells as substrate for adaptive timing of the classically conditioned eye blink response. *The Journal of Neuroscience*, 16, 3760–3774.

Fischer, M. H., and Zwaan, R. A. (2008). Embodied language: A review of the role of the motor system in language comprehension. *Quarterly Journal of Experimental Psychology*, 61, 825–850.

Francis, G., and Grossberg, S. (1996a). Cortical dynamics of boundary segmentation and reset: Persistence, afterimages, and residual traces. *Perception*, 35, 543–567.

Francis, G., and Grossberg, S. (1996b). Cortical dynamics of form and motion integration: Persistence, apparent motion, and illusory contours. *Vision Research*, 36, 149–173.

Francis, G., Grossberg, S., and Mingolla, E. (1994). Cortical dynamics of feature binding and reset: Control of visual persistence. *Vision Research*, 34, 1089–1104.

Franklin, D. J., and Grossberg, S. (2017). A neural model of normal and abnormal learning and memory consolidation: Adaptively timed conditioning, hippocampus, amnesia, neurotrophins, and consciousness. *Cognitive, Affective, and Behavioral Neuroscience*, 17, 24–76.

Frischen, A., Bayliss, A. P., and Tipper, S. P. (2007). Gaze cueing of attention: Visual attention, social cognition, and individual differences. *Psychological Bulletin*, 133, 694–724.

Fyshe, A., Sudre, G., Wehbe, L., Rafidi, N., and Mitchell, T. M. (2016). The semantics of adjective noun phrases in the human brain. *bioRxiv*, doi: https://doi.org/10.1101/089615.

Fyshe, A., Sudre, G., Wehbe, L., Rafidi, N., and Mitchell, T. M. (2019). The lexical semantics of adjective–noun phrases in the human brain. *Human Brain Mapping*, 40, 4457–4469.

Gallese, V., and Lakoff, G. (2005). The brain's concepts: The role of the sensory-motor system in conceptual knowledge. *Cognitive Neuropsychology*, 22, 455–479.

Gaudiano, P., and Grossberg, S. (1991). Vector associative maps: Unsupervised real-time error-based learning and control of movement trajectories. *Neural Networks*, 4, 147–183.

Gaudiano, P., and Grossberg, S. (1992). Adaptive vector integration to endpoint: Self-organizing neural circuits for control of planned movement trajectories. *Human Movement Science*, 11, 141–155.

Gegenfurtner, K. R. (2003). Cortical mechanisms of colour vision. *Nature Reviews Neuroscience*, 4, 563–572.

Gibbs, R. W. (1994). Metaphor interpretation as embodied simulation. *Mind and Language*, 21, 434–458.

Gibbs, R. W. (2006). Metaphor interpretation as embodied simulation. *Mind and Language*, 21, 434–458. doi: 10.1111/j.1468-0017.2006.00285.x

Gleitman, L. R., and Gleitman, C. (2022). Recollecting what we once knew: My life in psycholinguistics. *Annual Review of Psychology*, 73, 1–23.

Glenberg, A. M. (2010). *Embodiment as a Unifying Perspective for Psychology*. New York: John Wiley & Sons, Inc.

Glenberg, A. M., and Kaschak, M. P. (2002). Grounding language in action. *Psychological Bulletin & Review*, 9, 558–565.

Grahn, J. A., Parkinson, J. A., and Owen, A. M. (2009). The role of the basal ganglia in learning and memory: Neuropsychological studies. *Behavioural Brain Research*, 199, 53–60.

Granger, E., Rubin, M. A., Grossberg, S., and Lavoie, P. (2000). Classification of incomplete data using the fuzzy ARTMAP neural network. *Proceedings of the International Joint Conference on Neural Networks (IJCNN)*, 4, 35–40.

Grossberg, S. (1971a). On the dynamics of operant conditioning. *Journal of Theoretical Biology*, 33, 225–255.

Grossberg, S. (1971b). Pavlovian pattern learning by nonlinear neural networks. *Proceedings of the National Academy of Sciences*, 68, 828–831.

Grossberg, S. (1972a). A neural theory of punishment and avoidance, I: Qualitative theory. *Mathematical Biosciences*, 15, 39–67.

Grossberg, S. (1972b). A neural theory of punishment and avoidance, II: Quantitative theory. *Mathematical Biosciences*, 15, 253–285.

Grossberg, S. (1973). Contour enhancement, short-term memory, and constancies in reverberating neural networks. *Studies in Applied Mathematics*, 52, 213–257.

Grossberg, S. (1974). Classical and instrumental learning by neural networks. In R. Rosen and F. Snell (Eds.), *Progress in Theoretical Biology*. New York: Academic Press, pp. 51–141.

Grossberg, S. (1975). A neural model of attention, reinforcement, and discrimination learning. *International Review of Neurobiology*, 18, 263–327.

Grossberg, S. (1978a). A theory of human memory: Self-organization and performance of sensory-motor codes, maps, and plans. In R. Rosen and F. Snell (Eds.), *Progress in Theoretical Biology*. New York: Academic Press, Vol. 5, pp. 233–374.

Grossberg, S. (1978b). Behavioral contrast in short-term memory: Serial binary memory models or parallel continuous memory models? *Journal of Mathematical Psychology*, 3, 199–219.

Grossberg, S. (1982). Processing of expected and unexpected events during conditioning and attention: A psychophysiological theory. *Psychological Review*, 89, 529–572.

Grossberg, S. (1984a). Some normal and abnormal behavioral syndromes due to transmitter gating of opponent processes. *Biological Psychiatry*, 19, 1075–1118.

Grossberg, S. (1984b). Some psychophysiological and pharmacological correlates of a developmental, cognitive, and motivational theory. In R. Karrer, J. Cohen, and P. Tueting (Eds.), *Brain and Information: Event Related Potentials*. New York: New York Academy of Sciences, pp. 58–142.

Grossberg, S. (1986). The adaptive self-organization of serial order in behavior: Speech, language, and motor control. In E. C. Schwab and H. C. Nusbaum (Eds.), *Pattern Recognition by Humans and Machines, Vol. 1: Speech Perception*. New York: Academic Press, pp. 187–294.

Grossberg, S. (1987a). Cortical dynamics of three-dimensional form, color, and brightness perception, I: Monocular theory. *Perception and Psychophysics*, 41, 87–116.

Grossberg, S. (1987b). Cortical dynamics of three-dimensional form, color, and brightness perception, II: Binocular theory. *Perception and Psychophysics*, 41, 117–158.

Grossberg, S. (1988). Nonlinear neural networks: Principles, mechanisms, and architectures. *Neural Networks*, 1, 17–61.

Grossberg, S. (1991). Why do parallel cortical systems exist for the perception of static form and moving form? *Perception and Psychophysics*, 49, 117–141.

Grossberg, S. (1994). 3-D vision and figure-ground separation by visual cortex. *Perception and Psychophysics*, 55, 48–120.

Grossberg, S. (1997). Cortical dynamics of three-dimensional figure-ground perception of two-dimensional figures. *Psychological Review*, 104, 618–658.

Grossberg, S. (1998). How is a moving target continuously tracked behind occluding cover? In T. Watanabe (Ed.), *High Level Motion Processing: Computational, Neurobiological, and Psychophysical Perspectives*. Cambridge, MA: MIT Press, pp. 3–52.

Grossberg, S. (2013). Recurrent neural networks, *Scholarpedia*. http://www.scholarpedia.org/article/Recurrent_neural_networks.

Grossberg, S. (2014). How visual illusions illuminate complementary brain processes: Illusory depth from brightness and apparent motion of illusory contours. *Frontiers in Human Neuroscience*, doi: 10.3389/fnhum.2014.00854. https://www.frontiersin.org/articles/10.3389/fnhum.2014.00854/full.

Grossberg, S. (2016). Cortical dynamics of figure-ground separation in response to 2D pictures and 3D scenes: How V2 combines border ownership, stereoscopic cues, and gestalt grouping rules. *Frontiers in Psychology*, January 26, 2016. https://www.frontiersin.org/articles/10.3389/fpsyg.2015.02054/full.

Grossberg, S. (2018). Desirability, availability, credit assignment, category learning, and attention: Cognitive-emotional and working memory dynamics of orbitofrontal, ventrolateral, and dorsolateral prefrontal cortices. *Brain and Neuroscience Advances*, May 8, 2018. https://journals.sagepub.com/doi/full/10.1177/2398212818772179.

Grossberg, S. (2019). The embodied brain of SOVEREIGN2: From space-variant conscious percepts during visual search and navigation to learning invariant object categories and cognitive-emotional plans for acquiring valued goals. *Frontiers in Computational Neuroscience*. Published online: June 25, 2019. https://www.frontiersin.org/articles/10.3389/fncom.2019.00036/full

Grossberg, S. (2021). *Conscious Mind, Resonant Brain: How Each Brain Makes a Mind*. New York: Oxford University Press.

Grossberg, S. (2022). Towards understanding the brain dynamics of music: Learning and conscious performance of lyrics and melodies with variable rhythms and beats. *Frontiers in Systems Neuroscience*, April 8, 2022. https://www.frontiersin.org/articles/10.3389/fnsys.2022.766239/full.

Grossberg, S., and Hong, S. (2006). A neural model of surface perception: Lightness, anchoring, and filling-in. *Spatial Vision*, 19, 263–321.

Grossberg, S., and Levine, D. S. (1987). Neural dynamics of attentionally modulated Pavlovian conditioning: Blocking, inter-stimulus interval, and secondary reinforcement. *Applied Optics*, 26, 5015–5030.

Grossberg, S., and Merrill, J. W. L. (1992). A neural network model of adaptively timed reinforcement learning and hippocampal dynamics. *Cognitive Brain Research*, 1, 3–38.

Grossberg, S., and Merrill, J. W. L. (1996). The hippocampus and cerebellum in adaptively timed learning, recognition, and movement. *Journal of Cognitive Neuroscience*, 8, 257–277.

Grossberg, S., and Mingolla, E. (1985a). Neural dynamics of form perception: Boundary completion, illusory figures, and neon color spreading. *Psychological Review*, 92, 173–211.

Grossberg, S., and Mingolla, E. (1985b). Neural dynamics of perceptual grouping: Textures, boundaries, and emergent segmentations. *Perception and Psychophysics*, 38, 141–171.

Grossberg, S., and Myers, C. (2000). The resonant dynamics of speech perception: Interword integration and duration-dependent backward effects, *Psychological Review*, 107, 735–767.

Grossberg, S. and Pessoa, L. (1998). Texture segregation, surface representation, and figure-ground separation. *Vision Research*, 38, 2657–2684.

Grossberg, S., and Pearson, L. (2008). Laminar cortical dynamics of cognitive and motor working memory, sequence learning and performance: Toward a unified theory of how the cerebral cortex works. *Psychological Review*, 115, 677–732.

Grossberg, S., and Pilly, P. (2008). Temporal dynamics of decision-making during motion perception in the visual cortex. *Vision Research*, 48, 1345–1373.

Grossberg, S., and Rudd, M. (1989). A neural architecture for visual motion perception: Group and element apparent motion. *Neural Networks*, 2, 421–450.

Grossberg, S., and Rudd, M. E. (1992). Cortical dynamics of visual motion perception: Short-range and long-range apparent motion. *Psychological Review*, 99, 78–121.

Grossberg, S., and Schmajuk, N. A. (1989). Neural dynamics of adaptive timing and temporal discrimination during associative learning. *Neural Networks*, 2, 79–102.

Grossberg, S., and Stone, G. O. (1986). Neural dynamics of word recognition and recall: Attentional priming, learning, and resonance. *Psychological Review*, 93, 46–74.

Grossberg, S., and Swaminathan, G. (2004). A laminar cortical model for 3D perception of slanted and curved surfaces and of 2D images: development, attention and bistability. *Vision Research*, 44, 1147–1187.

Grossberg, S., and Todorovic, D. (1988). Neural dynamics of 1-D and 2-D brightness perception: A unified model of classical and recent phenomena. *Perception and Psychophysics*, 43, 241–277.

Grossberg, S., and Vladusich, T. (2010). How do children learn to follow gaze, share joint attention, imitate their teachers, and use tools during social interactions? *Neural Networks*, 23, 940–965.

Grossberg, S., and Williamson, J. R. (1999). A self-organizing neural system for learning to recognize textured scenes. *Vision Research*, 39, 1385–1406.

Grossberg, S., and Wyse, L. (1992). Figure-ground separation of connected scenic figures: Boundaries, filling-in, and opponent processing. In G. A. Carpenter and S. Grossberg (Eds.), *Neural Networks for Vision and Image Processing*. Cambridge, MA: MIT Press, pp. 161–194.

Grossberg, S., Bullock, D., and Dranias, M. (2008). Neural dynamics underlying impaired autonomic and conditioned responses following amygdala and orbitofrontal lesions. *Behavioral Neuroscience*, 122, 1100–1125.

Grossberg, S., Leveille, J., and Versace, M. (2011). How do object reference frames and motion vector decomposition emerge in laminar cortical circuits? *Attention, Perception, and Psychophysics*, 73, 1147–1170.

Grossberg, S., Markowitz, J., and Cao, Y. (2011). On the road to invariant recognition: Explaining tradeoff and morph properties of cells in inferotemporal cortex using multiple-scale task-sensitive attentive learning. *Neural Networks*, 24, 1036–1049.

Grossberg, S., Mingolla, E., and Pack, C. C. (1999). A neural model of motion processing and visual navigation by cortical area MST. *Cerebral Cortex*, 9, 878–895.

Grossberg, S., Srihasam, K., and Bullock, D. (2012). Neural dynamics of saccadic and smooth pursuit eye movement coordination during visual tracking of unpredictably moving targets. *Neural Networks*, 27, 1–20.

Grossberg, S., Srinivasan, K., and Yazdabakhsh, A. (2011). On the road to invariant object recognition: How cortical area V2 transforms absolute to relative disparity during 3D vision. *Neural Networks*, 24, 686–692.

Grossberg, S., Srinivasan, K., and Yazdanbakhsh, A. (2014). Binocular fusion and invariant category learning due to predictive remapping during scanning of a depthful scene with eye movements. *Frontiers in Psychology: Perception Science*, 3 January 2015, Volume 5. https://doi.org/10.3389/fpsyg.2014.01457, http://journal.frontiersin.org/Journal/10.3389/fpsyg.2014.01457/full

Guenther, F. H. (1995). Speech sound acquisition, coarticulation, and rate effects in a neural network model of speech production. *Psychological Review*, 102, 594–621.

Guenther, F. H., Ghosh, S. S., and Tourville, J. A. (2006). Neural modeling and imaging of the cortical interactions underlying syllable production. *Brain and Language*, 96, 280–301.

Guilford, J. P. (1929). Illusory movement from a rotating barber pole. *American Journal of Psychology*, 41, 686.

Hamberger, M. J., Seidel, W. T., Mckhann, G. M., Perrine, K., and Goodman, R. R. (2005). Brain stimulation reveals critical auditory naming cortex. *Brain*, 128, 2742–2749.

Hauk, O., and Pulvemuller, F. (2004). Neurophysiological distinction of action words in the fronto-central cortex. *Human Brain Mapping*, 21, 191–201.

Haxby, J. V., Gobbini, M. I., Furey, M. L., Ishai, A., Schouten, J. L., and Pietrini, P. (2001). Distributed and overlapping representations of faces and object in ventral temporal cortex. *Science*, 293, 2425–2430.

Hikosaka, O., and Wurtz, R. H. (1983). Visual and oculomotor functions of monkey substantia nigra pars reticulata. IV. Relation of substantia nigra to superior colliculus. *Journal of Neurophysiology*, 49, 1285–1301.

Hikosaka, O., and Wurtz, R. H. (1989). The basal ganglia. In R. Wurtz and M. Goldberg (Eds.), *The Neurobiology of Saccadic Eye Movements*. Amsterdam: Elsevier, pp. 257–281.

Hong, S., and Grossberg, S. (2004). A neuromorphic model for achromatic and chromatic surface representation of natural images. *Neural Networks*, 2004, 17, 787–808.

Hubel, D. H., and Wiesel, T. N. (1962). Receptive fields, binocular interaction and functional architecture in the cat's visual cortex. *Journal of Physiology*, 160, 106–154.

Hubel, D. H., and Wiesel, T. N. (1968). Receptive fields and functional architectures of monkey striate cortex. *Journal of Physiology*, 195, 215–243.

Hubel, D. H., and Wiesel, T. N. (1977). Functional architecture of macaque monkey visual cortex. *Proceedings of the Royal Society of London, B*, 198, 1–59.

Huth, A. G., Nishimoto, S., Vu, A. T., and Gallant, J. L. (2012). A continuous semantic space describes the representation of thousands of object and action categories across the human brain. *Neuron*, 76, 1210–1224.

Jackendoff, R. (2006). On conceptual semantics. *Intercultural Pragmatics*, 3. https://doi.org/10.1515/IP.2006.020.

Johansson, G. (1950). *Configurations in Event Perception*. Uppsala: Almqvist and Wiksell.

Kable, J. W., Kan, I. P., Wilson, A., Thompson-Schill, S. L., and Chatterjee, A. (2005). Conceptual representations of action in the lateral temporal cortex. *Journal of Cognitive Neuroscience*, 17, 1855–1870.

Kable, J. W., Lease-Spellmeyer, J., and Chatterjee, A. (2002). Neural substrates of action event knowledge. *Journal of Cognitive Neuroscience*, 14, 795–805.

Kamin, L. J. (1968). Attention-like processes in classical conditioning. In M. R. Jones (Ed.), *Miami Symposium on the Prediction of Behavior: Aversive Stimulation*. Miami: University of Miami Press, pp. 1–33.

Kamin, L. J. (1969). Predictability, surprise, attention, and conditioning. In B. A. Campbell and R. M. Church (Eds.), *Punishment and Aversive Behavior*. New York: Appleton-Century-Crofts, pp. 279–296.

Kashino, M. (2006). Phonemic restoration: The brain creates missing speech sounds. *Acoustical Science and Technology*, 27, 318–321.

Kelly, F. J., and Grossberg, S. (2000). Neural dynamics of 3-D surface perception: Figure-ground separation and lightness perception. *Perception & Psychophysics*, 62, 1596–1619.

Kemmerer, D., Castillo, J. G., Talavage, T., Patterson, S., and Wiley, C. (2008). Neuroanatomical distribution of five semantic components of verbs: Evidence from fMRI. *Brain and Language*, 107, 16–43.

Kimppa, L., Kujala, T., Leminen, A., Vainion, Ma., and Shtyrov, Y. (2015). Rapide and automatic speech-specific learning mechanism in human neocortex. *NeuroImage*, 118, 282–291.

Kochari, A. R., Lewis, A. G., Schoffelen, J.-M., and Schriefers, H. (2021). Semantic and syntactice composition of minimal adjective-noun phrases in Dutch: An MEG study. *Neuropsychologia*, 155, 1–24.

Kolers, P. (1972). *Aspects of Motion Perception*. Oxford, UK: Pergamon Press.

Kovács, G. (2020). Getting to know someone: Familiarity, person recognition, and identification in the human brain. *Journal of Cognitive Neuroscience*, 32, 2205–2225.

Lange, V. M., Perret, C., and Laganaro, M. (2015). Comparison of single-word and adjective-noun phrase production using event-related brain potentials. *Cortex*, 67, 15–29.

Mahon, B. Z., and Caramazza, A. (2008). A critical look at the embodied cognition hypothesis and a new proposal for grounding conceptual content. *Journal of Physiology*, 102, 59–70.

Materna, S., Dicke, P. W., and Their, P. (2008). Dissociable roles of the superior temporal sulcus and the intraparietal sulcus in joint attention: a functional magnetic resonance imaging study. *Journal of Cognitive Neuroscience*, 20, 108–119.

Middleton, F., and Strick, P. (2000). Basal ganglia and cerebellar loops: Motor and cognitive circuits. *Brain Research Reviews*, 31, 236–250.

Motter, B. C. (1993). Focal attention produces spatially selective processing in visual cortical areas V1, V2, and V4 in the presence of competing stimuli. *Journal of Neurophysiology*, 70, 909–919.

Newport, C. (2023). What kind of mind does ChatGPT have? *The New Yorker*, April 13, 2023.

Pagel, M. (2017). Q&A: What is human language, when did it evolve and why should we care? *BMC Biology*, 15, 64. https://doi.org/10.1186/s12915-017-0405-3

Pavlov, I. P. (1927). *Conditioned Reflexes*. London: Constable and Company (Reprinted by Dover Publications, 1960).

Pecher, D., and Zwaan, R. A. (Eds.) (2014) *Grounding Cognition: The Role of Perception and Action in Memory, Language, and Thinking*. Cambridge: Cambridge University Press.

Piaget, J. (1945). *La Formation du Symbole Chez L'enfant*. Paris: Delachaux Niestle, S.A.

Piaget, J. (1951). *Play, Dreams and Imitation in Childhood*. C. Gattegno and C. F. M. Hodgson (Trans.). London: Routledge and Kegan Paul.

Piaget, J. (1952). *The Origins of Intelligence in Children*. New York: International Universities Press.

Posner, M. I., Rafal, R. D., Choate, L. S., and Vaughan, J. (1985). Inhibition of return: Neural basis and function. *Cognitive Neuropsychology*, 2, 211–228.

Rajimehr, R., Young, J. C., and Tootell, R. B. H. (2009). An anterior temporal face patch in human cortex, predicted by macaque maps. *Proceedings of the National Academy of Sciences*, 106, 1995–2000. doi: 10.1073/pnas.0807304106

Ramon, M., and Gobbini, M. I. (2017). Familiarity matters: A review on prioritized processing of personally familiar faces. *Visual Cognition*, 26, 179–195.

Raposo, A., Moss, J. E., Stamatakis, E. A., and Tyler, L. K. (2009). Modulation of motor and premotor cortices by actions, action words, and action sentences. *Neuropsychologia*, 47, 388–396.

Rauschecker, J. P., and Scott, S. K. (2009). Maps and streams in the auditory cortex: Nonhuman primates illuminate human speech processing. *Nature Neuroscience*, 12, 718–724.

Repp, B., Liberman, A., Eccardt, T., and Pesetsky, D. (1978). Perceptual integration of acoustic cues for stop, fricative, and affricate manner. *Journal of Experimental Psychology: Human Perception and Performance*, 4, 621–637.

Richardson, D. C., Spivey, M. J., Barsalou, L. W., and McRae, K. (2003). Spatial representations activated during real-time comprehension of verbs. *Cognitive Science*, 27, 767–780.

Rodman, H. R., and Albright, T. D. (1987). Coding of visual stimulus velocity in area MT of the macaque. *Vision Research*, 27, 2035–2048.

Samuel, A. (1981a). Phonemic restoration: Insights from a new methodoloty. *Journal of Experimental Psychology: Human Perception & Performance*, 4, 474–494.

Samuel, A. (1981b). The role of bottom-up confirmation in the phonemic restoration illusion. *Journal of Experimental Psychology: Human Perception & Performance*, 7, 1124–1131.

Sandhofer, C., and Smith, Linda B. (2007). Learning adjectives in the real world: How learning nouns impedes learning adjectives, *Language Learning and Development*, 3, 233–267.

Saygin, A., McCullough, S., Alac, M., and Emmorey, K. (2010). Modulation of BOLD response in motion-sensitive lateral temporal cortex by real and fictive motion sentences. *Journal of Cognitive Neuroscience*, 22, 2480–2490.

Sereno, M. I., Dale, A. M., Reppas, J. B., Kwong, K. K., Belliveau, J. W., Brady, T. J., Rosen, B. R., and Tootell, R. B. H. (1995). Borders of multiple visual areas in humans revealed by functional MRI. *Science*, 268, 889–893.

Sugiura, M., Mano, Y., Sasaki, A., and Sadato, N. (2011). Beyond the memory mechanism: Person-selective and nonselective processes in recognition of personally familiar faces. *Journal of Cognitive Neuroscience*, 23, 699–715.

Silver, M. R., Grossberg, S., Bullock, D., Histed, M. H., and Miller, E. K. (2011). A neural model of sequential movement planning and control of eye movements: Item-order-rank working memory and saccade selection by the supplementary eye fields. *Neural Networks*, 26, 29–58.

Singer, D. G., and Revenson, T. A. (1997). *A Piaget Primer: How a Child Thinks*. Revised Edition. Madison, CT: International Universities Press. Inc.

Srivastava, R. K., Greff, K., and Schmidhuber, J. (2015). Training very deep networks. *Advances in Neural Information Processing Systems* 28 *(NIPS)*. https://proceedings.neurips.cc/paper/2015/file/215a71a12769b056c3c32e7299f1c5ed-Paper.pdf

Tanaka, K. (1996). Inferotemporal cortex and object vision. *Annual Review of Neuroscience*, 19, 109–139.

Tanaka, K., Saito, H., Fukada, Y., and Moriya, M. (1991). Coding visual images of objects in the inferotemporal cortex of the macaque monkey. *Journal of Neurophysiology*, 66, 170–189.

Tomasello, M., and Farrar, M. J. (1986). Joint attention and early language. *Child Development*, 57, 1454–1463.

Tronick, E. (2007). *The Neurobehavioral and Social-Emotional Development of Infants and Children*. New York: W. W. Norton & Company.

Tyler, C. W., and Kontsevich, L. L. (1995). Mechanisms of stereoscopic processing: stereoattention and surface perception in depth reconstruction. *Perception*, 24, 127–153.

Villiers, J. G. de., and Villiers, P. A. de. (1973). A cross-sectional study of the acquisition of grammatical morphemes inchild speech. *Journal of Psycholinguistic Research*, 2, 267–278.

Wallentin, M., Nielsen, A. H., Vuust, P., Dohn, A., Roepstorff, A., and Lund, T. E. (2011). BOLD response to motion verbs in left posterior middle temporal gyrus during story comprehension. *Brain and Language*, 119, 221–225.

Wallentin, M. Östergaard, S., Lund, T. E., Östergaard, L., and Roepstorff, A. (2005). Concrete spatial language: See what I mean? *Brain and Language*, 119, 221–233.

Wallach, H. (1935). On the visually perceived direction of motion. *Psychologische Forschung*, 20, 325–380.

Watson, C. E., Cardillo, E. R., Ianni, G. R., and Chatterjee, A. (2013). Action concepts in the brain: An activation likelihood estimation meta-analysis. *Journal of Cognitive Neuroscience*, 25, 1191–1205.

Wittgenstein, L. (1922). *Tractatus Logico-Philosophicus*. New York: Harcourt, Brace & Company, Inc.

Wuerger, S., Shapley, R., and Rubin, N. (1996). On the visually perceived direction of motion, by Hans Wallach: 60 years later. *Perception*, 25, 1317–1367.

Zwaan, R. A., and Taylor, L. J. (2006). Seeing, acting, understanding: Motor resonance in language comprehension, *Journal of Experimental Psychology: General*, 135, 1–11.

Chapter 5

Amari, S.-I. (1972). Characteristics of random nets of analog neuron-like elements. *Transactions on Systems, Man, and Cybernetics*, 2, 643–657.

Bellmann, A., Meuli, R., and Clarke, S. (2001). Two types of auditory neglect. *Brain*, 124, 676–687.

Bregman, A. S. (1990). *Auditory Scene Analysis: The Perceptual Organization of Sound*. Cambridge, MA: MIT Press.

Brown, J. W, Bullock, D., and Grossberg, S. (1999). How the basal ganglia use parallel excitatory and inhibitory learning pathways to selectively respond to unexpected rewarding cues. *Journal of Neuroscience*, 19, 10502–10511.

Brown, J. W., Bullock, D., and Grossberg, S. (2004). How laminar frontal cortex and basal ganglia circuits interact to control planned and reactive saccades. *Neural Networks*, 17, 471–510.

Bullock, D., and Grossberg, S. (1988). Neural dynamics of planned arm movements: Emergent invariants and speed-accuracy properties during trajectory formation. *Psychological Review*, 95, 49–90.

Bullock, D., and Grossberg, S. (1989). VITE and FLETE: Neural modules for trajectory formation and postural control. In W. Hershberger (Ed.), *Volitional Action*. Amsterdam: North-Holland, pp. 253–297.

Bullock, D., Grossberg, S., and Guenther, F. H. (1993). A self-organizing neural model of motor equivalent reaching and tool use by a multijoint arm. *Journal of Cognitive Neuroscience*, 5, 408–435.

Cao, Y., and Grossberg, S. (2005). A laminar cortical model of stereopsis and 3D surface perception: Closure and da Vinci stereopsis. *Spatial Vision*, 18, 515–578.

Cao, Y., and Grossberg, S. (2012). Stereopsis and 3D surface perception by spiking neurons in laminar cortical circuits: A method of converting neural rate models into spiking models. *Neural Networks*, 26, 75–98.

Cao, Y., Grossberg, S., and Markowitz, J. (2011). How does the brain rapidly learn and reorganize view- and positionally-invariant object representations in inferior temporal cortex? *Neural Networks*, 24, 1050–1061.

Carpenter, G. A. (1989). Neural network models for pattern recognition and associative memory. *Neural Networks*, 2, 243–257.

Carpenter, G. A., and Grossberg, S. (1987). A massively parallel architecture for a self organizing neural pattern recognition machine. *Computer Vision, Graphics, and Image Processing*, 37, 54–115.

Carpenter, G. A., and Grossberg, S. (1988). The ART of adaptive pattern recognition by a self-organizing neural network. *Computer*, 21, 77–88.

Carpenter, G. A., and Grossberg, S. (1992). A self-organizing neural network for supervised learning, recognition, and prediction. *IEEE Communications Magazine*, 30, 38–49.

Carpenter, G. A., and Grossberg, S. (1993). Normal and amnesic learning, recognition, and memory by a neural model of cortico-hippocampal interactions. *Trends in Neurosciences*, 16, 131–137.

Carpenter, G. A. and Grossberg, S. (2003). Adaptive resonance theory. In M.A. Arbib (Ed.), *The Handbook of Brain Theory and Neural Networks*, Second Edition, Cambridge, MA: MIT Press, 87-90.

Carpenter, G. A., Grossberg, S., Markuzon, N., Reynolds, J. H., and Rosen, D. B. (1992). Fuzzy ARTMAP: A neural network architecture for incremental supervised learning of analog multidimensional maps. *IEEE Transactions on Neural Networks*, 3, 698–713.

Carpenter, G. A., Grossberg, S., and Reynolds, J. H. (1991). ARTMAP: Supervised real-time learning and classification of nonstationary data by a self-organizing neural network. *Neural Networks*, 4, 565–588.

Caudell, T. P., Smith, S. D. G., Escobedo, R., and Anderson, M. (1994). NIRS: Large scale ART-1 neural architectures for engineering design retrieval. *Neural Networks*, 7, 1339–1350.

Caudell, T., Smith, S., Johnson, C., Wunsch, D. C. II., and Escobedo, R. (1990). A data compressed ART1 neural network algorithms. *Proceedings of the SPIE Conference on Aerospace Sensing*, April.

Caudell, T., Smith, S., Johnson, C., Wunsch, D. C. II., and Escobedo, R. (1991). An industrial application of neural networks to reusable design. *Proceedings of the International Joint Conference on Neural Networks*, 2, 919.

Cavanagh, P., Labianca, A. T., & Thornton, I. M. (2001). Attention-based visual routines: sprites. *Cognition*, 80, 47–60.

Chang, H.-C., Grossberg, S., and Cao, Y. (2014). Where's Waldo? How perceptual cognitive, and emotional brain processes cooperate during learning to categorize and find desired objects in a cluttered scene. *Frontiers in Integrative Neuroscience*, doi: 10.3389/fnint.2014.0043, https://www.frontiersin.org/articles/10.3389/fnint.2014.00043/full

Cherry, E. C. (1953). Some experiments on the recognition of speech, with one and two ears. *Journal of the Acoustical Society of America*, 25, 975–979.

Childress, D. (2008). *Johannes Gutenberg and the Printing Press*. Twenty-First Century Books/Lerner Publishing Group, Inc.: Minneapolis, MN.

Clune, J., Mouret, J. B., and Lipson, H. (2013). The evolutionary origins of modularity. *Proceedings of the Royal Society B*, 280(1755), 20122863. https://royalsocietypublishing.org/doi/10.1098/rspb.2012.2863

Darwin, C. (1859). *The Origin of Species by Means of Natural Selection*. London, England: Murray.

Da Silva, L. E. B., Elnabarawy, I., and Wunsch, D. C, II. (2019). A survey of Adaptive Resonance Theory neural network models for engineering applications. *Neural Networks*, 120, 167–203.

Dranias, M., Grossberg, S., and Bullock, D. (2008). Dopaminergic and non-dopaminergic value systems in conditioning and outcome-specific revaluation. *Brain Research*, 1238, 239–287.

Driver, J., and Mattingley, J. B. (1998). Parietal neglect and visual awareness. *Nature Neuroscience*, 1, 17–22.

Escobedo, R., Smith, S. D. G., and Caudell, T. P. (1993). A neural information retrieval system. *The International Journal of Advanced Manufacturing Technology*, 8, 269–273.

Fazl, A., Grossberg, S., and Mingolla, E. (2009). View-invariant object category learning, recognition, and search: How spatial and object attention are coordinated using surface-based attentional shrouds. *Cognitive Psychology*, 58, 1–48.

Foley, N. C., Grossberg, S., and Mingolla, E. (2012). Neural dynamics of object-based multifocal visual spatial attention and priming: Object cueing, useful-field-of-view, and crowding. *Cognitive Psychology*, 65, 77–117.

French, R. M. (1999). Catastrophic forgetting in connectionist networks. *Trends in Cognitive Sciences*, 3, 128–135.

Gaudiano P., and Grossberg S. (1991). Vector associative maps: Unsupervised real-time error-based learning and control of movement trajectories. *Neural Networks*, 4, 147–183.

Gaudiano, P., and Grossberg, S. (1992). Adaptive vector integration to endpoint: Self-organizing neural circuits for control of planned movement trajectories. *Human Movement Science*, 11, 141–155.

Gnadt, W., and Grossberg, S. (2008). SOVEREIGN: An autonomous neural system for incrementally learning planned action sequences to navigate towards a rewarded goal. *Neural Networks*, 21, 699–758.

Grossberg, S., and Marshall, J. (1989). Stereo boundary fusion by cortical complex cells: A system of maps, filters, and feedback networks for multiplexing distributed data. *Neural Networks*, 2, 29–51.

Goodale, M. A., and Milner, A. D. (1992). Separate visual pathways for perception and action. *Trends in Neurosciences*, 15, 20–15.

Goodale, M. A., Milner, A. D., Jakobson, L. S., and Carey, D. P. (1991). A neurological dissociation between perceiving objects and grasping them. *Nature*, 349, 154–156.

Grossberg, S. (1974). Contour enhancement, short-term memory, and constancies in reverberating neural networks. *Studies in Applied Mathematics*, 52, 213–257.

Grossberg, S. (1975). A neural model of attention, reinforcement, and discrimination learning. *International Review of Neurobiology*, 18, 263–327.

Grossberg, S. (1976a). Adaptive pattern classification and universal recoding, I: Parallel development and coding of neural feature detectors. *Biological Cybernetics*, 23, 121–134.

Grossberg, S. (1976b). Adaptive pattern classification and universal recoding, II: Feedback, expectation, olfaction, and illusions. *Biological Cybernetics*, 23, 187–202.

Grossberg, S. (1978). A theory of human memory: Self-organization and performance of sensory-motor codes, maps, and plans. In R. Rosen and F. Snell (Eds.), *Progress in Theoretical Biology*. New York: Academic Press, Vol. 5, pp. 233–374.

Grossberg, S. (1980). How does a brain build a cognitive code? *Psychological Review*, 87, 1–51.

Grossberg, S. (1984). Some psychophysiological and pharmacological correlates of a developmental, cognitive, and motivational theory. In R. Karrer, J. Cohen, and P. Tueting (Eds.), *Brain and Information: Event Related Potentials*. New York: New York Academy of Sciences, pp. 58–142.

Grossberg, S. (1987a). Cortical dynamics of three-dimensional form, color, and brightness perception, I: Monocular theory. *Perception and Psychophysics*, 41, 87–116.

Grossberg, S. (1987b). Cortical dynamics of three-dimensional form, color, and brightness perception, II: Binocular theory. *Perception and Psychophysics*, 41, 117–158.

Grossberg, S. (1994). 3-D vision and figure-ground separation by visual cortex. *Perception and Psychophysics*, 55, 48–120.

Grossberg, S. (1997). Cortical dynamics of three-dimensional figure-ground perception of two-dimensional figures. *Psychological Review*, 104, 618–658.

Grossberg, S. (1988). Nonlinear neural networks: Principles, mechanisms, and architectures. *Neural Networks*, 1, 17–61.

Grossberg, S. (2000). The complementary brain: Unifying brain dynamics and modularity. *Trends in Cognitive Sciences*, 4, 233–246.

Grossberg, S. (2009). Cortical and subcortical predictive dynamics and learning during perception, cognition, emotion and action. *Philosophical Transactions of the Royal Society of London B Biological Sciences*, 364, 1223–1234.

Grossberg, S. (2013). Adaptive resonance theory: How a brain learns to consciously attend, learn, and recognize a changing world. *Neural Networks*, 37, 1–47.

Grossberg, S. (2017a). Acetylcholine neuromodulation in normal and abnormal learning and memory: Vigilance control in waking, sleep, autism, amnesia, and Alzheimer's disease. *Frontiers in Neural Circuits*, November 2, 2017. https://www.frontiersin.org/articles/10.3389/fncir.2017.00082/full

Grossberg, S. (2017b). Towards solving the Hard Problem of Consciousness: The varieties of brain resonances and the conscious experiences that they support. *Neural Networks*, 87, 38–95.

Grossberg, S. (2018). Desirability, availability, credit assignment, category learning, and attention: Cognitive-emotional and working memory dynamics of orbitofrontal, ventrolateral, and dorsolateral prefrontal cortices. *Brain and Neuroscience Advances*, May 8, 2018. http://journals.sagepub.com/doi/full/10.1177/2398212818772179

Grossberg, S. (2019a). The embodied brain of SOVEREIGN2: From space-variant conscious percepts during visual search and navigation to learning invariant object categories and cognitive-emotional plans for acquiring valued goals. *Frontiers in Computational Neuroscience*. https://www.frontiersin.org/articles/10.3389/fncom.2019.00036/full

Grossberg, S. (2019b). The resonant brain: How attentive conscious seeing regulates action sequences that interact with attentive cognitive learning, recognition, and prediction. *Attention, Perception & Psychophysics*. Published online: June 19, 2019. https://link.springer.com/article/10.3758/s13414-019-01789-2

Grossberg, S. (2020). *Conscious Mind/Resonant Brain: How Each Brain Makes a Mind*. New York: Oxford University Press.

Grossberg, S., and Hong, S. (2006). A neural model of surface perception: Lightness, anchoring, and filling-in. *Spatial Vision*, 19, 263–321.

Grossberg, S., and Howe, P. D. L. (2003). A laminar cortical model of stereopsis and three-dimensional surface perception. *Vision Research*, 43, 801–829.

Grossberg, S., and Kishnan, D. (2018). Neural dynamics of autistic repetitive behaviors and Fragile X syndrome: Basal ganglia movement gating and mGluR-modulated adaptively timed learning. *Frontiers in Psychology, Psychopathology*. https://www.frontiersin.org/articles/10.3389/fpsyg.2018.00269/full

Grossberg, S., and Pilly, P. K. (2014). Coordinated learning of grid cell and place cell spatial and temporal properties: multiple scales, attention, and oscillations. *Philosophical Transactions of the Royal Society B.*, 369, 20120524. https://royalsocietypublishing.org/doi/pdf/10.1098/rstb.2012.0524

Grossberg, S., and Raizada, R. (2000). Contrast-sensitive perceptual grouping and object-based attention in the laminar circuits of primary visual cortex. *Vision Research*, 40, 1413–1432.

Grossberg, S., and Swaminathan, G. (2004). A laminar cortical model for 3D perception of slanted and curved surfaces and of 2D images: development, attention and bistability. *Vision Research*, 44, 1147–1187.

Grossberg, S., and Versace, M. (2008). Spikes, synchrony, and attentive learning by laminar thalamocortical circuits. *Brain Research*, 1218, 278–312.

Grossberg, S., and Vladusich, T. (2010). How do children learn to follow gaze, share joint attention, imitate their teachers, and use tools during social interactions? *Neural Networks*, 23, 940–965.

Grossberg, S., and Yazdanbakhsh, A. (2005). Laminar cortical dynamics of 3D surface perception: Stratification, transparency, and neon color spreading. *Vision Research*, 45, 1725–1743.

Grossberg, S., and Zajac, L. (2017). How humans consciously see paintings and paintings illuminate how humans see. *Art & Perception*, 5, 1–95.

Grossberg, S., Bullock, D., and Dranias, M. (2008). Neural dynamics underlying impaired autonomic and conditioned responses following amygdala and orbitofrontal lesions. *Behavioral Neuroscience*, 122, 1100–1125.

Grossberg, S., Hwang, S., and Mingolla, E. (2002). Thalamocortical dynamics of the McCollough effect: Boundary-surface alignment through perceptual learning. *Vision Research*, 42, 1259–1286.

Grossberg, S., Kuhlmann, L., and Mingolla, E. (2007). A neural model of 3D shape-from-texture: Multiple-scale filtering, boundary grouping, and surface filling-in. *Vision Research*, 47, 634–672.

Grossberg, S., Srinivasan, K., and Yazdanbakhsh, A. (2014). Binocular fusion and invariant category learning due to predictive remapping during scanning of a depthful scene with eye movements. *Frontiers in Psychology: Perception Science*, doi: 10.3389/fpsyg.2014.01457 https://www.frontiersin.org/articles/10.3389/fpsyg.2014.01457/full

Grossberg, S., Govindarajan, K. K., Wyse, L. L., and Cohen, M. A. (2004). ARTSTREAM: A neural network model of auditory scene analysis and source segregation. *Neural Networks*, 17, 511–536.

Guenther, F. H. (1995). Speech sound acquisition, coarticulation, and rate effects in a neural network model of speech production. *Psychological Review*, 102, 594–621.

Guenther, F. H., Ghosh, S. S., and Tourville, J. A. (2006). Neural modeling and imaging of the cortical interactions underlying syllable production. *Brain and Language*, 96, 280–301.

Heeger, D. J. (1992). Normalization of cell responses in cat striate cortex. *Visual Neuroscience*, 9, 181–197.

Heilman, K. M., Bowers, D., Coslett, H. B., Whelan, H., and Watson, R. T. (1985). Directional hypokinesia. *Neurology*, 35, 855–859.

Hinton, G., Deng, L., Yu, D., Dahl, G. E., Mohamed, A-r., Jaitly, N., Senior, A., Vanhoucke, V., Nguyen, P., Sainath, T. N., and Kingsbury, B. (2012). Deep neural networks for acoustic modeling in speech recognition: The shared views of four research groups. *IEEE Signal Processing Magazine*, 29, 82–97.

Kelly, F. J., and Grossberg, S. (2000). Neural dynamics of 3-D surface perception: Figure-ground separation and lightness perception. *Perception & Psychophysics*, 62, 1596–1619.

Kirkpatrick, J., Pancanu, R., Rabinowitz, N., Veness, J., Desjarkins, G., Rusu, A. A., Milan, K., Zuan, J., Ramalho, T., Grabska-Barwinska, A., Hassabis, D., Clopath, C., Kumaran, D., and Hadsell, R. (2017). Overcoming catastrophic forgetting in neural networks. *Proceedings of the National Academy of Sciences USA*, 114, 3521–3526.

LeCun, Y., Bengio, Y., and Hinton, G. (2015). Deep learning. *Nature*, 521, 436–444.

LeVine, S. (2017). Artificial intelligence pioneer says we need to start over. *Axios*, September 15, 2017.

Leonard, B., and McNaughton, B. L. (1990). Spatial representation in the rat: Conceptual, behavioral, and neurophysiological perspectives. In R. P. Kesner and D. S. Olton (Eds.), *Neurobiology of Comparative Cognition*. Hillsdale, NJ: Lawrence Erlbaum Associates, Chapter 13, pp. 363–422.

Lincoff, G. H. (1981). *The Audubon Society Field Guide to North American Mushrooms*. New York: Alfred A Knopf.

MacNeilage, P. R. (1998). The frame/content theory of evolution of speech. *Behavioral and Brain Sciences*, 21, 499–546.

Marshall, J. C. (2001). Auditory neglect and right parietal cortex. *Brain*, 124, 645–646.

Mattingley, J. B., Husain, M., Rorden, C., Kennard, C., and Driver, J. (1998). Motor role of human inferior parietal lobe revealed in unilateral neglect patients. *Nature*, 392, 179–182.

McCloskey, M., and Cohen, N. (1989). Catastrophic interference in connectionist networks: The sequential learning problem. In G. H. Bower (Ed.), *The Psychology of Learning and Motivation*, Vol. 24, pp. 109–164.

Mesulam, M.-M. (1999). Spatial attention and neglect: Parietal, frontal and cingulate contributions to the mental representation and attentional targeting of salient extrapersonal events. *Philosophical Transactions of the Royal Society B*, 354, 1325–1346.

Mishkin, M. (1982). A memory system in the monkey. *Philosophical Transactions Royal Society of London B*, 298, 85–95.

Mishkin, M., Ungerleider, L. G., and Macko, K. A. (1983). Object vision and spatial vision: Two cortical pathways. *Trends in Neurosciences*, 6, 414–417.

Muller, R. A. (1996). A quarter of a century of place cells. *Neuron*, 17, 813–822.

Näätänen, R., Simpson, M., and Loveless, N. E. (1982). Stimulus deviance and evoked potentials. *Biological Psychiatry*, 14, 53–98.

Näätänen, R. (1982). Processing negativity: An evoked-potential reflection. *Psychological Bulletin*, 92, 605–640.

O'Keefe, J., and Dostrovsky, J. (1971). The hippocampus as a spatial map. Preliminary evidence from unit activity in the freely-moving rat. *Brain Research*, 34, 171–175.

O'Keefe, J., and Nadel, L. (1978). *The Hippocampus as a Cognitive Map*. Oxford, UK: Clarendon Press.

Omer, D. B., Maimon, S. R., Las, L., and Ulanovsky, N. (2018). Social place-cells in the bat hippocampus. *Science*, 359, 218–224.

Parker, D. B. (1985). Learning-Logic. Technical Report TR-47, Center for Computational Research in Economics and Management Science, MIT.

Parker, D. B. (1986). A comparison of algorithms for neuron-like cells. In J. Denker (Ed.), *Proceedings of the Second Annual Conference on Neural Networks for Computing.* Proceedings. New York: American Institute of Physics, Vol. 151, pp. 327–332.

Parker, D. B. (1987). Optimal algorithms for adaptive networks: Second order back propagation, second order direct propagation, and second order Hebbian learning. *Proceedings of the 1987 IEEE International Conference on Neural Networks.* New York: IEEE Press, Vol. II, pp. 593–600.

Pavlov, I. P. (1927). *Conditioned Reflexes.* London: Constable and Company. (Reprinted by Dover Publications, 1960).

Piaget, J. (1945). *La Formation du Symbole Chez L'enfant.* Paris: Delachaux Niestle, S.A.

Piaget, J. (1951). *Play, Dreams and Imitation in Childhood.* C. Gattegno and C. F. M. Hodgson (Trans.). London: Routledge and Kegan Paul.

Piaget, J. (1952). *The Origins of Intelligence in Children.* New York: International Universities Press.

Pribe, C., Grossberg, S., and Cohen, M. A. (1997). Neural control of interlimb oscillations, II: Biped and quadruped gaits and bifurcations. *Biological Cybernetics*, 77, 141–152.

Pylyshyn, Z. (1989). The role of location indexes in spatial perception: A sketch of the FINST spatial- index model. *Cognition*, 32, 65–97.

Ratcliff, R. (1990). Connectionist models of recognition memory: Constraints imposed by learning and forgetting functions. *Psychological Review*, 97, 285–308.

Robertson, I. H., Manly, T., Beschin, N., Daini, R., Haeske-Dewick, H., Homberg, V., Jehkonen, M., Pizzamiglio, G., Shiel, A., and Weber, E. (1997). Auditory sustained attention is a marker of unilateral spatial neglect. *Neuropsychologia*, 35, 1527–1532.

Rosenblatt, F. (1958). The perceptron: A probabilistic model for information storage and organization in the brain. *Psychological Review*, 65, 386–408.

Rosenblatt, F. (1962). *Principles of Neurodynamics: Perceptrons and the Theory of Brain Mechanisms.* Washington, DC: Spartan Books.

Rueckert, L., and Grafman, J. (1988). Sustained attention deficits in patients with lesions of parietal cortex. *Neuropsychologia*, 36, 653–660.

Rumelhart, D. E., Hinton, G. E., and Williams, R. J. (1986). Learning representations by back-propagating errors. *Nature*, 323, 533–536.

Sams, M., Paavilainen, P., Alho, K., and Näätänen, R. (1985). Auditory frequency discrimination and event-related potentials. *Electroencephalography and Clinical Neurophysiology/ Evoked Potentials Section*, 62, 437–448.

Schafer, M., and Schiller, D. (2018). Navigating social space. *Neuron*, 100, 476–489.

Schafer, M., and Schiller, D. (2020). The brain's social road maps. *Scientific American*, February, pp. 31–35.

Schmidhuber, J. (2020). Critique of Honda Prize for Dr. Hinton. http://people.idsia.ch/~juergen/critique-honda-prize-hinton.html#reply

Silver, M. R., Grossberg, S., Bullock, D., Histed, M. H., and Miller, E. K. (2011). A neural model of sequential movement planning and control of eye movements: Item-order-rank working memory and saccade selection by the supplementary eye fields. *Neural* Networks, 26, 29–58.

Skinner, B. F. (1938). *The Behavior of Organisms: An Experimental Analysis.* Cambridge, MA: B. F. Skinner Foundation.

Suga, N., Xiao, Z., Ma, X., and Ji, W. (2002). Plasticity and corticofugal modulation for hearing in adult animals. *Neuron*, 36, 1, 9–18.

Suga, N. (2008). Role of corticofugal feedback in hearing. *Journal of Comparative Physiology A*, 194, 169–183.

Velez, R., and Clune, J. (2017). Diffusion-based neuromodulation can eliminate catastrophic forgetting in simple neural networks. *PLoS One*, 12(11), e0187736. https://www.ncbi.nlm.nih.gov/pmc/articles/PMC5690421/

Werbos, P. (1974). *Beyond Regression: New Tools for Prediction and Analysis in the Behavioral Sciences.* Unpublished Doctoral Dissertation, Harvard University.

Werbos, P. (1994). *The Roots of Backpropagation: From Ordered Derivatives to Neural Networks and Political Forecasting.* New York: John Wiley & Sons, Inc.

Wunsch, D. C. II. (2019). Admiring the Great Mountain: A celebration special issue in honor of Stephen Grossberg's 80th birthday. *Neural Networks*, 120, 1–4.

Yao, X. (1999). Evolving artificial neural networks. *Proceedings of the IEEE*, 87, 1423–1447.

Name Index

Subject Index